HarperCollins*Publishers*
哈珀·柯林斯出版集团

双子星科普文库

THE CARTOON GUIDE TO CHEMISTRY

爆笑科学漫画

化学妙想

汉英对照

[美] 拉里·高尼克　克雷格·克里德尔◎著　那日松◎译　刘夙◎校订
Larry Gonick & Craig Criddle

科学普及出版社

·北京·

图书在版编目（CIP）数据

化学妙想：汉英对照/(美)高尼克，(美)克里德尔著；那日松译. —北京：科学普及出版社，2014.9（2023.11重印）
（爆笑科学漫画）
ISBN 978-7-110-08487-8

Ⅰ. ①化… Ⅱ. ①高… ②克… ③那… Ⅲ. ①化学—青年读物—汉、英 ②化学—少年读物—汉、英 Ⅳ. ①O6-49

中国版本图书馆CIP数据核字（2014）第002695号

策划编辑　郑洪炜
责任编辑　郑洪炜
责任校对　王勤杰
责任印制　徐　飞
装帧设计　中文天地

出版发行　科学普及出版社
地　　址　北京市海淀区中关村南大街16号
邮　　编　100081
发行电话　010-62173865
传　　真　010-62173081
网　　址　http://www.cspbooks.com.cn

开　　本　710mm×1000mm　1/16
字　　数　400千字
印　　张　20.25
版　　次　2014年9月第1版
印　　次　2023年11月第10次印刷
印　　刷　北京盛通印刷股份有限公司

书　　号　ISBN 978-7-110-08487-8/O・143
定　　价　58.80元

CONTENTS 目录

CHAPTER 1. HIDDEN INGREDIENTS 隐藏的原料 1

CHAPTER 2. MATTER BECOMES ELECTRIC 万物皆带电 21

CHAPTER 3. TOGETHERNESS 团结 59

CHAPTER 4. CHEMICAL REACTIONS 化学反应 88

CHAPTER 5. HEAT OF REACTION 反应热 111

CHAPTER 6. MATTER IN A STATE 物质的状态 138

CHAPTER 7. SOLUTIONS 溶液 169

CHAPTER 8. REACTION RATE AND EQUILIBRIUM 反应速率和平衡 186

CHAPTER 9. ACID BASICS 酸碱 219

CHAPTER 10. CHEMICAL THERMODYNAMICS 化学热力学 …… 252

CHAPTER 11. ELECTROCHEMISTRY 电化学 …… 271

CHAPTER 12. ORGANIC CHEMISTRY 有机化学 …… 291

索引 …… 313

Chapter 1 Hidden Ingredients

第一章 隐藏的原料

THE VERY FIRST CHEMICAL REACTION TO IMPRESS OUR ANCESTORS WAS **FIRE.**

人类祖先对化学反应的第一印象来自“火”。

WELL, MAYBE FIRE WAS THE **SECOND** ONE, THEN...

哦~那就说成火是人类对化学反应的**第二**印象吧……

ANYWAY, FIRE WAS **REALLY, REALLY IMPORTANT.**

总之,火是**非常,非常重要**的!

FIRE—AND THOSE OTHER PROCESSES—REVEALED **HIDDEN FEATURES OF MATTER.** IF YOU HEAT A PIECE OF WOOD, ALL YOU GET IS A HOT PIECE OF WOOD, AT FIRST... BUT SUDDENLY, AT SOME POINT, THE WOOD BURSTS INTO FLAME. WHERE DID **THAT** COME FROM?

火，还有前面提到的其他现象——都揭示了那些原本**看不见摸不着的物质的秘密。**例如，你烧一块木头，起初木头只是变热了而已，但是到了某个时刻，它突然就自己着火了。那火焰是从哪儿来的呢?

CHEMISTRY IS THE SCIENCE THAT ANSWERS THAT QUESTION, AND **CHEMICAL REACTIONS** ARE THE STRANGE TRANSFORMATIONS THAT REVEAL MATTER'S **HIDDEN PROPERTIES.**

化学就是回答这些问题的一门科学。**化学反应**就是上述这些奇怪的变化，这些变化揭示了物质的**秘密性质**。

CHEMISTRY IS A SCIENCE ABOUT THE OCCULT, THE HIDDEN, THE INVISIBLE. NO WONDER IT TOOK SO LONG FOR CHEMICAL SECRETS TO COME OUT... AND IT ALL STARTED WITH **FIRE.**

化学是一门神秘的科学，它专门研究那些隐藏着的、看不见的世界。难怪化学秘密的揭示花费了人类几千年时间……而这一切都始于**火**。

PROBABLY THE BEST THING ABOUT FIRE WAS THAT IT COULD BE USED TO CONTROL **OTHER** CHEMICAL REACTIONS: COOKING, FOR EXAMPLE!

也许火最大的用处是调控**其他**化学反应的进行：例如煮东西。

YOU KNOW HOW SCIENTISTS ARE: IF THEY CAN COOK ONE THING, THEY'LL COOK ANOTHER. PRETTY SOON, THEY WERE COOKING ROCKS.

正如你所知道的，科学家都是些胡闹鬼：他们一旦烤熟了一种东西，就会尝试去烤各种其他的东西。很快，他们就开始烤石头了。

SOUNDS CRAZY, BUT ONE OF THOSE GREEN, CRUMBLY ROCKS MELTED, CHANGED, AND BECAME AN ORANGE LIQUID THAT COOLED INTO SHINY, METALLIC **COPPER.**

尽管这事儿听起来很疯狂，但是有一种绿色的、脆脆的石头一烧就熔化了，并且逐渐变成橙色的液体，冷却之后又变成了闪闪发亮的金属——**铜**。

THIS ENCOURAGED THEM TO SMELT RED ROCKS INTO IRON... BAKE MUD INTO BRICKS... SAUTE FAT AND ASHES INTO SOAP... AND (WITHOUT FIRE) TO CURDLE MILK INTO YOGURT... FERMENT GRAIN INTO BEER... AND CABBAGE INTO KIMCHEE. THE NEXT THING YOU KNEW, CHEMISTRY HAD CAUSED **CIVILIZATION!**

铜的发现，鼓励人们创造了更多的东西：红色的石头通过煅烧形成铁……烤泥巴得到砖块……炒脂肪加骨灰制成肥皂；还有不用火的：牛奶凝固制成了乳酪……小麦发酵酿成啤酒……卷心菜腌成泡菜。故事进行下去无非是——化学创造了**人类文明**！

WHAT ACCOUNTS FOR MATTER'S SECRETS? THE ANCIENT GREEKS CAME UP WITH AT LEAST THREE DIFFERENT THEORIES.

什么解释了物质的秘密？古希腊先哲们提出了至少三种不同的理论。

THE **ATOMISTS,** LED BY **DEMOCRITUS,** THOUGHT THAT MATTER WAS MADE OUT OF TINY, INDIVISIBLE PARTICLES, OR **ATOMS** (A-TOM = "NO CUT"). IF YOU CUT AND CUT AND CUT AND CUT, THEY REASONED, THE PROCESS HAD TO STOP SOMEWHERE.

由**德谟克利特***提出的**原子论**指出，物质由看不见的、微小的、不可分割的粒子——原子组成。“原子”的英文单词atom由a-tom构成，意为不可分割。他们指出，当你不断地分割一个东西时，最终会到达一个不可继续分割的地步。

* 编者注：德谟克利特，古希腊伟大的唯物主义哲学家，原子唯物论学说创始人之一。

ANOTHER PHILOSOPHER, **HERACLITUS,** SUGGESTED THAT EVERYTHING WAS MADE OUT OF **FIRE.**

另一位哲学家**赫拉克利特***则认为所有东西都由**火**构成。

* 编者注：赫拉克利特，一位富有传奇色彩的古希腊哲学家，他认为火与万物可相互转化。

BUT ATOMS COULDN'T BE SEEN, AND... FIRE? I MEAN, REALLY! THE GREAT **ARISTOTLE** ANNOUNCED THAT THERE WERE REALLY **FOUR ELEMENTS**, OR BASIC SUBSTANCES, FROM WHICH ALL ELSE WAS COMPOSED. THESE WERE **AIR, EARTH, FIRE,** AND **WATER.** OTHER STUFF, HE OPINED, WAS A BLEND OF THESE FOUR.

原子尚且不可见，但……火?开玩笑!（它应该是明显可见的，怎么会组成万物?）

伟大的**亚里士多德***宣布，最基本的元素——组成万物的基本单元一共有四种：**空气、土、火**和**水**。地球上的一切物质都由它们混合而成。

* 编者注：亚里士多德，世界古代史上最伟大的哲学家、科学家和教育家之一。

OF THE THREE IDEAS, FOR SOME REASON, IT WAS ARISTOTLE'S THAT MOST INFLUENCED MEDIEVAL SCIENCE. IT WAS SO **OPTIMISTIC!** IF EVERYTHING WAS A MIXTURE OF FOUR ELEMENTS, THEN YOU SHOULD BE ABLE TO TURN ANYTHING INTO ANYTHING ELSE JUST BY TWEAKING THE INGREDIENTS!

在以上三种说法中，由于种种原因，亚里士多德的说法最终对中世纪的科学产生了最深远的影响。如果所有东西都只由这四种元素组成，那么你就能用这四种元素调配出任何东西啦！够**爽**吧。

THIS HOPELESS QUEST WAS TAKEN UP IN PERSIA BY **JABIR** (EIGHTH CENTURY) AND **AL-RAZI** (TENTH CENTURY), WHO INVENTED ALL SORTS OF USEFUL LAB EQUIPMENT AND PROCEDURES IN THE PROCESS. THIS PROVES YOU CAN MAKE TREMENDOUS PRACTICAL PROGRESS WITH STUPID IDEAS.

这个注定无望的探索首先由公元8世纪的波斯人**贾比尔**和公元10世纪的**拉奇**尝试。他们发明了各种各样的实用实验装置和实验流程。他们的事例说明：哪怕是个愚蠢的想法也能让人在实践上取得巨大的进步。

MEDIEVAL EUROPE BORROWED THE ISLAMIC SCIENCE—AND ITS NAME, **ALCHEMY** (="THE CHEMISTRY" IN ARABIC)—AND ITS HUNGER FOR TRANSMUTED GOLD. THE GERMAN ALCHEMIST **HENNIG BRAND,** FOR EXAMPLE, TRIED TO MAKE GOLD BY DISTILLING 60 BUCKETS OF URINE.

在伊斯兰科学中兴起的化学连同它当时的名称——“**炼金术**”（炼金术的英语“ALCHEMY”在阿拉伯语中就是“化学”的意思）传到了中世纪的欧洲。炼金术的终极渴求是炼出金子，这一愿望也被一起传到了欧洲。例如，德国炼金师**亨尼希·布兰德**曾经试图用60桶尿来提炼金子。

IN THE END, BRAND'S EQUIPMENT GLOWED IN THE DARK. HE HAD DISCOVERED PHOSPHORUS—BUT NO GOLD...

最后，布兰德的盛尿桶居然在夜晚会发光。只不过，他最终得到的是磷，而不是金子……

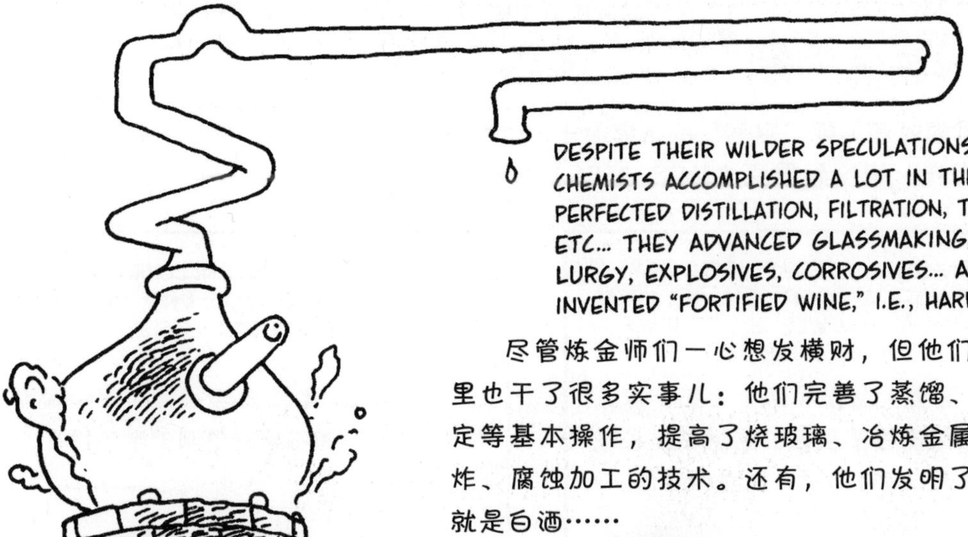

DESPITE THEIR WILDER SPECULATIONS, THE ALCHEMISTS ACCOMPLISHED A LOT IN THE LAB: THEY PERFECTED DISTILLATION, FILTRATION, TITRATION, ETC... THEY ADVANCED GLASSMAKING, METALLURGY, EXPLOSIVES, CORROSIVES... AND THEY INVENTED "FORTIFIED WINE," I.E., HARD LIQUOR...

尽管炼金师们一心想发横财，但他们在实验室里也干了很多实事儿：他们完善了蒸馏、过滤、滴定等基本操作，提高了烧玻璃、冶炼金属，还有爆炸、腐蚀加工的技术。还有，他们发明了烈酒，也就是白酒……

BUT THEIR LAB TECHNIQUE MISSED ONE BIG THING: THEY FAILED TO COLLECT **GASES.** IF A REACTION CONSUMED GAS, THE ALCHEMISTS HAD NO WAY OF KNOWING. IF IT GAVE OFF GAS, THEY LET IT ESCAPE.

但是他们在实验技术上忽略了一个重要的方面：未能收集**气体**。那时候，炼金师们完全不知道化学反应会消耗气体，如果有化学反应生成了气体，他们就直接让它跑掉。

THIS MEANT THEY COULD NEVER FULLY ACCOUNT FOR THE **INGREDIENTS** OR **PRODUCTS** OF CHEMICAL REACTIONS.

这意味着，炼金师们总是不能完整地解释清楚化学反应的所有**原料**和**产物**。

THE MODERN STUDY OF GASES OR "AIRS" BEGAN IN THE 1600s, WITH SOME INVESTIGATIONS INTO THE EFFECTS OF AIR PRESSURE. CONSIDER THIS DEMONSTRATION BY **OTTO VON GUERICKE** (1602–1686).

现代科学对气体或“空气”的研究始于17世纪，最初研究的是大气压的性质。让我们来看看由**奥托·冯·居里克**（1602—1686）展示的实验。

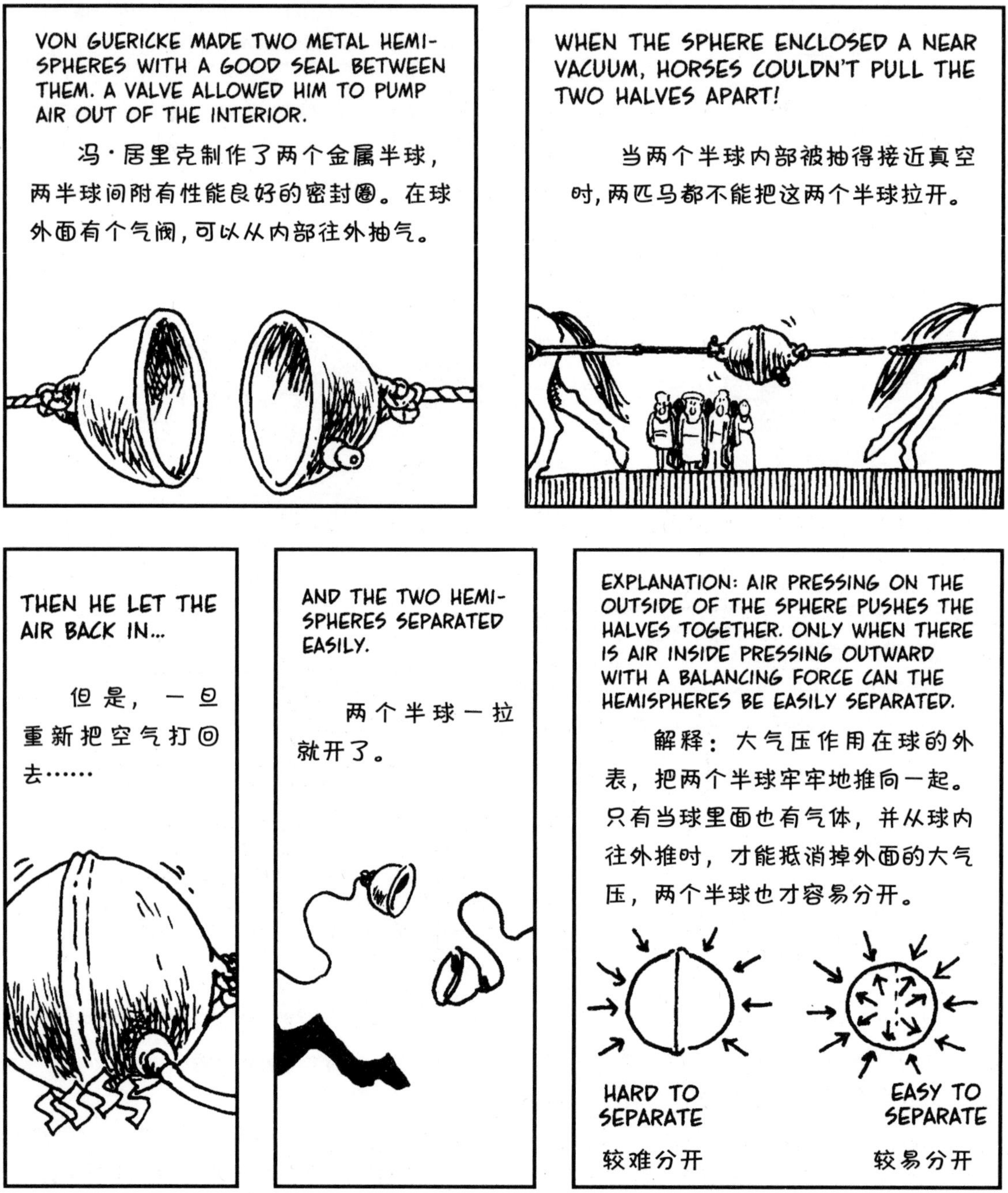

我们在家里也能通过简单的实验证明同样的原理：

将一个瓶子装满水并拧紧瓶盖，将瓶子倒置并将有盖的一端浸入水盆中（例如厨房的水池），瓶盖拿掉之后，瓶子里的水依然能保持充满状态，不会漏下来。

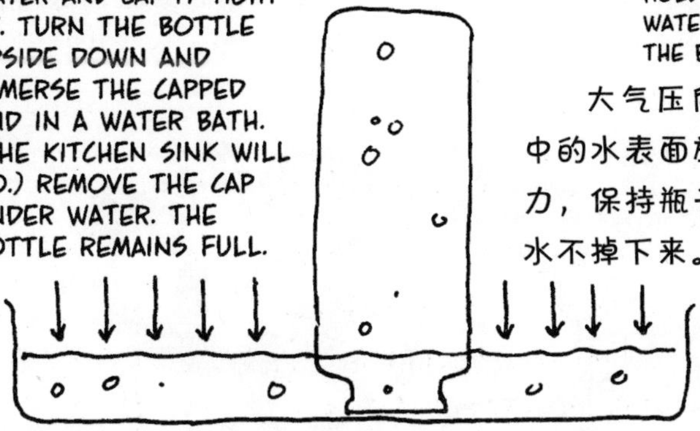

大气压向水盆中的水表面施加压力，保持瓶子里的水不掉下来。

这个倒置的瓶子后来被一位牧师**约瑟夫·普利斯特列**（1733—1804）改装成了收集气体的工具。他在厨房建立了一个化学实验室。

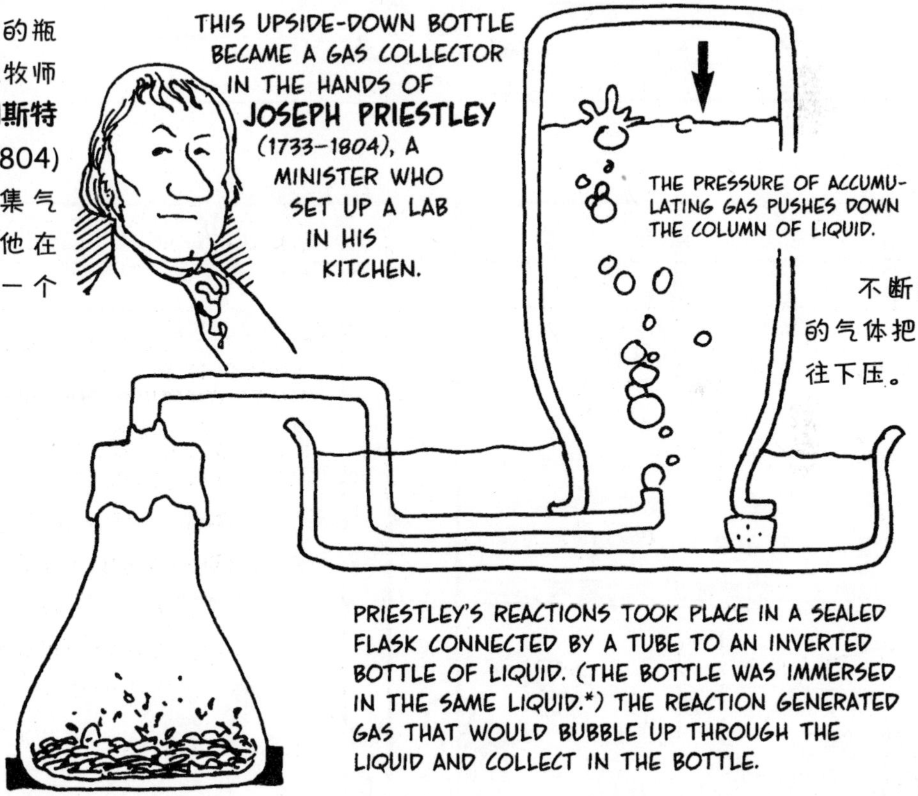

不断积累的气体把液柱往下压。

普利斯特列的化学实验在一个密封的烧瓶里进行，烧瓶通过一个管子与一个倒置的装满液体的瓶子相连（瓶子与水盆里装有同样的液体*）。反应生成的气体就会在倒置的瓶子中冒出泡泡，上升到瓶底，并被收集在瓶中。

*WATER, UNLESS THE GAS WAS WATER SOLUBLE, IN WHICH CASE PRIESTLEY USED MERCURY.

*一般使用水作为实验用液体。如果某种气体会溶于水，那么，普利斯特列就用水银作为实验用液体。

PRIESTLEY STORED THE GASES IN PIG BLADDERS HE HAPPENED TO HAVE LYING AROUND THE HOUSE.

普利斯特列把气体收集在一个个猪膀胱里，散落在家中。

FOR EXAMPLE, WHEN HE COMBINED A STRONG ACID WITH IRON FILINGS, THE REACTION PRODUCED A GAS, OR "INFLAMMABLE AIR," THAT BURNED EXPLOSIVELY. WE KNOW IT AS **HYDROGEN.**

例如，他把强酸和铁屑混合，（一经点燃）反应生成的气体会剧烈燃烧，这种被称为“燃气”的气体其实就是我们现在所说的**氢气**。

ANOTHER EXPERIMENT HEATED A RED MINERAL CALLED "CALX OF MERCURY." AS THE "CALX" MELTED, DROPLETS OF PURE MERCURY CONDENSED ON THE WALLS OF THE VESSEL, WHILE GAS ACCUMULATED IN THE WATER BOTTLE.

在另一个实验中，他加热了一种叫作“汞石灰”的红色矿物。石灰熔化后，一滴滴的纯汞（即水银）在瓶壁凝结了出来。同时，生成的气体也被收集在了倒置的瓶子里。

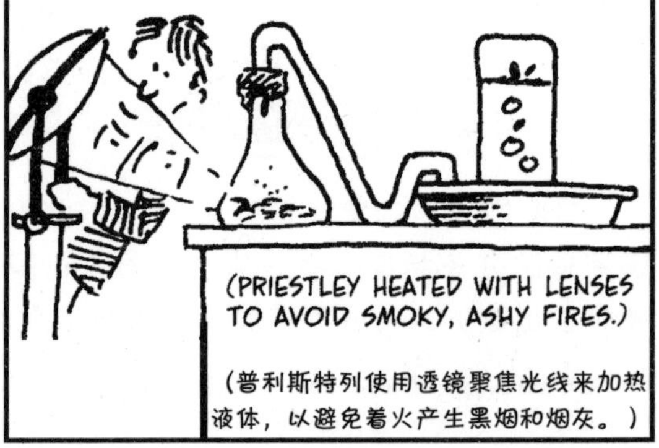

PRIESTLEY NOTICED THAT A FLAME BURNED EXTRA BRIGHTLY WHEN SURROUNDED BY THIS NEW GAS.

普利斯特列还注意到，当被这种新的气体环绕时，火焰燃烧得特别旺盛。

SINCE HE KNEW THAT FLAMES BURN WELL IN "GOOD" (I.E., BREATHABLE) AIR AND SNUFF OUT IN BAD AIR (AS IN A COAL MINE), HE DECIDED TO TAKE A WHIFF.

当他知道火焰在“好气”（能维持呼吸）里燃烧旺盛，在“坏气”（使人窒息）里会熄灭之后，他决定自己吸一口。

AFTERWARD, HE WROTE:

"THE FEELING OF IT TO MY LUNGS WAS NOT SENSIBLY DIFFERENT FROM THAT OF COMMON AIR. BUT I FANCIED THAT MY BREATH FELT PARTICULARLY LIGHT AND EASY FOR SOME TIME AFTERWARD. WHO CAN TELL BUT THAT, IN TIME, THIS PURE AIR MAY BECOME A FASHIONABLE ARTICLE IN LUXURY? HITHERTO ONLY TWO MICE AND MYSELF HAVE HAD THE PRIVILEGE OF BREATHING IT."

吸完之后，他写道：

“从肺的感觉来看，与吸入一般的空气没什么区别，但是不久之后就会感到呼吸和运动变得非常轻松。谁也说不准这种气体将来会不会成为时尚的奢侈品，反正现在只有我和两只老鼠有享受此的福分。”

FOR OXYGEN IT WAS...

氧气是……

与此同时，法国化学家**安托万·拉瓦锡**（1743—1794）也做了类似的实验，只是实验步骤截然不同。

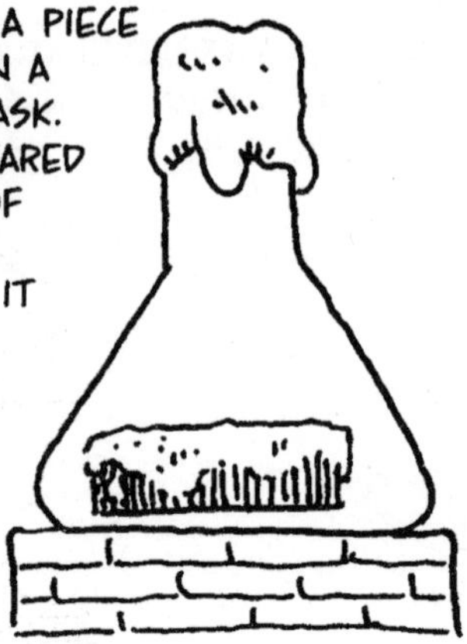

拉瓦锡在一个密封的烧瓶中放置了一块金属锡并对其加热。金属表面逐渐生成了一层灰。拉瓦锡继续加热了一天半，直到再也没有新的灰生成为止。

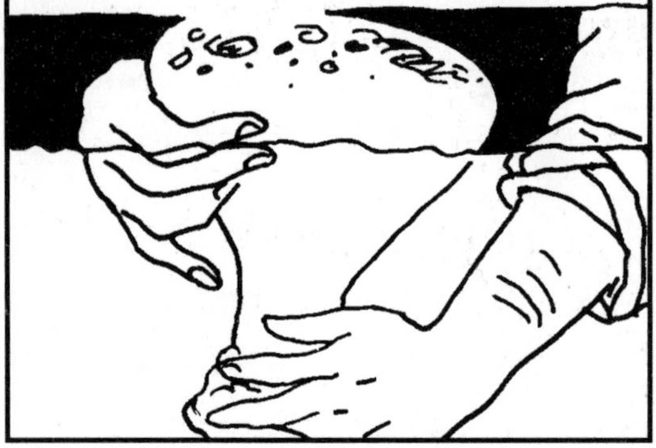

待烧瓶冷却后，他把烧瓶倒置在水中并把封口揭开。

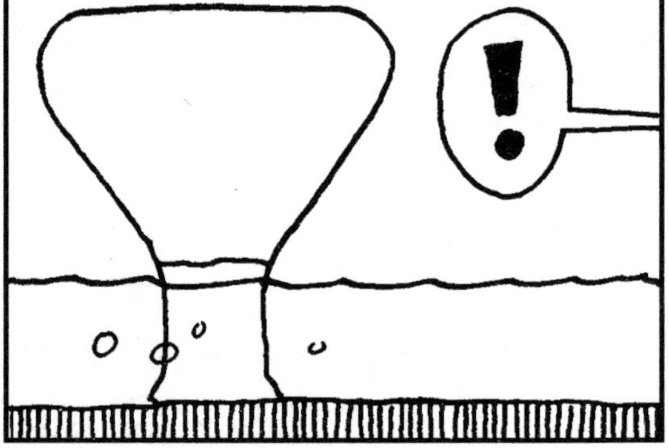

他发现，烧瓶内的水面上升至瓶身的五分之一处。

CONCLUSION: ONE-FIFTH OF THE AIR ORIGINALLY IN THE FLASK WAS REMOVED BY THE REACTION. THIS GAS MUST HAVE COMBINED WITH THE TIN TO FORM THE ASHY SUBSTANCE.

结论：烧瓶里原有的空气被加热时发生的反应消耗掉了五分之一。这部分气体一定是与锡块结合而生成了那些灰一样的物质。

AIR, SAID LAVOISIER, MUST BE A **MIXTURE** OF TWO DIFFERENT GASES. ONE OF THEM, WHICH MAKES UP ONE-FIFTH OF THE TOTAL VOLUME, COMBINED WITH THE TIN, WHILE THE OTHER DID NOT.

拉瓦锡说，空气一定是两种气体的**混合物**。其中一种气体占总体的五分之一，并且能与锡结合；而另一种气体则不与锡发生反应。

NEXT LAVOISIER REPEATED THE EXPERIMENT USING MERCURY INSTEAD OF TIN. OVER HIGH HEAT, MERCURY ALSO FORMED AN ASH (OR "CALX") AND REMOVED GAS FROM THE AIR. THEN, WHEN HEATED GENTLY, THE ASH GAVE BACK THE GAS AND ALL THE ORIGINAL MERCURY, A LA PRIESTLEY.

接着，拉瓦锡用水银代替锡重复了一次实验。在高温作用下，水银也转变成了像灰一样的物质（称为“金属灰”），并且消耗掉一部分空气。然后，在较低的温度加热下，这些像灰一样的物质又能全部恢复为水银，消耗的气体也重新释放出来，这与普利斯特列的实验差不多。

IN OTHER WORDS, PRIESTLEY'S "GOOD AIR" WAS THE SAME GAS THAT LAVOISIER HAD FOUND TO MAKE UP 20% OF THE ATMOSPHERE. THE FRENCH CHEMIST GAVE IT A NEW NAME: **OXYGEN.**

换句话说，普利斯特列所谓的“好气”，与拉瓦锡所发现的占空气成分20%的气体是同一种物质。这位法国化学家给这种气体起了一个新名称：**氧气**。

INTERPRETATION: THE ASH WAS A **COMPOUND** OF THE METAL AND OXYGEN (A METALLIC OXIDE, WE WOULD SAY).

解释：这些灰是金属和氧气所形成的**化合物**（或称为金属氧化物）。

LAVOISIER CONFIRMED THIS BY WEIGHING: THE WEIGHT OF THE REMAINING (UNREACTED) METAL PLUS THE WEIGHT OF ASH WAS **GREATER** THAN THE WEIGHT OF THE ORIGINAL METAL.

为了求证，拉瓦锡用天平称量了反应后的金属和金属灰，发现它比反应前的完好金属质量重。

LAVOISIER DREW A GENERAL CONCLUSION: **COMBUSTION** WAS A PROCESS WHEREBY FUEL COMBINED WITH OXYGEN. IN OTHER WORDS, **FIRE IS NOT AN ELEMENT**; IT'S A CHEMICAL REACTION THAT GOBBLES UP OXYGEN AND GIVES OFF HEAT AND LIGHT.

拉瓦锡由此得出一个一般性结论：**燃烧**是燃料与氧气化合的过程。换句话说，**“火”不是一种元素**；它只是一类能消耗氧气并且能发热发光的化学反应而已。

AND MORE: LAVOISIER ALSO FOUND THAT THE TOTAL WEIGHT OF THE SEALED FLASK PLUS CONTENTS WAS THE SAME BEFORE AND AFTER THE REACTION.

此外，拉瓦锡还发现，密封烧瓶与反应物的总重量在化学反应前后是相同的。

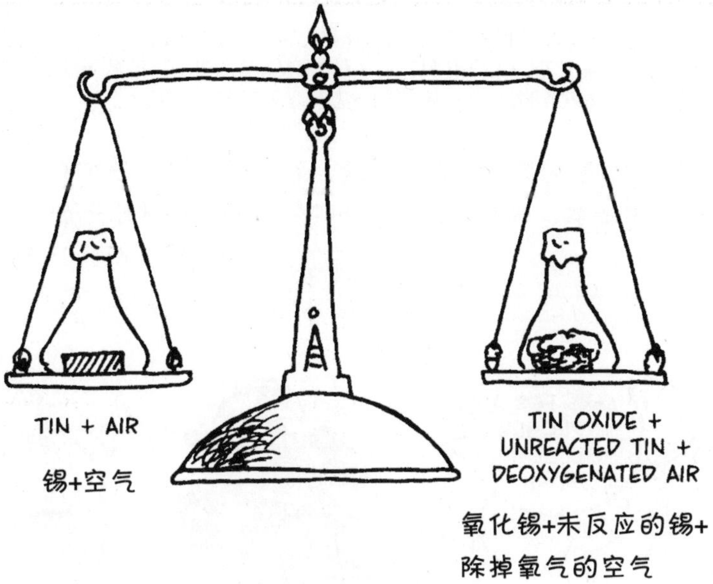

AND SO HE LAID DOWN THE LAW OF **CONSERVATION OF MATTER.**

于是，拉瓦锡得到了**质量守恒定律**。

In chemical reactions, nothing is created or destroyed. The elements are merely rearranged in new combinations.

在化学反应中，既没有新的物质生成，也没有物质被毁灭，而仅仅是各化学元素以新的方式重新组合起来。

LAVOISIER PROPOSED A PROGRAM FOR CHEMISTRY: FIND THE ELEMENTS, THEIR WEIGHTS, AND THEIR RULES OF COMBINATION. THEN HE LOST HIS HEAD IN THE FRENCH REVOLUTION, AND THE PROGRAM, LIKE HIS HEAD, HAD TO BE CARRIED OUT BY OTHERS.

拉瓦锡还提出了一种化学研究的程序：鉴别出化学元素，得到它们的重量和化合反应的规律。后来，拉瓦锡自己在法国大革命中丢了脑袋。他提倡的化学研究的程序就和他的脑袋一样——落在了后人的肩上。

CHEMISTS FOLLOWED THROUGH WITH ENTHUSIASM, AND BY 1800 HAD DISCOVERED ABOUT THIRTY ELEMENTS—AND NONE OF THEM WAS **WATER.** IT TURNED OUT TO BE A COMPOUND OF HYDROGEN AND OXYGEN.

化学家们热火朝天地探索，到了1800年，已经有大概30种元素被发现，其中没有一个是**水**。后来发现，水其实是由氢和氧两种元素组成的化合物。

AND **COMPOUNDS,** THEY FOUND, WERE NO MERE ARISTOTELIAN MISH-MASH. INSTEAD, COMPOUNDS ALWAYS COMBINED ELEMENTS IN **FIXED PROPORTIONS.** WATER, FOR EXAMPLE, WAS ALWAYS MADE OF EXACTLY **TWO** VOLUMES OF HYDROGEN GAS AND **ONE** VOLUME OF OXYGEN.

化学家们还发现，**化合物**并非像亚里士多德说的那样是元素大杂烩。实际上，化合物中的元素是以**固定的比例**结合的。例如，水总是准确地由两份体积的氢气和一份体积的氧气组成。

SUCH DISCOVERIES LED **JOHN DALTON** (1766–1844) TO REVIVE THE **ATOMIC THEORY OF MATTER.** EACH ELEMENT, HE REASONED, WAS MADE OF TINY, INDIVISIBLE ATOMS. THE ATOMS OF ANY ONE ELEMENT ARE ALL ALIKE, BUT DIFFER FROM THE ATOMS OF OTHER ELEMENTS.

这一发现启发了**约翰·道尔顿**（1766—1844），他提出了**原子学说。**他认为，每种元素都由微小的、不可分割的原子组成。同属一种元素的原子都很相似，而属于不同元素的原子则各不相同。

COMPOUND SUBSTANCES, SAID DALTON, WERE COMPOSED OF FIXED GROUPINGS OF ATOMS CALLED **MOLECULES.**

道尔顿说，化合物是由原子的固定组合体**分子**直接构成的。

ALTHOUGH ATOMS WERE INVISIBLY SMALL, SCIENTISTS ACCEPTED THE ATOMIC THEORY ANYWAY, BECAUSE IT EXPLAINED WHAT THEY COULD SEE...

虽然原子小到肉眼无法看见，但由于原子学说能解释当时所观察到的现象，因此化学家们便接受了这一学说。

MEANWHILE, THEY KEPT UP THE HUNT FOR NEW ELEMENTS, FINDING NEARLY **SEVENTY** BY THE 1860s—AND WHAT A LIST IT WAS! ELEMENTS MIGHT BE SOLID, LIQUID, OR GASEOUS; YELLOW, GREEN, BLACK, WHITE, OR COLORLESS; CRUMBLY OR BENDY; WILDLY REACTIVE OR RELATIVELY INERT.

同时，他们继续寻找新的元素。到19世纪60年代，科学家们已经发现了近70种元素。多么壮观的列表啊！在这些元素（译者注：这里指的是元素单质）中，有固体、液体和气体；有黄的、绿的、黑的、白的或无色的；有的脆、有的韧；有的反应激烈、有的反应迟钝。

ONE MORNING IN 1869, A RUSSIAN NAMED **DMITRI MENDELEEV** (1834–1907) WOKE UP WITH AN IDEA: LIST THE ELEMENTS IN ORDER OF INCREASING ATOMIC WEIGHT AND DO A "TEXT WRAP" AT REGULAR INTERVALS.

1869年的一个早上，俄国科学家**德米特里·门捷列夫**（1834—1907）被梦里的一个想法惊醒：他把所有元素按原子的重量大小进行排列，然后以相同的间隔“换行”。

THE RESULT WAS A SORT OF TABLE, WITH THE ELEMENTS ARRANGED IN ROWS. HERE'S A BABY VERSION OF MENDELEEV'S TABLE. (YOU'LL SEE THE REAL THING NEXT CHAPTER.)

结果出现了一个由排列成行的元素组成的表格。以下是门捷列夫表格的节选版（完整版见下一章）。

HYDROGEN 氢							
LITHIUM 锂	BERYLLIUM 铍	BORON 硼	CARBON 碳	NITROGEN 氮	OXYGEN 氧	FLUORINE 氟
SODIUM 钠	MAGNESIUM 镁	ALUMINUM 铝	SILICON 硅	PHOSPHORUS 磷	SULFUR 硫	CHLORINE 氯
POTASSIUM 钾	CALCIUM 钙					

THE ELEMENTS SHOWED A **PERIODIC PATTERN:** EACH VERTICAL COLUMN CONTAINED CHEMICALLY SIMILAR ELEMENTS. IN FACT, MENDELEEV NOTED GAPS FARTHER DOWN THE TABLE AND SUCCESSFULLY PREDICTED **NEW ELEMENTS** THAT WOULD FILL THEM!

元素具有**周期律**：每一列元素的化学性质相似。门捷列夫还将表格更往下的地方留了空缺，他预测将来会有**新元素**来填补这些空白。

TO FIND THE ANSWER, SCIENTISTS FOLLOWED THE SAME LINE OF THOUGHT THEY'D BEEN USING ALL ALONG: IF SUBSTANCES ARE MADE OF ELEMENTS, AND ELEMENTS ARE MADE OUT OF ATOMS, THEN WHAT, THEY ASKED, ARE ATOMS MADE OUT OF?

为了找到这一问题的答案，科学家沿着他们的探索之路继续前行：如果物质由不同元素构成，而元素由原子构成，那么，原子又是由什么构成的呢？

Chapter 2
Matter Becomes Electric

第二章
万物皆带电

NATURE HAD ANOTHER SECRET BESIDES FIRE... AT LEAST, IT LOOKED LIKE ANOTHER SECRET AT FIRST...

除了“火”之外，自然界还有另一个秘密……至少看起来像是另一个……

THIS ONE INVOLVED **AMBER...** OR AS THE GREEKS CALLED IT, **ELEKTRA.**

这个秘密与琥珀有关……琥珀的希腊语是ELEKTRA（在希腊语中琥珀还有相吸引的意思）。

WHEN THEY RUBBED THIS STUFF WITH FUR, IT ACTED STRANGELY, ATTRACTING FLUFF AND FEATHERS AND THE HAIR ON THE BACK OF YOUR ARM.

用一块琥珀摩擦毛皮，琥珀就变得怪里怪气的——可以吸附绒毛、羽毛和手臂上的汗毛。

CENTURIES LATER, AN ENGLISHMAN NAMED WILLIAM GILBERT FOUND OTHER MATERIALS WITH THE SAME PROPERTY. HE SAID THEY ALL HAD "ELEKTRA."

几百年之后，一个叫威廉·吉尔伯特的英国人发现许多其他物体具有与琥珀同样的性质，他统称这些物体为“跟琥珀一样”。

* Elektra也是希腊神话中一位公主的名字，她联合其弟兄杀死了母亲和继父。

THEN PEOPLE NOTICED THAT THERE WERE REALLY TWO KINDS OF "ELECTRIC" MATERIALS: ONE REPELLED WHAT THE OTHER ATTRACTED, AND VICE VERSA.

随后，人们发现了两类能生“电”的物体：其中一类能吸引什么，另一类就排斥什么；反之亦然。

AROUND 1750, **BENJAMIN FRANKLIN** (1706–1790) FIRST CALLED THESE TWO KINDS OF ELECTRICITY **POSITIVE** AND **NEGATIVE.**

POSITIVE, SAID FRANKLIN, REPELS POSITIVE; NEGATIVE REPELS NEGATIVE; AND POSITIVE AND NEGATIVE ATTRACT EACH OTHER AND CANCEL EACH OTHER OUT. IN ORDINARY, **NEUTRAL** MATTER, OPPOSITE CHARGES ARE PRESENT IN EQUAL AMOUNT.

1750年左右，**本杰明·富兰克林**（1706—1790）首次把这两种电称为“**正电荷**”和“**负电荷**”。

他说，正电荷排斥正电荷；负电荷排斥负电荷；正电荷与负电荷相互吸引和抵消。在普通的**电中性**物体中，正电荷与负电荷的数量相等。

NEGATIVE CHARGES CAN SOMETIMES FLOW OUT OF A SUBSTANCE, CREATING A CHARGE **IMBALANCE**—AN EXCESS OF NEGATIVITY HERE AND POSITIVITY THERE...

负电荷有时可以从物体身上流走，导致**电荷不平衡**——一处带有富余的正电荷，另一处带有富余的负电荷。

BUT BECAUSE OF THE MUTUAL ATTRACTION, THE NEGATIVES MAY SUDDENLY FLOW BACK TO THE POSITIVE CHARGE WITH A SPARK.

但由于相互吸引，富余的负电荷有时会突然往富余的正电荷流动，打出一个电火花。

"TWO NIGHTS AGO, BEING ABOUT TO KILL A TURKEY BY THE SHOCK FROM TWO LARGE GLASS JARS,* CONTAINING AS MUCH ELECTRICAL FIRE AS FORTY COMMON PHIALS, I INADVERTENTLY TOOK THE WHOLE THROUGH MY OWN ARMS AND BODY, BY RECEIVING THE FIRE FROM THE UNITED TOP WIRES WITH ONE HAND WHILE THE OTHER HELD A CHAIN CONNECTED WITH THE OUTSIDE OF BOTH JARS."

—BENJAMIN FRANKLIN, 1750

“前天晚上，我正准备用两个大玻璃瓶储存的电来杀死一只火鸡，这两个大瓶子储存的电量相当于40个常见的小药瓶的储量*。当时，我双手不小心分别碰到了两个电叉的连接线和瓶身外表的铁链，随着一个大火花，所有的电从我的手臂通过了整个身体。”

——本杰明·富兰克林，1750

*JUST ONE OF THE WAYS THE FUN-LOVING FOUNDING FATHER LIKED TO AMUSE HIMSELF!

*“国父”总爱用自己的兴趣来自娱自乐！

WITH THE INVENTION OF THE ELECTRIC BATTERY (BY VOLTA IN 1800), ONE COULD RUN A STEADY STREAM OF NEGATIVE CHARGE—A **CURRENT**—THROUGH A COPPER WIRE, AND MAYBE THROUGH OTHER MATERIALS AS WELL.

电池发明之后（1800年由伏特发明），人们便可以在铜线或其他材料上实现负电荷的平稳流动——**电流**。

CHEMISTS TRIED RUNNING ELECTRICITY THROUGH ORDINARY WATER. TWO METAL STRIPS, OR ELECTRODES, WERE CONNECTED TO A BATTERY AND IMMERSED IN WATER.

化学家们尝试让电流通过普通的清水。在此实验中，用两个金属条作为电极浸入水中，并连接在电池的两极。

AS CHARGE BUILT UP ON THE ELECTRODES, BUBBLES OF **HYDROGEN GAS** APPEARED AT THE NEGATIVE STRIP, OR **CATHODE**. BUBBLES OF **OXYGEN** FORMED AT THE POSITIVE STRIP, OR ANODE.

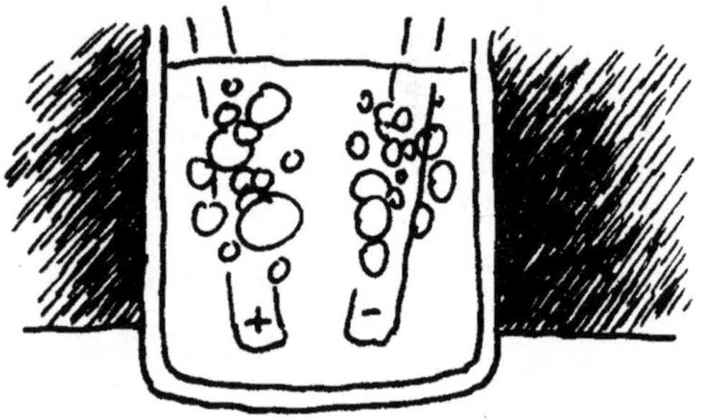

当电荷集聚在两电极处时，带负电的金属条——**阴极**，产生了大量**氢气**气泡；而带正电的电极——**阳极**，则产生**氧气**气泡。

ELECTRICITY SPLITS WATER! SCIENTISTS SOON TRIED THIS **ELECTROLYSIS** (ELECTRIC SPLITTING) ON OTHER SUBSTANCES. MELTED TABLE SALT, THEY FOUND, YIELDS METALLIC **SODIUM** AT THE CATHODE AND GREEN, TOXIC **CHLORINE GAS** AT THE ANODE.

电可以分解水！科学家马上试着用电来分解其他物质——这就叫**电解**。他们发现，对熔化的食盐进行电解，能在阴极得到金属**钠**，并在阳极得到绿色的有毒气体——**氯气**。

IT'S A BIG LEAP FROM FINDING ELECTRICITY IN A FEW PLACES TO SEEING IT EVERYWHERE, BUT THAT'S SCIENCE FOR YOU!

从偶然发现电现象，到承认电的普遍存在，跨越了一个布满荆棘的大鸿沟，但这恰是为我们所称道的科学！

BY THE END OF THE 19TH CENTURY, SCIENTISTS WERE CONVINCED THAT ATOMS WERE MADE OF **ELECTRIC INGREDIENTS.**

到19世纪末，事实已经让科学家们相信原子是由**带电成分**组成的。

AND SO THEY ARE. HERE'S THE IDEA:

而原子确实由带电的成分组成：

ATOMS ARE MADE UP OF SMALLER, ELECTRICALLY CHARGED PARTICLES (AND SOME NEUTRAL PARTICLES TOO). EACH ATOM HAS AN EQUAL NUMBER OF POSITIVE AND NEGATIVE CHARGES. THE NEGATIVELY CHARGED PARTICLES, CALLED **ELECTRONS,** WEIGH LITTLE AND MOVE AROUND EASILY.

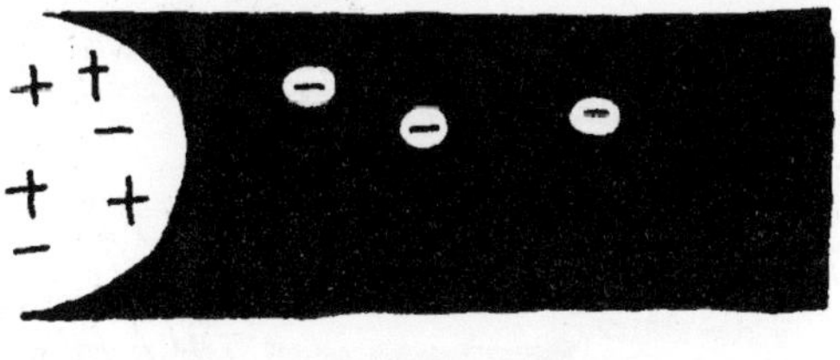

原子由更小的带电粒子（当然还有电中性粒子）所组成。每个原子都含有相同数量的正电荷和负电荷。其中，带负电荷的粒子称为**电子**，它们非常轻，移动迅速。

A DEPARTING ELECTRON LEAVES BEHIND A POSITIVELY CHARGED ATOM, OR **POSITIVE ION.** SUCH IONS, ATTRACTED TO CATHODES (WHICH ARE NEGATIVE), ARE CALLED **CATIONS** (PRONOUNCED "CAT-EYE-ONZ").

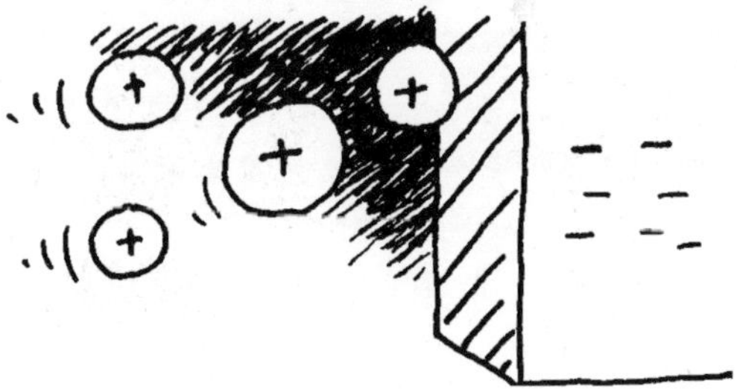

如果电子跑丢了，就留下原子中带正电荷的部分，即**带正电的离子**。这个离子会吸附在阴极（阴极带负电荷），称为**阳离子**。

OTHER KINDS OF ATOMS **ACQUIRE** ELECTRONS TO BECOME NEGATIVELY CHARGED IONS, OR **ANIONS,** ATTRACTED TO ANODES.

有些原子容易**捕获**额外的电子变成带负电的离子——**阴离子**。它们会被阳极所吸引。

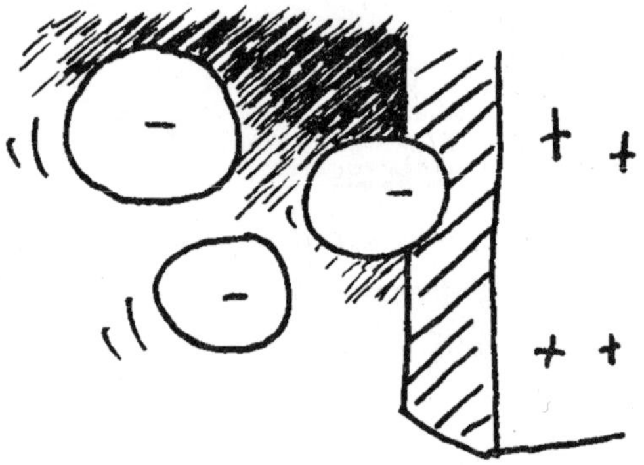

IN TABLE SALT, FOR EXAMPLE, SODIUM CATIONS ARE ATTRACTED TO CHLORIDE ANIONS AND ARRANGE THEMSELVES INTO A CRYSTAL, **SODIUM CHLORIDE.**

以食盐为例，大量钠阳离子与氯阴离子互相吸引，并按照一定的方式排列，组成**氯化钠**晶体。

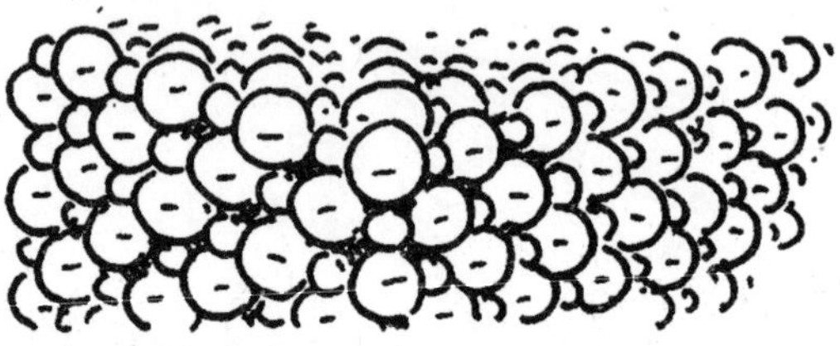

DURING ELECTROLYSIS, THESE IONS MIGRATE TOWARD THE ELECTRODES, AND THE SALT DISSOCIATES.

在电解时，这些离子迁移到两个电极处，食盐也就被分解了。

All-Important Fact: 重要提示：

ATOMS COMBINE CHEMICALLY BY SHARING OR TRANSFERRING ELECTRONS.

原子与原子之间通过共享或得失电子而化合。

SO—TO UNDERSTAND CHEMISTRY, WE NEED TO SEE HOW ELECTRONS BEHAVE WITHIN EACH ATOM.

因此，要理解化学就得知道每个原子中的电子是怎么一回事。

HOW SMALL IS THE SMALL PICTURE? LET'S TRY SHRINKING DOWN **ONE MILLION TIMES.** A HUMAN HAIR IS NOW THIRTY STORIES THICK... BACTERIA ARE THE SIZE OF TORPEDOES... AND ATOMS ARE JUST BARELY VISIBLE AS TINY SPECKS.

原子到底有多小呢？先试着把我们自己缩小**100万倍**。一根头发就变成30层楼那么粗……细菌就跟鱼雷一样……但是原子却仍像灰尘一样勉强可见。

SHRINKING ANOTHER THOUSAND TIMES BRINGS US TO **NANOMETER** (= 10^{-9} METER) SCALE. I'M JUST SHY OF 2 nm TALL. THE ATOMS ARE NOW ABOUT ONE-TENTH MY SIZE. WE'RE IN A VERY ENERGETIC ENVIRONMENT: LIGHT WAVES ARE ZOOMING AROUND, AND ALL THE ATOMS ARE JIGGLING.

再缩小1000倍，我们就来到了**纳米**（=10^{-9}米）级别。我的身高变成只有不到2纳米。现在，原子的尺寸大概是我身体的十分之一。我们置身于一个非常活跃的氛围里：光波在游走，所有原子都不安分地在推推攘攘。

THIS IS **GRAPHITE** FROM SOME PENCIL SHAVINGS. THE CARBON ATOMS ARE ARRANGED IN SHEETS THAT CAN SLIDE OVER EACH OTHER EASILY. THIS EXPLAINS WHY GRAPHITE IS A GOOD LUBRICANT.*

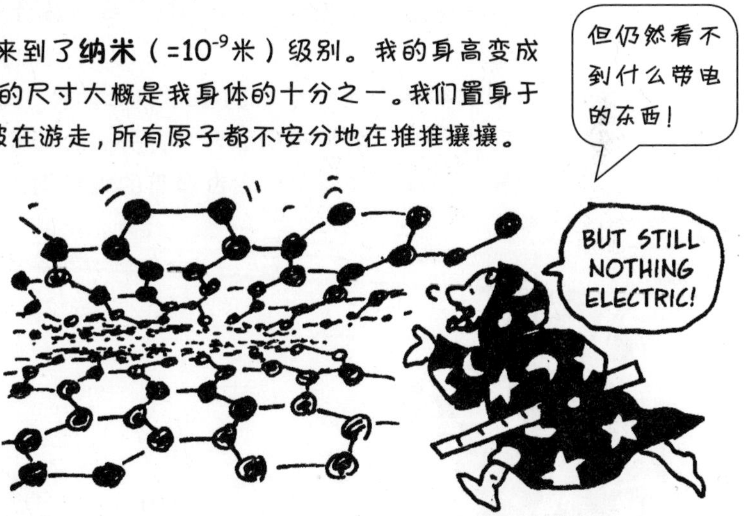

这是存在于铅笔屑中的**石墨**。它的碳原子排列成一层层的薄片，薄片之间很容易滑移。这也是石墨能作为优良润滑剂的缘故。*

*IN PURE FORM. PENCIL LEAD IS A MIXTURE OF GRAPHITE AND CLAY.
*铅笔芯实际上由石墨和黏土组成。

LET'S SHRINK TEN MORE TIMES TO ATOMIC SIZE—10^{-10} METER—AND LOOK AT A SINGLE CARBON ATOM. I CAN VAGUELY SENSE SOME ELECTRONS HUMMING AROUND, ALTHOUGH THEY'RE AWFULLY HARD TO PIN DOWN.

再缩小10倍，就达到了原子的尺寸——10^{-10}米。我们把焦点放在一个单独的碳原子上，可以模糊地看到一团像云雾般的电子，但依然很难分辨彼此。

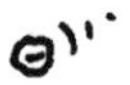

NOW I'M A HUNDRED TIMES SMALLER, AT **PICOMETER** SCALE. THAT'S A MILLIONTH OF A MILLIONTH, OR 10^{-12} ACTUAL SIZE. THERE AT LAST ARE THE POSITIVE CHARGES, ALL LUMPED TOGETHER AT THE VERY CENTER OF THE ATOM IN A TINY CORE OR **NUCLEUS.** IF THE DIAMETER OF THE ATOM WERE THE LENGTH OF A FOOTBALL FIELD, THEN THE NUCLEUS WOULD BE SMALLER THAN A PEA. THE ATOM IS MOSTLY EMPTY SPACE!

现在，我又缩小了100倍，处于**皮米**级。这相当于平常的一百万分之一的一百万分之一，记为10^{-12}。我们终于找到了正电荷的所在，它们拥挤在原子中心的小小核心部位，称为**原子核**。如果一个原子的直径像一个足球场那么大，那么，原子核比一粒豌豆还要小。整个原子基本上是一个什么也没有的纯空间！

ORDINARILY, THE CARBON NUCLEUS CONSISTS OF TWELVE PARTICLES: SIX **PROTONS** WITH A POSITIVE CHARGE AND SIX **NEUTRONS** WITH NO CHARGE AT ALL. THE PROTONS' CHARGE IS BALANCED BY THE SIX HOVERING NEGATIVE ELECTRONS, SO THE ATOM IS NEUTRAL OVERALL.

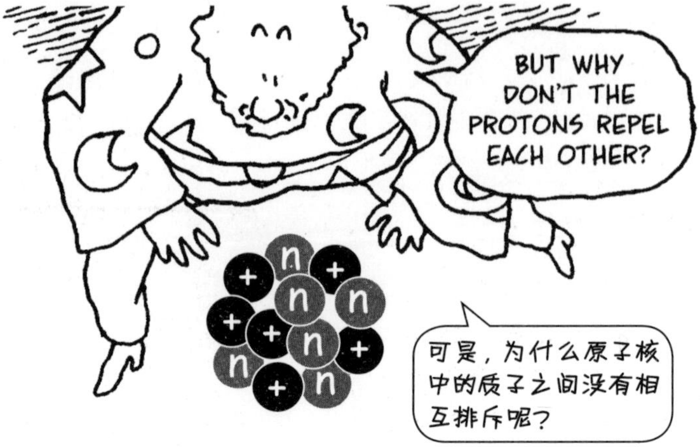

通常，碳原子核由12个粒子组成：6个带正电的**质子**和6个不带电的**中子**。质子所带的正电荷被6颗徘徊在周围的带负电的电子所抵消，因此整个原子是电中性的。

THE NUCLEUS IS HELD TOGETHER BY A POWERFUL, SHORT-RANGE ATTRACTION CALLED THE **STRONG FORCE,*** WHICH OVERCOMES ELECTRICAL REPULSION. THIS INTENSE PULL MAKES MOST NUCLEI VIRTUALLY INDESTRUCTIBLE. THIS VERY SAME CARBON ATOM HAS BEEN ROAMING THE EARTH FOR BILLIONS OF YEARS.

原子核被一种叫作**强相互作用***的高能、短程吸引力结合在一起。这种力能克服静电排斥力。这一强烈吸引力的存在使得几乎所有原子核都坚不可摧。这种碳原子已经在地球上漫游几十亿年了。

*SCIENTISTS DON'T INVENT NEARLY SUCH COLORFUL NAMES AS THEY USED TO.

*科学家们没有像过去那样，给它们起花里胡哨的名字。

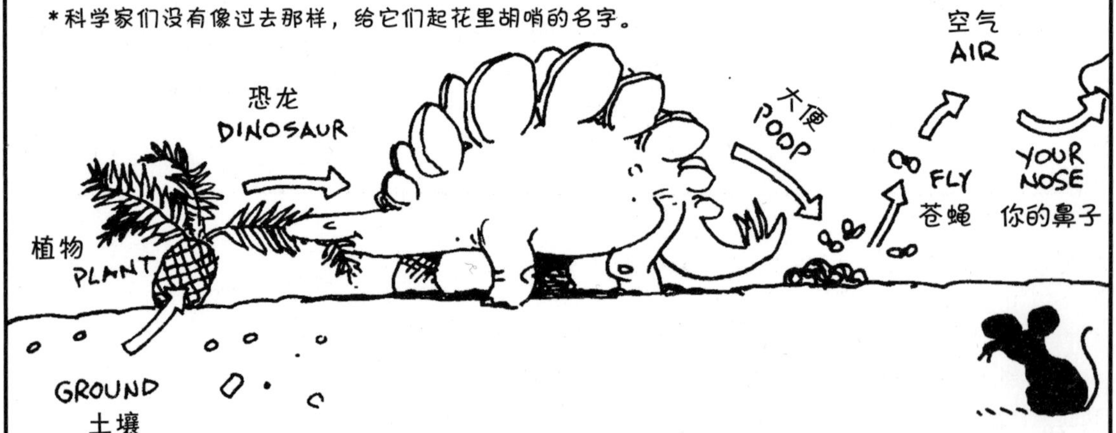

NEARLY ALL THE ATOM'S MASS IS CONCENTRATED IN THE TINY NUCLEUS. EACH PROTON AND NEUTRON (THEY HAVE ALMOST EXACTLY THE SAME WEIGHT) HAS 1840 TIMES THE MASS OF AN ELECTRON.

一个原子的绝大部分质量都集中在小小的原子核处。一个质子或中子的质量都是一个电子的1840倍（原子和中子的质量相当）。

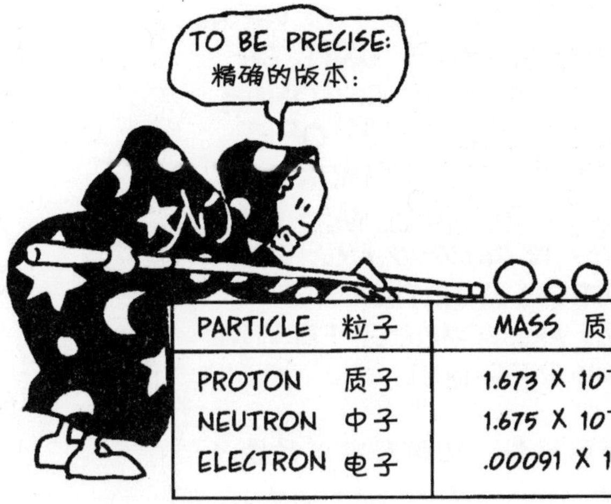

PARTICLE 粒子	MASS 质量
PROTON 质子	1.673×10^{-24}g
NEUTRON 中子	1.675×10^{-24}g
ELECTRON 电子	$.00091 \times 10^{-24}$g

NOW FOR A FEW HELPFUL

definitions:

AN ELEMENT'S **ATOMIC NUMBER** IS THE NUMBER OF PROTONS IN ITS NUCLEUS. CARBON'S ATOMIC NUMBER IS 6.

下面是一些有用的**定义**：

一个元素的**原子序数**是该元素原子核中所含质子的数量。例如，碳的原子序数为6。

ALMOST 99% OF ALL CARBON ATOMS ON EARTH HAVE SIX NEUTRONS ALONG WITH THEIR SIX PROTONS. WE CALL THIS CARBON-12 (AND SOMETIMES WRITE ^{12}C), SINCE ITS MASS IS SO CLOSE TO THAT OF TWELVE NUCLEAR PARTICLES.

地球上近99%的碳原子除含有6个质子以外，还含有6个中子。因此，我们把它们称为碳-12（有时也写成^{12}C），因为它的质量近似于12个核粒子。

MORE PRECISELY, CHEMISTS DEFINE AN **ATOMIC MASS UNIT,** OR **AMU,** TO BE PRECISELY **ONE-TWELFTH THE MASS OF A ^{12}C ATOM.** THE COMMON CARBON ATOM HAS A MASS OF EXACTLY 12.000000 AMU, BY DEFINITION. ALL OTHER ATOMIC MASSES ARE COMPUTED RELATIVE TO THIS REFERENCE.

严格地说，化学家定义了**原子质量单位**(Atomic Mass Unit, AMU)，它等于^{12}C**原子质量的十二分之一**。于是根据定义，常见的碳原子质量就精确地等于12.000000 AMU。所有其他的原子质量都参照这一标准来换算。

THE OTHER 1.1% OF CARBON ATOMS HAVE SEVEN NEUTRONS. THERE MUST STILL BE SIX PROTONS (OTHERWISE IT'S NOT CARBON!), BUT THIS **CARBON-13** ATOM WEIGHS APPRECIABLY MORE THAN CARBON-12.

另外还有1.1%的碳原子含有7个中子，同时它必然含有6个质子（否则它就不是碳原子了！），但这个**碳**-13原子就要比**碳**-12原子重一点。

^{12}C, ^{13}C, AND A VERY RARE FORM, ^{14}C, WITH EIGHT NEUTRONS, ARE CALLED **ISOTOPES** OF CARBON. THE ISOTOPES OF AN ELEMENT HAVE THE SAME NUMBER OF PROTONS, BUT DIFFERENT NUMBERS OF NEUTRONS.

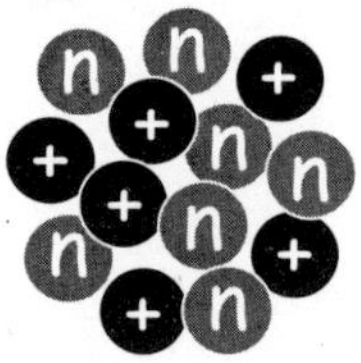

^{13}C NUCLEUS

^{13}C原子核

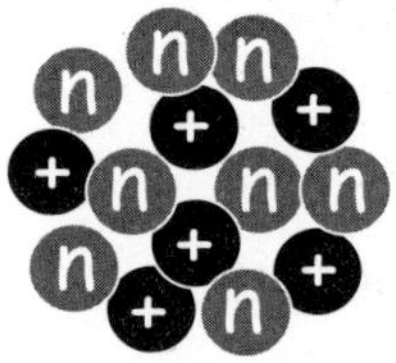

^{14}C NUCLEUS

^{14}C原子核

^{12}C、^{13}C，还有极其罕见的^{14}C（含有8个中子），它们统称为碳的**同位素**。一个元素的同位素含有相同数量的质子和不同数量的中子。

THE SIMPLEST ATOM OF ALL IS **HYDROGEN,** SYMBOL H, WITH AN ATOMIC NUMBER OF ONE. IN NEARLY ALL HYDROGEN ATOMS, A SINGLE ELECTRON ORBITS A SINGLE PROTON, BUT ISOTOPES WITH ONE AND TWO NEUTRONS ALSO EXIST.

最简单的原子是**氢原子**，符号为H，原子序数是1。绝大部分氢原子除了一个质子和一个电子轨道之外什么都没有，但是也存在含有一个或两个中子的氢同位素。

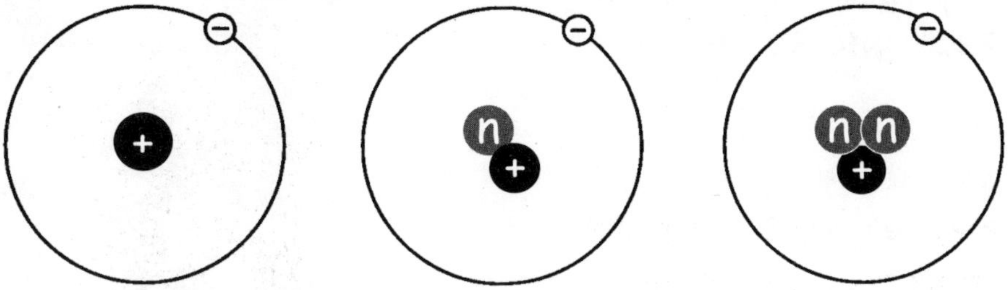

^{1}H　　^{2}H ("DEUTERIUM")　　^{3}H ("TRITIUM")

^{1}H　　^{2}H（氘）　　^{3}H（氚）

ANOTHER FAMILIAR ELEMENT IS **OXYGEN,** SYMBOL O. ITS ATOMIC NUMBER IS 8. ITS MOST COMMON ISOTOPE HAS EIGHT NEUTRONS FOR AN ATOMIC WEIGHT OF APPROXIMATELY 16.* OTHER ISOTOPES INCLUDE ^{17}O AND ^{18}O.

另一个为人所熟悉的元素是**氧**，符号是O，原子序数是8。它最常见的同位素含有8个中子，原子质量约为16*。其他同位素包括^{17}O和^{18}O。

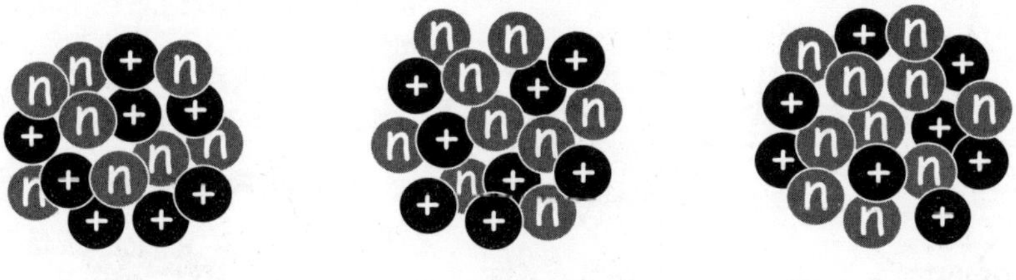

^{16}O NUCLEUS　　^{17}O NUCLEUS　　^{18}O NUCLEUS

^{16}O原子核　　^{17}O原子核　　^{18}O原子核

*THE ACTUAL MASS OF ^{16}O IS 15.9949 AMU. THE "MISSING MASS" IS CONVERTED TO THE **ENERGY** OF THE STRONG FORCE THAT BINDS THE NUCLEUS TOGETHER. OTHER ATOMS HAVE SIMILAR FRACTIONAL WEIGHTS.

*^{16}O的实际原子质量是15.9949AMU。"丢失的原子质量"被转换成强相互作用的**能量**以便原子核聚在一起。其他原子也具有类似的情况，只占有部分重量。

NOW YOU MIGHT ASK, IF EVERY ELEMENT HAS AN ATOMIC NUMBER, DOES EVERY NUMBER HAVE AN ELEMENT? IS THERE AN ELEMENT WITH 37 PROTONS? 52? 92?

这时你可能要问，如果所有的元素都有一个原子序数，那是否每个数字都对应着一种元素呢？有没有一种元素含37、52或92个质子呢？

NATURE, IT TURNS OUT, MAKES ATOMS WITH EVERY NUMBER FROM 1 (HYDROGEN) TO 92 (URANIUM), ALTHOUGH A FEW ELEMENTS IN THERE ARE VERY SCARCE.

大自然果然是从1（氢）到92（铀）为每个数字都准备了一种元素，尽管其中有些元素非常稀少。

THE SEQUENCE STOPS THERE BECAUSE LARGE NUCLEI (THOSE ABOVE 83, BISMUTH) ARE UNSTABLE. BEYOND URANIUM, 92, THEY FALL APART SO QUICKLY THAT WE DON'T SEE THEM IN NATURE. PHYSICISTS CAN MAKE NUCLEI WITH MORE THAN 92 PROTONS, BUT THEY DON'T SURVIVE LONG.

元素排到92号之后就差不多了，因为过大的原子核是不稳定的（那些大于83号，铋）。超过92号铀以上的元素，原子分解太快了，因此我们很难看到它们自然存在。物理学家可以人工造出含有超过92个质子的原子核，但它们都不能长期存在。

HERE IS A LIST OF THE 92 NATURALLY OCCURRING ELEMENTS:

以下是92种自然存在的元素的列表：

1. Hydrogen, H 氢，H
2. Helium, He 氦，He
3. Lithium, Li 锂，Li
4. Beryllium, Be 铍，Be
5. Boron, B 硼，B
6. Carbon, C 碳，C
7. Nitrogen, N 氮，N
8. Oxygen, O 氧，O
9. Fluorine, F 氟，F
10. Neon, Ne 氖，Ne
11. Sodium, Na 钠，Na
12. Magnesium, Mg 镁，Mg
13. Aluminum, Al 铝，Al
14. Silicon, Si 硅，Si
15. Phosphorus, P 磷，P
16. Sulfur, S 硫，S
17. Chlorine, Cl 氯，Cl
18. Argon, Ar 氩，Ar
19. Potassium, K 钾，K
20. Calcium, Ca 钙，Ca
21. Scandium, Sc 钪，Sc
22. Titanium, Ti 钛，Ti
23. Vanadium, V 钒，V
24. Chromium, Cr 铬，Cr
25. Manganese, Mn 锰，Mn
26. Iron, Fe 铁，Fe
27. Cobalt, Co 钴，Co
28. Nickel, Ni 镍，Ni
29. Copper, Cu 铜，Cu
30. Zinc, Zn 锌，Zn
31. Gallium, Ga 镓，Ga
32. Germanium, Ge 锗，Ge
33. Arsenic, As 砷，As
34. Selenium, Se 硒，Se
35. Bromine, Br 溴，Br
36. Krypton, Kr 氪，Kr
37. Rubidium, Rb 铷，Rb
38. Strontium, Sr 锶，Sr
39. Yttrium, Y 钇，Y
40. Zirconium, Zr 锆，Zr
41. Niobium, Nb 铌，Nb
42. Molybdenum, Mo 钼，Mo
43. Technetium, Tc 锝，Tc
44. Ruthenium, Ru 钌，Ru
45. Rhodium, Rh 铑，Rh
46. Palladium, Pd 钯，Pd
47. Silver, Ag 银，Ag
48. Cadmium, Cd 镉，Cd
49. Indium, In 铟，In
50. Tin, Sn 锡，Sn
51. Antimony, Sb 锑，Sb
52. Tellurium, Te 碲，Te
53. Iodine, I 碘，I
54. Xenon, Xe 氙，Xe
55. Cesium, Cs 铯，Cs
56. Barium, Ba 钡，Ba
57. Lanthanum, La 镧，La
58-71—Never mind these! 58-71——先别管了！
72. Hafnium, Hf 铪，Hf
73. Tantalum, Ta 钽，Ta
74. Tungsten, W 钨，W
75. Rhenium, Re 铼，Re
76. Osmium, Os 锇，Os
77. Iridium, Ir 铱，Ir
78. Platinum, Pt 铂，Pt
79. Gold, Au 金，Au
80. Mercury, Hg 汞，Hg
81. Thallium, Tl 铊，Tl
82. Lead, Pb 铅，Pb
83. Bismuth, Bi 铋，Bi
84. Polonium, Po 钋，Po
85. Astatine, At 砹，At
86. Radon, Rn 氡，Rn
87. Francium, Fr 钫，Fr
88. Radium, Ra 镭，Ra
89. Actinium, Ac 锕，Ac
90. Thorium, Th 钍，Th
91. Protactinium, Pa 镤，Pa
92. Uranium, U 铀，U

(93, 94, AND ABOVE ARE ARTIFICIAL AND UNSTABLE.)

（93号、94号及以上都是人造元素而且很不稳定）

The Elusive Electron
难以捉摸的电子

TO TURN THAT RATHER STARK LIST INTO A PERIODIC TABLE—FOR THAT IS OUR GOAL—WE NOW TURN TO THE ATOM'S OTHER MAIN INGREDIENT, ITS ELECTRONS. THESE, WE SHOULD WARN YOU, DEFY COMMON SENSE, BECAUSE ELECTRONS, YOU SEE, OBEY THE BIZARRE RULES OF MODERN PHYSICS CALLED **QUANTUM MECHANICS.**

我们的目标是将这个晦涩难懂的列表变成一个有规律的表格。因此，现在我们要研究原子的另一个要素——电子。事先警告一下，这玩意儿的性子违反人的常识，因为电子遵循的是奇妙的现代物理学说——**量子力学**。

WRAP YOUR MIND AROUND THIS: AN ELECTRON IS A **PARTICLE,** LIKE A MARBLE, BUT ALSO A **WAVE,** LIKE A BEAM OF LIGHT. AS A PARTICLE, IT HAS A DEFINITE **MASS, CHARGE,** AND **SPIN,** BUT IT ALSO HAS A **WAVELENGTH.** IT'S "SMEARED OUT" IN SOME WAY. ITS PRECISE POSITION IS ALWAYS A BIT UNCERTAIN. MAKE SENSE? WE DIDN'T THINK SO!

打起精神来，试着理解以下这些话：一个电子是一个**粒子**——就像一颗弹珠，但又是一束**波**，就好像一道光。作为一个粒子，它有确定的**质量**、**电荷**和**自旋**，但它同时又有一定的**波长**。它的存在就好像被抹淡了一样，总是让人弄不清其确切位置所在。这个合理吗？我们可不这么认为。

IN ITS GUISE AS A PARTICLE, AN ELECTRON INHABITS A SORT OF "PROBABILITY CLOUD"—**NOT** A CIRCULAR ORBIT. THE DENSEST PARTS OF THE CLOUD ARE WHERE THE ELECTRON IS LIKELIEST TO "BE"—IF IT CAN BE SAID TO BE ANYWHERE, WHICH IT CAN'T EXACTLY. THESE CLOUDS NEED NOT BE ROUND, BY THE WAY.

在"粒子"的伪装下，电子存在于一个"概率云"当中，**而不是**在一个圆形的轨道中。如果说电子可以出现在任何地方，那么，概率云最密集的部分就是它最可能存在的地方，但也不可能非常准确地定位它。另外，这个"概率云"也可能不是圆的。

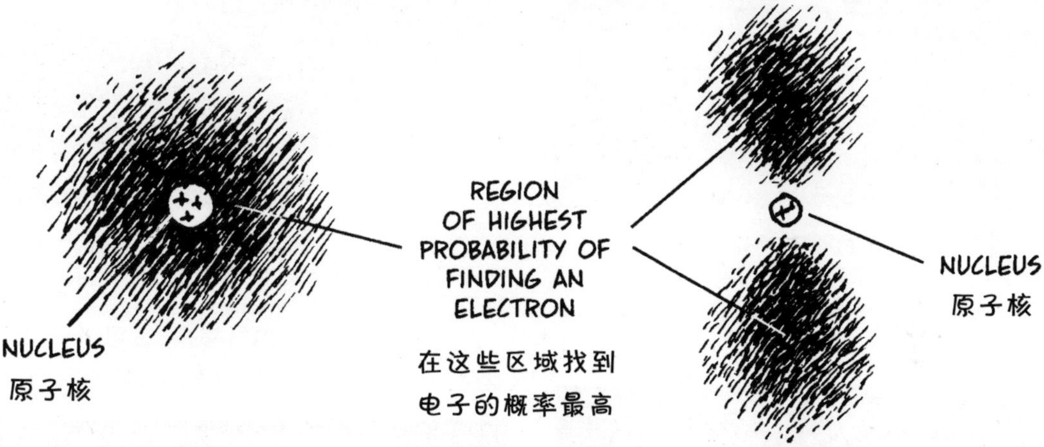

WE CAN ALSO VISUALIZE THE ELECTRON AS A WAVE, BEAMING AROUND THE NUCLEUS. IN THIS PICTURE, QUANTUM MECHANICS TELLS US THAT THE ELECTRON IS ALWAYS A "STANDING WAVE." THAT IS, IT "GOES AROUND" THE NUCLEUS A **WHOLE NUMBER OF WAVELENGTHS**: 1, 2, 3, 4, ETC., BUT NEVER A FRACTIONAL VALUE.

我们也可以把电子看成是闪耀在原子核周围的一束波。如果这样看的话，根据量子力学，电子总是一个"驻波"。即它"围绕"着原子核显示**整数**波长：1、2、3、4等，永远不会有分数值。

IN OTHER WORDS, ONLY CERTAIN DISCRETE "ORBITS" ARE AVAILABLE TO AN ELECTRON IN AN ATOM.

换句话说，对于一个原子里的电子来说，它只能出现在某些特定的离散"轨道"上。

LET'S CONTRAST THIS WITH A MORE FAMILIAR SYSTEM: A PLANET ORBITING A STAR.

让我们拿这个与一个更熟悉的系统对比一下：行星绕着恒星运转。

IMAGINE THAT SOMETHING GIVES THE PLANET A NUDGE, ADDING ENERGY TO IT.

想象一下：有人捅了一下这个行星，给其增加了能量。

THE EXTRA ENERGY PUSHES THE PLANET INTO AN ORBIT FARTHER FROM THE STAR.

额外的能量将行星推到离恒星更远的轨道上去。

IN FACT, WITH A BIG ENOUGH JOLT, THE PLANET WILL ESCAPE THE STAR'S GRAVITATIONAL PULL COMPLETELY.

实际上,如果你力气够大,你可以一棍子把这个行星打飞，让其彻底离开恒星的引力。

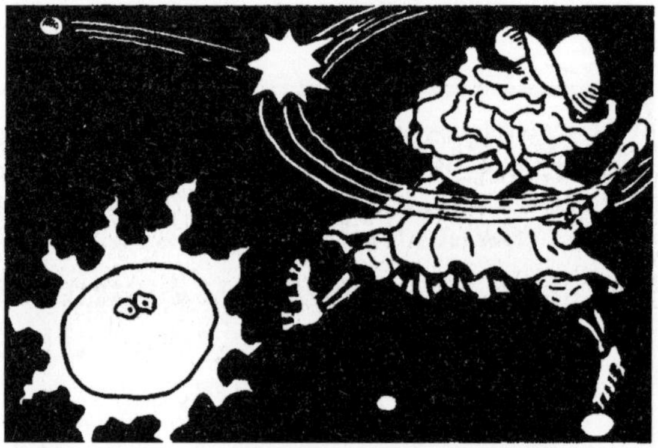

AN ORBITING ELECTRON IS SIMILAR: IT MAY ABSORB A JOLT OF ENERGY, TOO, IN THE FORM OF A BEAM OF LIGHT, FOR EXAMPLE.

电子的轨道也差不多：它也可以吸收一个能量波，例如，一个形式为光束的能量波。

BUT THE ELECTRON MUST JUMP TO AN ORBIT CONSISTENT WITH A WHOLE NUMBER OF WAVELENGTHS.

但是电子必须跳到另一个拥有连续的整数个波长的轨道上去。

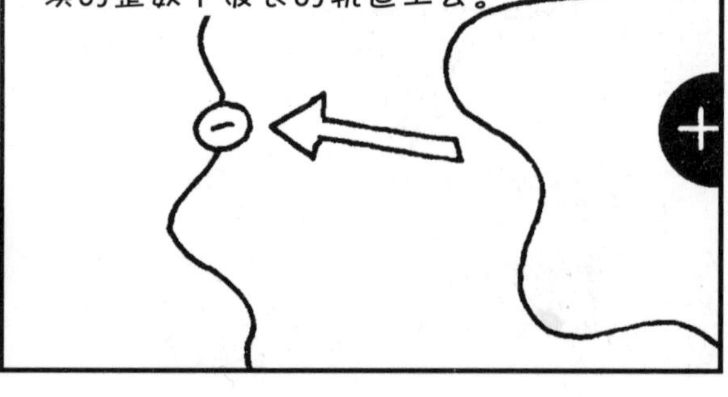

THIS MEANS IT CAN ABSORB ONLY CERTAIN **FIXED AMOUNTS OF ENERGY:** JUST ENOUGH TO JUMP THE ELECTRON TO ONE OF THE HIGHER AVAILABLE ORBITS. UNLIKE A PLANET, WHICH CAN ABSORB ENERGY GRADUALLY AND ORBIT AT ANY DISTANCE, AN ELECTRON'S ENERGY IS LIMITED TO CERTAIN VALUES.

这意味着电子只能吸收某些**固定大小的能量**：不多不少，恰好足够其跳到另一个更高能量的轨道上去。行星则不同，它可以逐步地吸收能量，在任意距离的轨道上绕行，而电子的能量只能取有限的几个值。

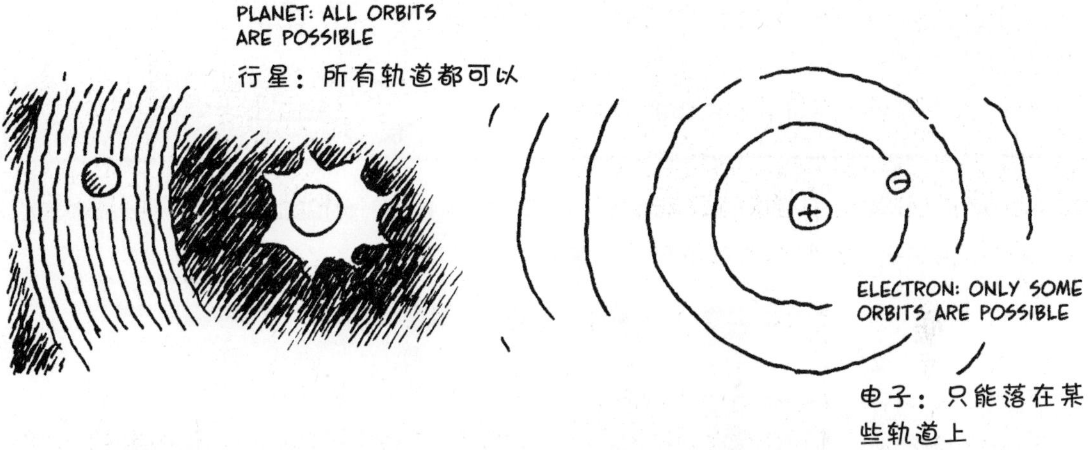

WE SAY THE ELECTRON'S ENERGY IS **QUANTIZED:** IN ANY GIVEN ATOM, THE ELECTRONS CAN ASSUME ONLY CERTAIN FIXED, DISCRETE ENERGY LEVELS.

因此，我们说电子的能量是**量子化**的：在任意给定的原子中，电子只能取某些固定的、离散的能级。

THE ELECTRON CONFIGURATIONS WITHIN EACH ENERGY LEVEL ARE CALLED **ORBITALS** (NAMED, NO DOUBT, BY NOSTALGIC PHYSICISTS DREAMING OF PLANETS).

电子在每个能级中的存在方式就称为**轨道**（这当然是那些满脑子想着行星的物理学家们起的名字）。

THE SIMPLEST EXAMPLE IS **HYDROGEN:** ONE ELECTRON PULLED BY A SINGLE PROTON. THE ELECTRON CAN INHABIT ANY ONE OF SEVEN DIFFERENT LEVELS, OR "SHELLS," MISLEADINGLY DEPICTED HERE AS CIRCULAR ORBITS.

最简单的例子是**氢**：一个电子受到一个质子的引力作用。电子能够待在7个不同的能级或“壳层”的任何一级上。上图把这些能级画成精确而均匀的圆形轨道其实有些误导读者。

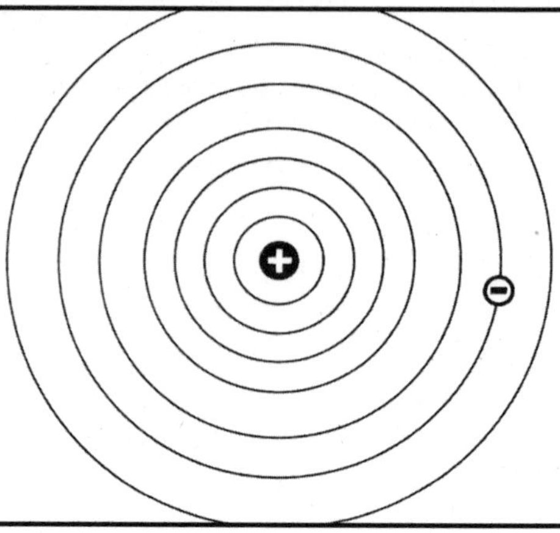

THIS GRAPH SHOWS THE ELECTRON'S ENERGY IN EACH SHELL.

下图给出了电子在每一能级上所带的能量。

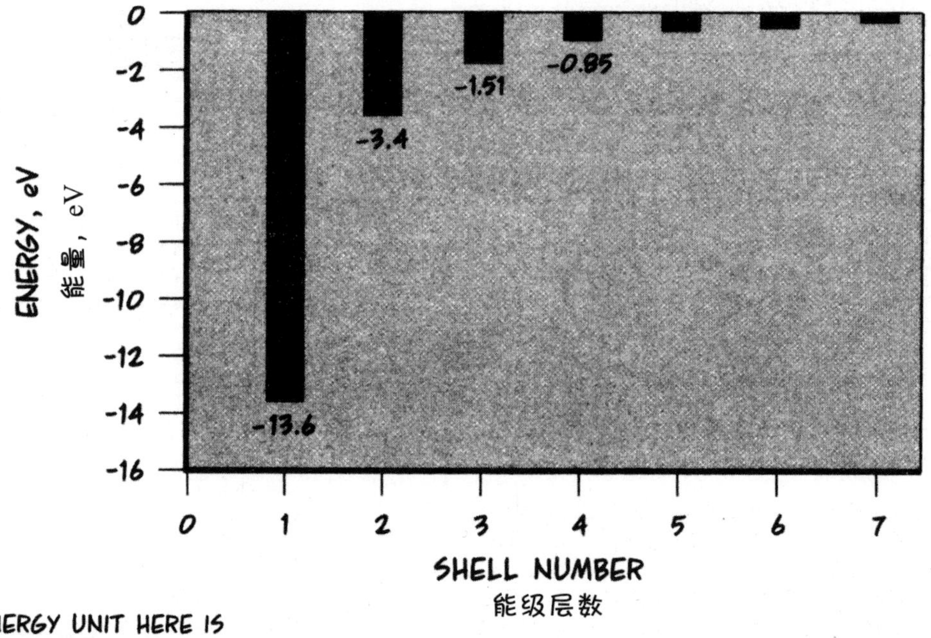

THE ENERGY UNIT HERE IS THE **ELECTRON VOLT** (eV). ONE eV IS THE ENERGY GAINED BY ONE ELECTRON PUSHED BY ONE VOLT. (NOTE: IN ATOMS, AN ELECTRON'S ENERGY IS NEGATIVE, SINCE ENERGY MUST BE ADDED TO PULL THE ELECTRON FREE OF THE NUCLEUS. THE FREE STATE IS TAKEN TO HAVE ENERGY = 0.)

图中能量的单位是**电子伏**（eV）。1电子伏的能量等于1个电子跨越1伏电势差所需的能量。（注意：在原子里面，电子所带的能量是负的，因为要把电子从原子核上拉开需要付出能量。同时，完全不受束缚状态的电子能量为0。）

TO RAISE AN ELECTRON FROM SHELL 1 TO SHELL 2 REQUIRES AN ENERGY EQUAL TO THE DIFFERENCE (-3.4) – (-13.6) = 13.6 – 3.4 = **10.2** eV.

要把一个电子从第1层搬到第2层，需要的能量是(-3.4) - (-13.6) = 13.6 - 3.4 = 10.2电子伏。

TO REMOVE THE ELECTRON COMPLETELY AND MAKE A HYDROGEN ION REQUIRES **13.6** eV. THIS IS CALLED THE ATOM'S **IONIZATION ENERGY.**

同理，要完全地把电子移走，只剩下一个氢离子，需要13.6电子伏。这叫作这个原子的**电离能**。

* 西方习惯，有人打喷嚏之后，其他人要对他说“祝你健康”，而英语中的“打喷嚏”（sneeze）的发音与“电离”（ionization）的发音近似。

LARGER ATOMS, LIKE HELIUM, LITHIUM, OR TIN, ALSO HAVE UP TO SEVEN ELECTRON SHELLS. BUT IN THESE ATOMS, THE "HIGHER" SHELLS CAN HOLD MORE ELECTRONS THAN LOWER SHELLS CAN.

大一点儿的原子，例如氦、锂或锡，也具有7个电子层。但是在这些原子里，高层可以容纳的电子数比低层要多。

HIGHER-SHELL ELECTRONS CAN ALSO HAVE MORE COMPLEX CONFIGURATIONS, OR **ORBITALS,** THAN LOWER-SHELL ELECTRONS. YOU CAN THINK OF THESE ORBITALS AS ENERGY SUBLEVELS. DIFFERENT SUBLEVELS ARE CALLED s, p, d, AND f, AND EACH ORBITAL CAN HOLD **UP TO TWO ELECTRONS.**

处于高层的电子还有比低层更复杂的结构，称为**原子轨道**。可以把这些轨道想象成一个大层下面的亚层，分别记为s、p、d和f轨道。每个轨道最多能容纳**两个电子**。

SHELL 1 HAS ONLY AN s ORBITAL, WHICH IS SPHERICAL. IT CAN HOLD ONE OR TWO ELECTRONS.

第一层只有一个s轨道，是圆形的，能容纳一个或两个电子。

SHELL 2 HAS ONE s AND THREE p ORBITALS, WHICH LOOK SOMETHING LIKE DUMBBELLS. WHEN FULL, THIS SHELL HOLDS EIGHT ELECTRONS.

第二层有一个s和三个p轨道。p轨道的形状像哑铃一样。全满的时候，这一层最多能容纳8个电子。

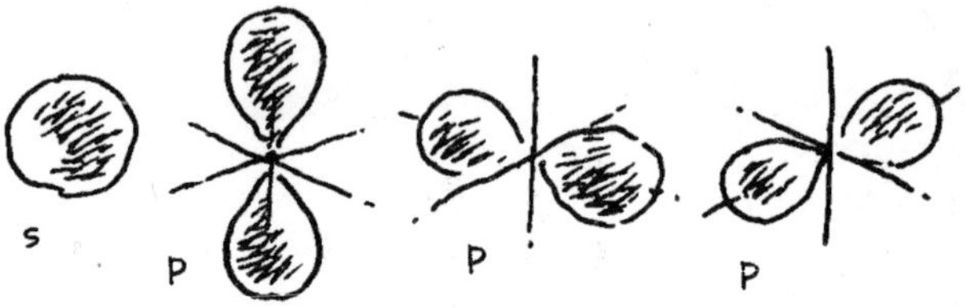

SHELL 3 HAS ONE s, THREE p, AND FIVE d ORBITALS (FORGET DRAWING THEM ALL!). WHEN FULL, IT HOLDS 18 ELECTRONS (2 X [1 + 3 + 5]).

第三层有一个s轨道、三个p轨道和5个d轨道（咱就不都画出来了！）。全满的时候可容纳18个电子（2×[1+3+5]）。

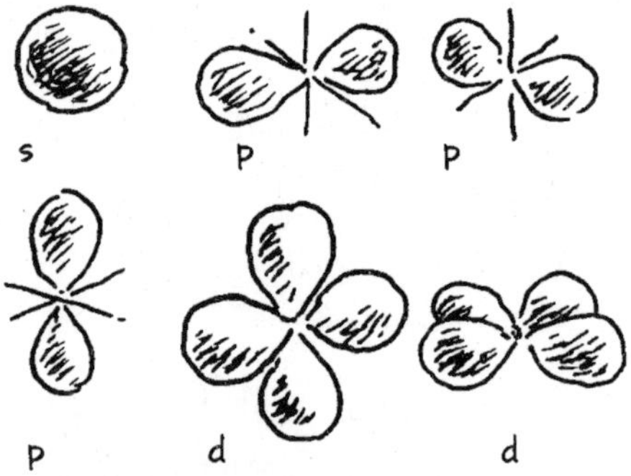

AND THREE MORE d ORBITALS

另外还有3个d轨道

SHELLS 4 AND HIGHER HAVE ALL OF THAT PLUS SEVEN f ORBITALS—UP TO 32 ELECTRONS TOTAL.

第四层以及更高层除了含有以上轨道之外，还包括7个f轨道，总共可以容纳32个电子。

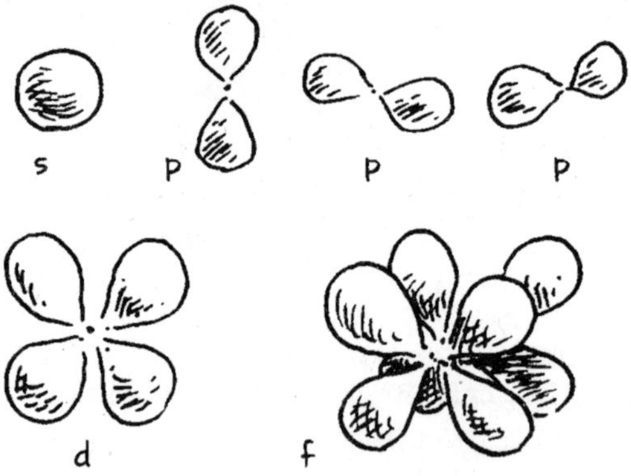

5 OF THESE

5个这样的轨道

7 OF THESE

7个这样的轨道

THIS DIAGRAM SHOWS THE ENERGY LEVELS OF THE DIFFERENT ORBITALS. THE FARTHER UP THE PAGE, THE HIGHER THE ENERGY.

右图按能级高低列出了所有原子轨道，从页底到页顶，能量越来越高。

NOTE THAT THE SHELLS HAVE **OVERLAPPING ENERGIES:** E.G., SOME ORBITALS IN SHELL 4 (4d AND 4f) HAVE HIGHER ENERGY THAN SOME ORBITALS IN SHELL 5 (5s), EVEN THOUGH 4 IS "LOWER" THAN 5.

很容易发现，不同层之间有**能量重叠**的现象。例如，尽管第四层低于第五层，但第四层的两条轨道（4d和4f）比第五层的轨道（5s）能量还要高。

NOTE: 2s MEANS THE s ORBITAL IN SHELL 2, 4d MEANS THE d ORBITAL IN SHELL 4, ETC. EACH ARROW LEADS TO THE ORBITAL WITH THE NEXT-HIGHEST ENERGY.

注意：右图中，2s代表第二层的s轨道，4d代表第四层的d轨道，依此类推。箭头指向下一个更高能量的轨道。

AS WE BUILD UP AN ATOM, EACH ELECTRON "WANTS" TO GO INTO THE LOWEST AVAILABLE ENERGY STATE. WE START AT THE LOWEST, THEN WHEN THAT FILLS UP, GO TO THE NEXT-LOWEST, ETC.

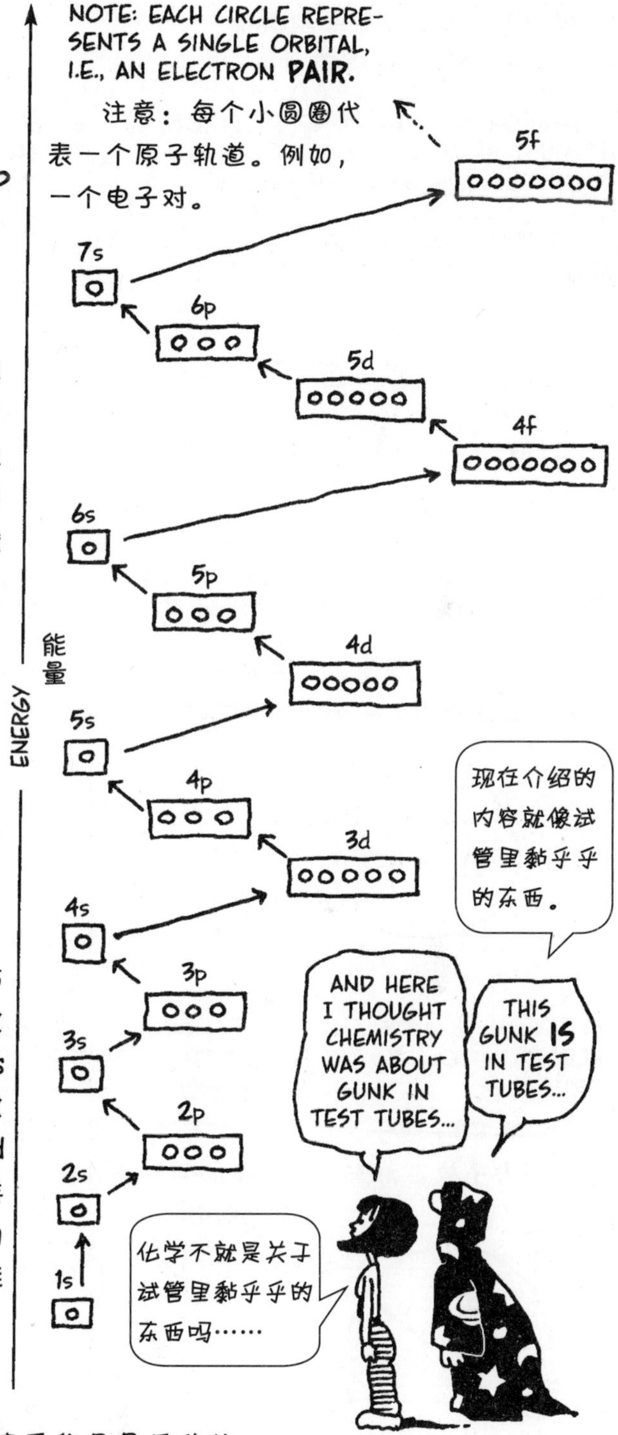

在构造原子的时候，每个电子都偏爱能量最低的状态。因此，我们必须从能量最低的地方开始放电子，当该处放满后就放在倒数第二低的地方，依此类推。

NOW LET'S BUILD SOME ATOMS.
现在可以试着构建一些原子了。

1. HYDROGEN, H, HAS ONE ELECTRON. IT MUST BE IN THE LOWEST SHELL'S s ORBITAL. WE WRITE THIS AS $1s^1$.

1. 氢（H）有一个电子，它必须放在最低层的s轨道上，记为$1s^1$。

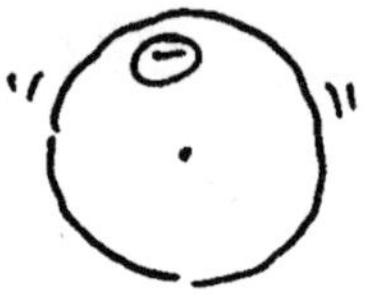

$1s^1$

2. HELIUM, He, ADDS A SECOND ELECTRON TO THIS s ORBITAL. NOW SHELL 1 IS FULL, AND WE WRITE $1s^2$.

2. 氦(He)增加了一个电子,将其放在s轨道上,恰好就把第一层的s轨道填满了,记为$1s^2$。

REMEMBER: TWO ELECTRONS PER ORBITAL, TOPS!

记住：每个原子轨道最多只能有两个电子！

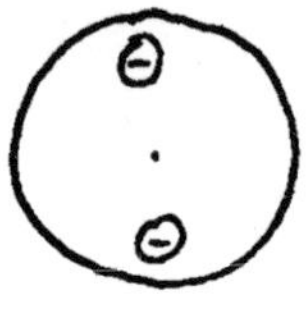

$1s^2$

3. LITHIUM, Li, HAS TO PUT THE THIRD ELECTRON IN A NEW SHELL, SHELL 2.

3. 锂（Li）原子，必须把第三个电子放到新的一层——第二层。

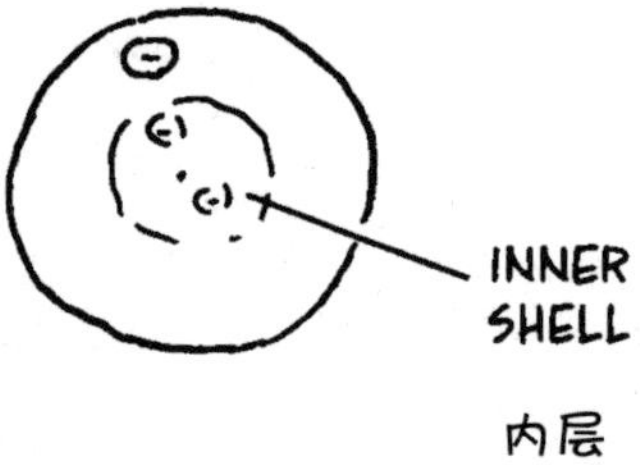

$1s^2 2s^1$

4. BERYLLIUM, Be, COMPLETES THE 2s ORBITAL.

4. 到了铍（Be），2s轨道也满了。

FROM HERE ON, WE OMIT THE INNER SHELL IN THE DRAWING.

往后内层都省略不画了。

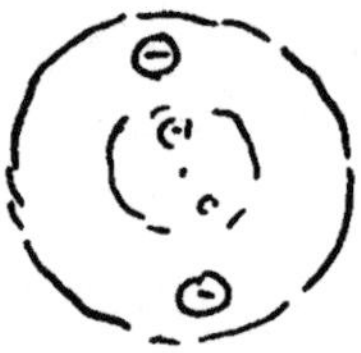

$1s^2 2s^2$

5. BORON, B, ADDS AN ELECTRON TO A 2p ORBITAL.

5. 硼（B）原子，新加的电子放在2p轨道上。

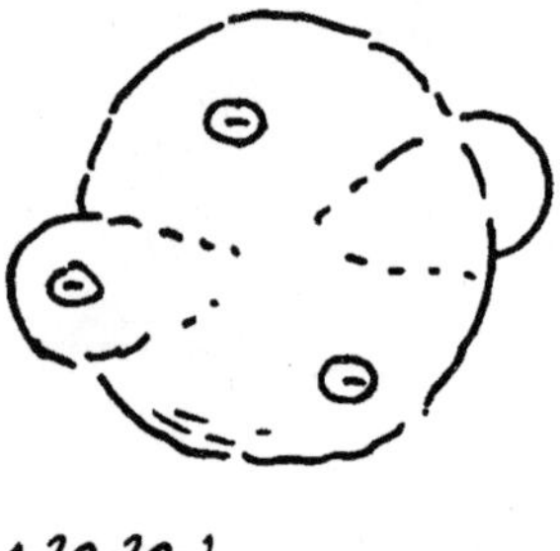

$1s^2 2s^2 2p^1$

6. CARBON, C, ADDS AN ELECTRON TO THE SECOND p ORBITAL.

6. 碳（C）原子，新加的电子放在另一个p轨道。

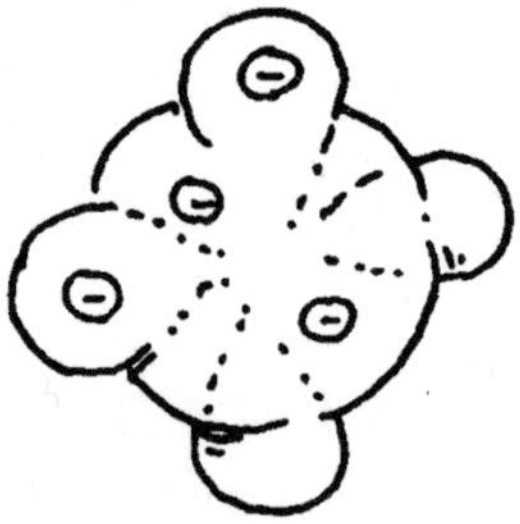

$1s^2 2s^2 2p^2$

7. NITROGEN, N, ADDS AN ELECTRON TO THE THIRD p ORBITAL.

7. 氮（N）原子，再加一个电子到第三个p轨道。

$1s^2 2s^2 2p^3$

8. OXYGEN, O

8. 氧（O）

$1s^22s^22p^4$

9. FLUORINE, F

9. 氟（F）

$1s^22s^22p^5$

10. NEON, Ne, COMPLETES SHELL 2.

10. 氖（Ne），整个第二层都满了。

$1s^22s^22p^6$

TO FIND OUT WHAT HAPPENS IN ELEMENT #11, LOOK AT THE CHART ON p 33. AFTER 2p FILLS UP, THE LOWEST-ENERGY AVAILABLE ORBITAL IS 3s, IN THE THIRD SHELL, FOLLOWED BY 3p. SO WE HAVE:

想了解第11号元素，可以重新参照第43页的图表。2p轨道全填满之后，剩下能量最低的原子轨道就是第三层的3s，然后是3p。我们继续：

11. SODIUM, Na. WE CAN WRITE THIS AS $Ne3s^1$, INDICATING ONE s ELECTRON ORBITING OUTSIDE A GROUP OF ELECTRONS JUST LIKE NEON'S.

11. 钠（Na）原子的电子可以写成$Ne3s^1$，表示在像氖一样的电子群外部存在一个单电子s轨道。

$Ne3s^1$

12. MAGNESIUM, Mg. SIMILARLY, WE CAN WRITE THIS AS $Ne3s^2$.

12. 镁（Mg），类似的，可以写成$Ne3s^2$。

$Ne3s^2$

13. ALUMINUM, Al

13. 铝（Al）

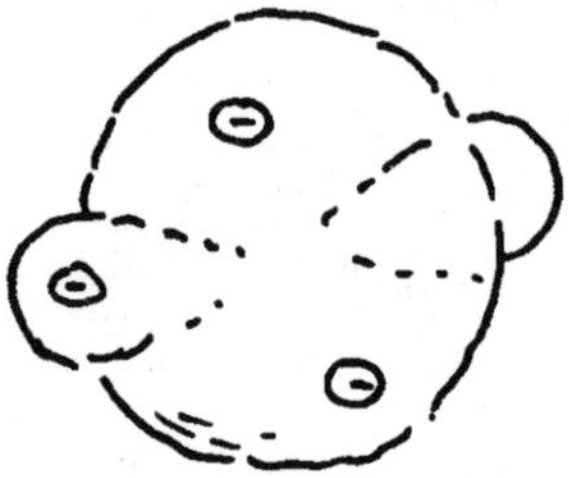

$Ne3s^23p^1$

14. SILICON, Si

14. 硅（Si）

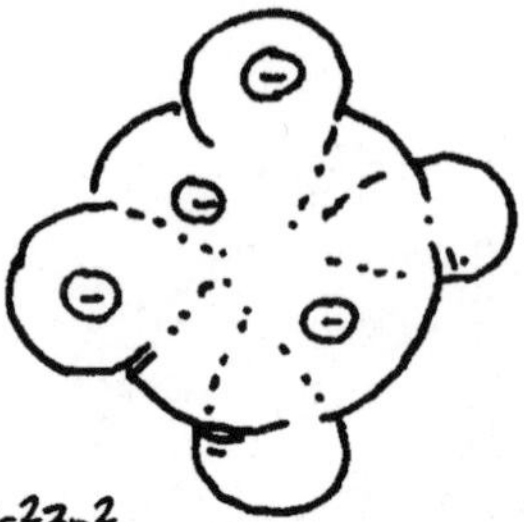

$Ne3s^23p^2$

15. PHOSPHORUS, P

15. 磷（P）

$Ne3s^23p^3$

16. SULFUR, S

16. 硫（S）

$Ne3s^23p^4$

17. CHLORINE, Cl

17. 氯（Cl）

$Ne3s^23p^5$

18. ARGON, Ar

18. 氩（Ar）

$Ne3s^23p^6$

IF YOU COMPARE THESE ATOMS WITH THOSE ON THE PREVIOUS PAGE, YOU WILL SEE THAT ELEMENTS 11-18 ARE LIKE "BIG SISTERS" TO ELEMENTS 3-10. EACH OF THE ATOMS ON THIS PAGE HAS AN **OUTER SHELL** IDENTICAL TO THAT OF THE ATOM JUST EIGHT ELEMENTS BEHIND IT!

比较后不难发现，第11-18号元素是第3-10号元素的放大版。第11-18号元素每个原子的**外层**电子与前8号原子是一样的。

WE WRITE THE FIRST EIGHTEEN ELEMENTS IN A TABLE. IN ANY COLUMN, ALL THE ATOMS HAVE THE SAME OUTER ELECTRON CONFIGURATION.

我们可以把前18号元素列在一个表中，使每列的元素都具有相同的外层电子结构。

1 H							2 He
3 Li	4 Be	5 B	6 C	7 N	8 O	9 F	10 Ne
11 Na	12 Mg	13 Al	14 Si	15 P	16 S	17 Cl	18 Ar

(EXCEPT HELIUM, WHICH GOES IN THE LAST COLUMN BECAUSE ITS OUTER SHELL IS FULL.)

（氦除外，因为它的外层电子是满的，把它放在最边上）

NEXT, ACCORDING TO THE CHART ON P. 43 THE 4s ORBITAL FILLS AS WE BEGIN THE FOURTH ROW OF THE TABLE. NEXT, SAYS THE CHART, ELECTRONS BEGIN TO OCCUPY THE 3d ORBITALS. BEFORE WE CAN CONTINUE IN THE FOURTH SHELL, TEN ELECTRONS MUST GO INTO THESE **INNER** ORBITALS. WE WRITE THESE TEN ELEMENTS ON A LOOP, SINCE WE'RE STALLED FILLING THE FOURTH SHELL.

下一步，根据第43页的图表，接下来先要把4s轨道填满，作为表中第四行的开始，接下来才轮到内层的3d轨道的10个空位，对应着10个不同的元素。填不完这10个空位，就不能继续填第4层的p轨道。因此我们把这10种元素全写在一条纸圈上。

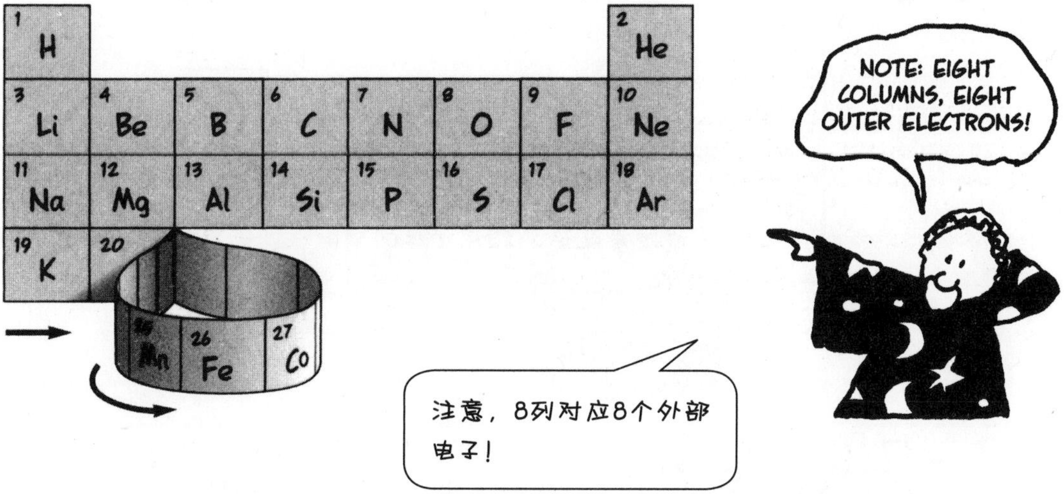

AFTER THOSE TEN, WE CAN RESUME PUTTING ELECTRONS IN THE FOURTH SHELL, UNTIL ALL THE 4s AND 4p ORBITALS ARE FULL AT ELEMENT 36, KRYPTON, Kr.

写完这10个元素之后，就可以继续往第4层填电子了，一直到所有4s和4p轨道全填满，来到第36号的氪（Kr）为止。

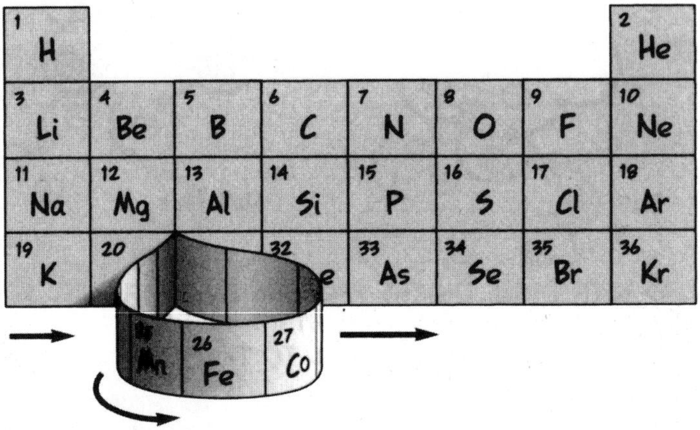

AGAIN, WITHIN EACH COLUMN THAT LIES "FLAT ON THE PAGE," ATOMS HAVE OUTER SHELLS THAT LOOK THE SAME.

填完第4行之后就会发现，表格每列元素仍然保持外层电子数相同的规律。

THE FIFTH ROW FILLS UP IN EXACTLY THE SAME WAY AS THE FOURTH: FIRST THE OUTER s, THEN THE INNER d, THEN THE OUTER p.

第五行元素的排列方法与第4行一样，首先是外层s轨道，再到内层d轨道，最后轮到外层的p轨道。

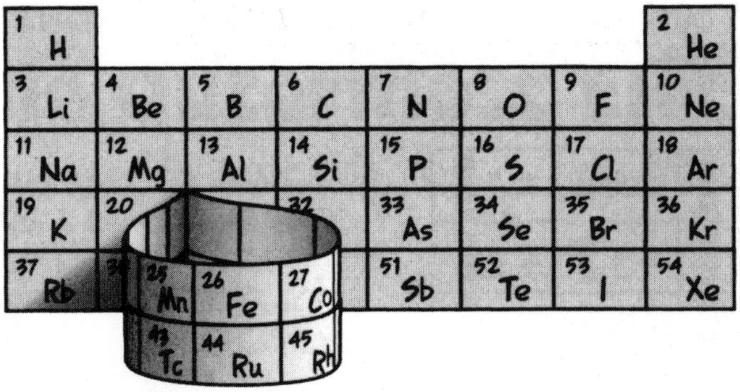

THE ELEMENTS THAT ARE "FLAT ON THE PAGE" ARE CALLED **MAIN-GROUP ELEMENTS.** THOSE IN THE LOOPS ARE CALLED **TRANSITION METALS.**

我们把正式填在表格平面上的元素称为**主族元素**，那些写在纸圈上的则称为**过渡金属**。

THE SIXTH ROW HAS A LOOP WITHIN A LOOP, AS 4f ORBITALS FILL BEFORE 5d. (SEE P. 43!) AS THERE ARE SEVEN 4f ORBITALS, THIS LOOP HAS 14 ELEMENTS. IT IS CALLED THE **LANTHANIDE SERIES,** AFTER ITS FIRST ELEMENT, LANTHANUM.

第六行是纸圈套纸圈，因为在填5d轨道前必须先填满4f轨道（参考第43页！），这个纸圈有7个4f轨道共14个元素。以这一系列的第一个元素镧命名，称为**镧系元素**。

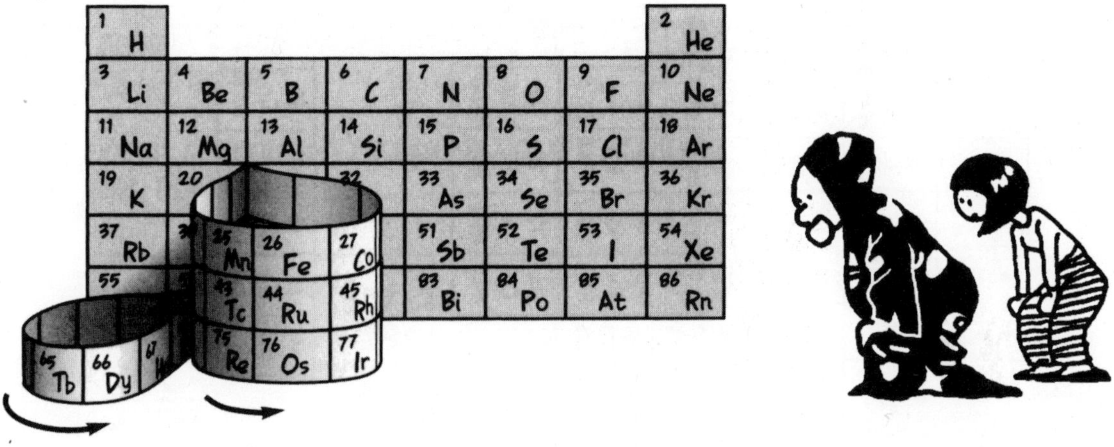

THE SEVENTH ROW PETERS OUT WHEN WE RUN OUT OF ELEMENTS.

第7行与第6行填法相同，但只能填到铀为止，填不完1整行。

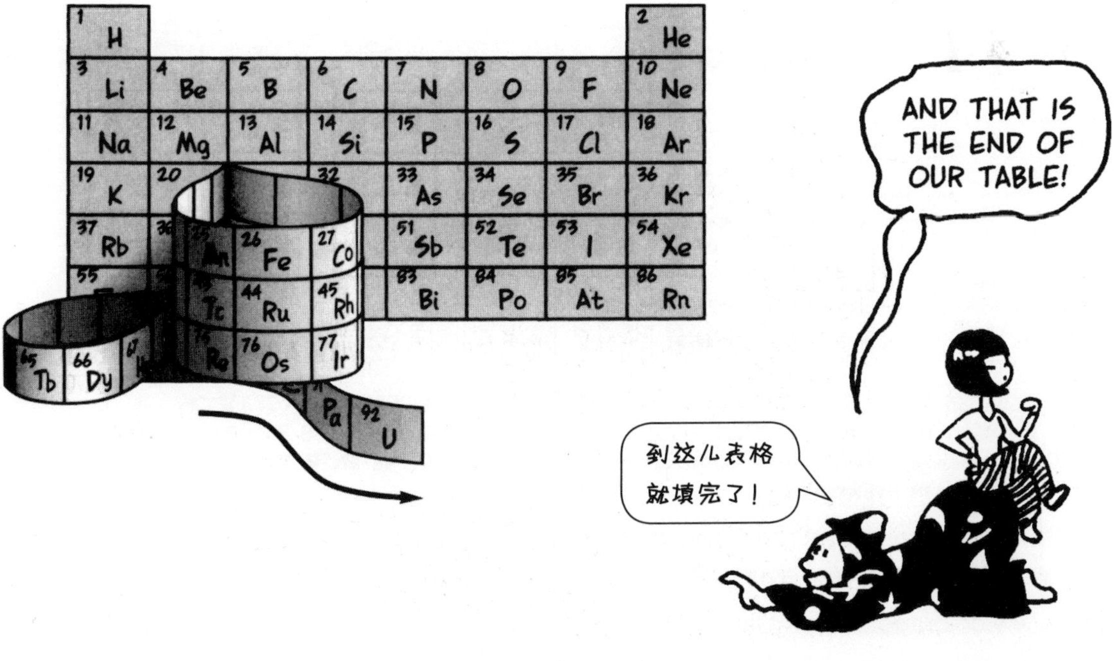

TURN THIS PAGE SIDEWAYS TO SEE THE PERIODIC TABLE AS IT IS USUALLY DISPLAYED. THE d-LOOPS ARE FLATTENED OUT TO SHOW EVERY ELEMENT. THE 14-ELEMENT f-LOOP, AFTER 57, LANTHANUM, IS CUT OUT AND PUT BELOW THE MAIN TABLE. THE TABLE'S "TAIL," THE ACTINIDE SERIES AFTER 89, IS ALSO AT THE BOTTOM.

横过来看此项，就能看到元素周期表的通常形式。那些由d-轨道系列形成的纸圈也被放平了，显示在一个表中。从57号镧开始的f-轨道的纸圈以及作为表格的“尾巴”的89号之后的锕系元素则另外放在独立的表格中。

1 H 1.01																	2 He 4.00
3 Li 6.94	4 Be 9.01											5 B 10.81	6 C 12.01	7 N 14.01	8 O 16.00	9 F 19.00	10 Ne 20.18
11 Na 22.99	12 Mg 24.31											13 Al 26.98	14 Si 28.09	15 P 30.97	16 S 32.07	17 Cl 35.45	18 Ar 39.95
19 K 39.10	20 Ca 40.08	21 Sc 44.96	22 Ti 47.88	23 V 50.94	24 Cr 52.00	25 Mn 54.94	26 Fe 55.85	27 Co 58.93	28 Ni 58.69	29 Cu 63.55	30 Zn 65.39	31 Ga 69.72	32 Ge 72.59	33 As 74.92	34 Se 78.96	35 Br 79.90	36 Kr 83.80
37 Rb 85.47	38 Sr 87.62	39 Y 88.91	40 Zr 91.22	41 Nb 92.91	42 Mo 95.94	43 Tc (98)	44 Ru 101.1	45 Rh 102.9	46 Pd 106.4	47 Ag 107.9	48 Cd 112.4	49 In 114.8	50 Sn 118.7	51 Sb 121.8	52 Te 127.6	53 I 126.9	54 Xe 131.3
55 Cs 132.9	56 Ba 137.3	57 La* 138.9	72 Hf 178.5	73 Ta 180.9	74 W 183.9	75 Re 186.2	76 Os 192.2	77 Ir 190.2	78 Pt 195.1	79 Au 197.0	80 Hg 200.5	81 Tl 204.4	82 Pb 207.2	83 Bi 209.0	84 Po (209)	85 At (210)	86 Rn (222)
87 Fr (223)	88 Ra (226)	89 Ac** (227)															

58 *Ce 140.1	59 Pr 140.9	60 Nd 144.2	61 Pm (145)	62 Sm 150.4	63 Eu 152.0	64 Gd 157.3	65 Tb 158.9	66 Dy 162.5	67 Ho 164.9	68 Er 167.3	69 Tm 168.9	70 Yb 173.0	71 Lu 175.0

90 **Th 232.0	91 Pa (231)	92 U (238)

EACH BOX CONTAINS AN ELEMENT'S ATOMIC NUMBER, SYMBOL, AND ATOMIC WEIGHT. WEIGHTS ARE NOT WHOLE NUMBERS BECAUSE THEY REPRESENT AN AVERAGE OF SEVERAL ISOTOPES.

每个格子中有元素的原子序数、元素符号和原子质量。这些原子质量是该元素所有同位素质量的平均值，因此都不是整数。

FOR A WONDERFULLY INFORMATION-RICH PERIODIC TABLE WITH A DETAILED PROFILE OF EVERY ELEMENT, SEE http://pearl1.lanl.gov/periodic/default.htm. ANOTHER WEB-BASED TABLE, AT www.colorado.edu/physics/2000/applets/a3.html, SHOWS THE ENERGIES OF ALL THE ELECTRONS IN EVERY ATOM.

这个网址http://pearl1.lanl.gov/periodic/default.htm有一个非常棒的交互式元素周期表，每个元素都附有详细介绍。另一个在线元素周期表还显示了每个原子中所有电子的能量：www.colorado.edu/physics/2000/applets/a3.html。

WHAT'S SO PERIODIC ABOUT THE PERIODIC TABLE? WHAT PROPERTIES REPEAT THEMSELVES IN THE COLUMNS? WHAT TRENDS DO WE TRACE ALONG THE ROWS?

元素周期表到底有什么周期？表中每列的元素有什么共同性质？而每行又有什么变化规律？

The Outermost Electrons 最外层电子

MOVING LEFT TO RIGHT ALONG A ROW OF MAIN-GROUP ELEMENTS, THE NUMBER OF OUTER ELECTRONS GOES UP STEADILY. GROUP 1 ELEMENTS ALL HAVE ONE OUTER ELECTRON, GROUP 2 ELEMENTS HAVE TWO, ETC., UNTIL THE LAST GROUP, WHICH ALL HAVE EIGHT. TRANSITION METALS HAVE EITHER ONE OR TWO OUTER ELECTRONS.*

每一行从左到右的各主族元素最外层电子个数依次加1。第一族的所有元素最外层电子数均为1，第二族元素最外层电子数均为2，依此类推，直到最后一族元素，最外层电子数均为8。而过渡金属元素则可能会有1或2个外层电子*。

*TRANSITION METALS' INNER ELECTRONS SOMETIMES HAVE HIGH ENOUGH ENERGY TO ACT LIKE OUTER ELECTRONS, HOWEVER.

*过渡元素的内层电子有时会有足够高的能量发挥和外层电子一样的作用。

NUMBER OF OUTER-SHELL ELECTRONS

外层电子个数

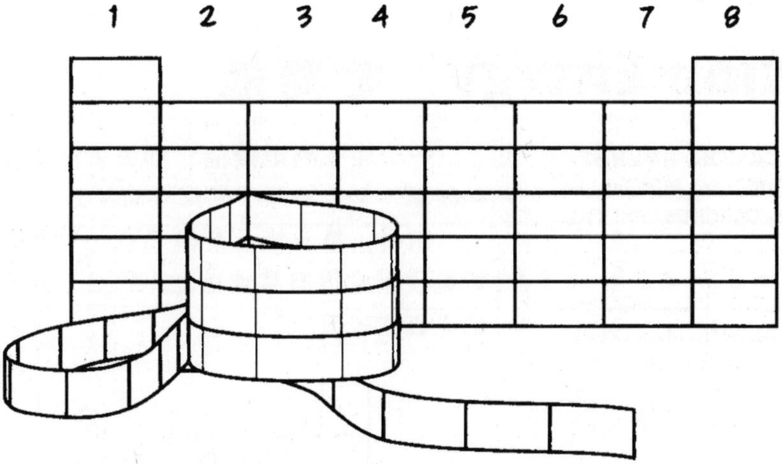

THE OUTER ELECTRONS, CALLED **VALENCE ELECTRONS,** ACCOUNT FOR MOST CHEMICAL REACTIONS.

我们把最外层的电子称为**价电子**，它们决定了该元素大部分的化学性质。

Atomic Size

GOING ALONG A ROW FROM LEFT TO RIGHT, ATOMS GET SMALLER, AND MOVING DOWN A COLUMN, THEY GET BIGGER.

REASON: MOVING TO THE RIGHT, THE BIGGER CHARGE OF THE NUCLEUS PULLS ELECTRONS CLOSER IN. GOING DOWN A COLUMN, THE OUTER ELECTRONS ARE IN HIGHER SHELLS, HENCE FARTHER AWAY FROM THE NUCLEUS.

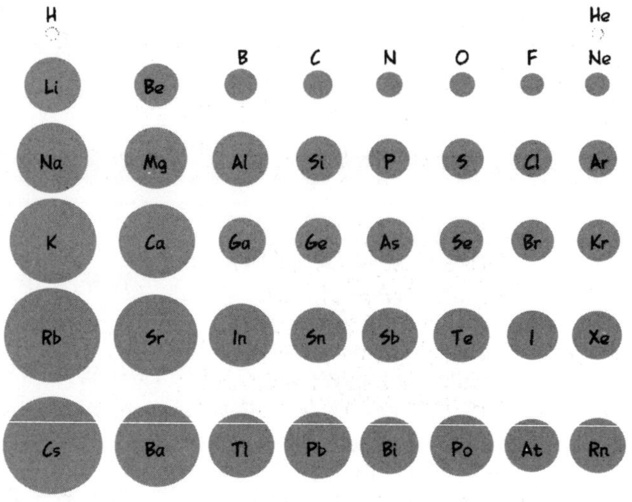

原子尺寸

每一行从左到右的元素，原子尺寸依次变小。而每一列从上到下，原子尺寸依次变大。为什么？

因为从左到右，原子核的正电荷越来越大，因此束缚电子的力量也越来越大，把电子绑紧了，整个原子尺寸就小了。而从上到下，由于原子层数增多，最外层电子离原子核越来越远，受原子核的束缚力也越来越小，原子尺寸就越来越大。

Ionization Energy 电离能

AN ATOM'S **IONIZATION ENERGY**—THE ENERGY NEEDED TO REMOVE AN OUTER ELECTRON—DEPENDS ON THE ATOM'S SIZE.

FOR EXAMPLE, GROUP 1 ELEMENTS HAVE A SINGLE VALENCE ELECTRON FAR AWAY FROM THE NUCLEUS. IT SHOULD BE EASY TO PRY OFF. THESE ELEMENTS SHOULD HAVE LOW IONIZATION ENERGIES.

一个原子的**电离能**，即失去一个外层电子所需的能量，与这一原子的尺寸有关。

例如，第一族元素都只有一个离原子核最远的价电子，它很容易被拐走。因此，我们说这些元素的电离能较低。

AND SO THEY DO. GROUP 1 ELEMENTS—LITHIUM, SODIUM, POTASSIUM, RUBIDIUM, AND CESIUM, THE **ALKALI METALS**—SHED ELECTRONS EASILY.

事实确实如此。第一族元素的锂、钠、钾、铷和铯都易失电子，它们被统称为**碱金属**。

IN FACT, THEY ARE SO REACTIVE THAT THEY ARE NEVER FOUND NATURALLY PURE, BUT ALWAYS IN COMBINATION WITH OTHER ELEMENTS.

这些元素的价电子实在太活跃了，因此在自然界中很难找到以完整原子形式存在的碱金属元素，它们基本上都是以与其他原子结合的状态存在。

MOVING RIGHTWARD ALONG A ROW, ELECTRONS ARE CLOSER TO THE NUCLEUS, WHICH HOLDS THEM MORE TIGHTLY, SO IONIZATION ENERGIES SHOULD RISE TO A MAXIMUM IN THE LAST COLUMN.

每一行从左到右的各元素，由于电子受原子核的束缚力依次增大，因此电离能也越来越大，位于最右一列元素的电离能最大。

AT THE START OF THE NEXT ROW, WITH A NEW OUTER SHELL, IONIZATION ENERGY DROPS AGAIN. THIS GRAPH SHOWS THE PERIODICITY OF IONIZATION ENERGY.

在下一行的开头，由于新增加了一个电子层，电离能突降，然后随着从左到右又重新增加。下图显示了电离能的这一周期律。

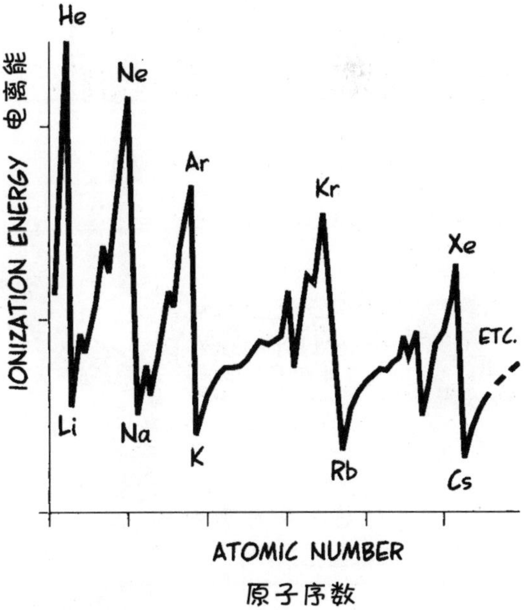

Electron Affinity

THIS PROPERTY, THE FLIP SIDE OF IONIZATION ENERGY, MEASURES AN ATOM'S "WILLINGNESS" TO BECOME AN ANION, I.E., TO ADD AN EXTRA ELECTRON.

STRAY ELECTRONS MAY FEEL THE NUCLEAR PULL AND ATTACH THEMSELVES TO ATOMS, ESPECIALLY IF AN UNFILLED OUTER ORBITAL IS AVAILABLE.

电子亲合能

这个性质与电离能恰好相反，它衡量一个原子有多喜欢夺取电子，即变成一个阴离子。

杂散电子会受原子核的吸引力的影响而被拉近，结合到原子上，尤其是当这原子的最外层电子有未填满的轨道可供容纳的时候。

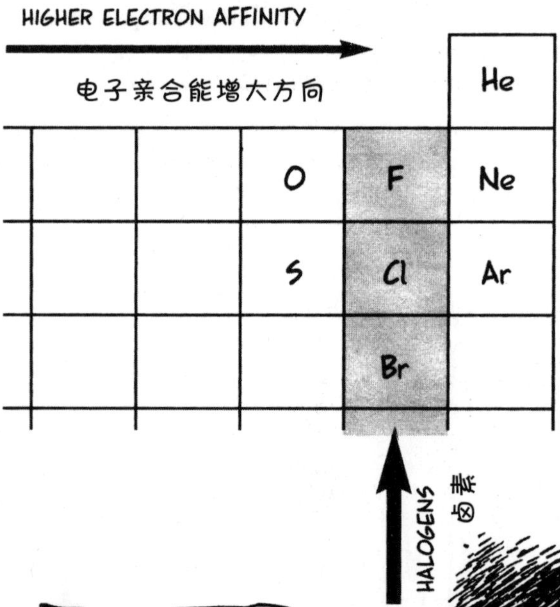

ATOMS TOWARD THE RIGHT SIDE OF THE PERIODIC TABLE TEND TO HAVE HIGHER ELECTRON AFFINITY: SMALL DIAMETER (SO ELECTRONS CAN GET CLOSER), BIG PULL FROM THE NUCLEUS, AND AN UNFILLED ORBITAL OR TWO.

在周期表中右边的这两列元素，电子亲合能越高、原子直径越小（电子越靠近原子核），原子核对电子的束缚力越大，同时有一到两条未填满的空轨道。

THE NEXT-TO-LAST GROUP IS ESPECIALLY ELECTRON HUNGRY. THESE ELEMENTS, THE **HALOGENS,** HAVE A SMALL DIAMETER AND ONE VACANT SPOT IN A p ORBITAL. AS YOU MIGHT IMAGINE, HALOGENS COMBINE WITH THE ELECTRON-SHEDDING ALKALI METALS OF GROUP 1. TABLE SALT, NaCl, IS A PRIME EXAMPLE OF AN ALKALI-HALOGEN COMPOUND.

倒数第二族元素是最能夺取电子的，称为**卤素**。它们具有小的原子直径和一条空的p轨道。可以设想，卤素和易失电子的碱金属能很好地结合起来。例如食盐——氯化钠，就是一个典型的碱卤化合物。

THE PERIODIC TABLE IS BROADLY DIVIDED ALONG A STAIRSTEP BORDER INTO METALS AND NONMETALS, WITH A FEW CONFUSED "METALLOIDS" STRADDLING THE FENCE. METALS, ON THE LEFT, VASTLY OUTNUMBER NONMETALS, THANKS TO ALL THE ELEMENTS IN THE "LOOPS".

整个元素周期表由一条阶梯型的边界大致划分为金属元素和非金属元素。除了恰好落在界线附近的少数"类金属"难以区分是金属还是非金属之外，整条界线以左的元素都是金属，界线以右的元素都是非金属。由于那些突出来的纸圈，金属元素的种类比非金属元素要多得多。

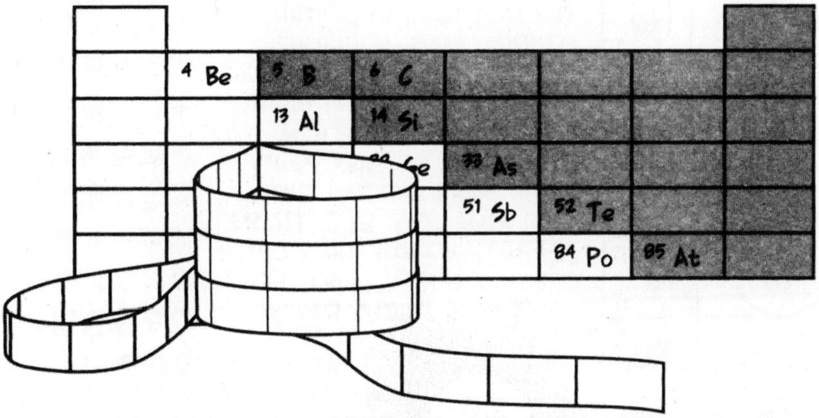

METALS TEND TO GIVE UP ELECTRONS FREELY, WHEREAS NONMETALS GENERALLY PREFER TO GAIN OR SHARE ELECTRONS. BUT METALS DO SHARE ELECTRONS AMONG THEMSELVES, FORMING TIGHTLY-PACKED, DENSE SOLIDS. NONMETALS USUALLY HAVE A LESS COHESIVE STRUCTURE.

金属元素一般容易失去自己的电子，而非金属元素则容易夺取或共享额外的电子。不过金属元素的原子之间倒可以互相共享电子，并形成紧密堆砌的高密度固体。相比之下，非金属元素的结构就比较松散。

Properties of metals 金属的性质

HIGH DENSITY	密度大
HIGH MELTING POINT AND BOILING POINT	熔点和沸点高
GOOD ELECTRICAL CONDUCTIVITY	导电性好
SHINY	有金属光泽
MALLEABLE (EASY TO SHAPE)	具有可塑性（容易塑形）
DUCTILE (EASY TO STRETCH INTO WIRES)	具有韧性（能拉细而不断）
REACTIVE WITH NONMETALS	易与非金属元素反应

Properties of nonmetals 非金属的性质

OFTEN LIQUID OR GASEOUS AT ROOM TEMPERATURE	在室温下往往是液体或气体
BRITTLE WHEN SOLID	固态时易碎
DULL-LOOKING	无光泽
POOR ELECTRICAL CONDUCTIVITY	导电性差
REACTIVE WITH METALS (EXCEPT FOR THE LAST GROUP)	易与金属元素反应（最后一族元素除外）

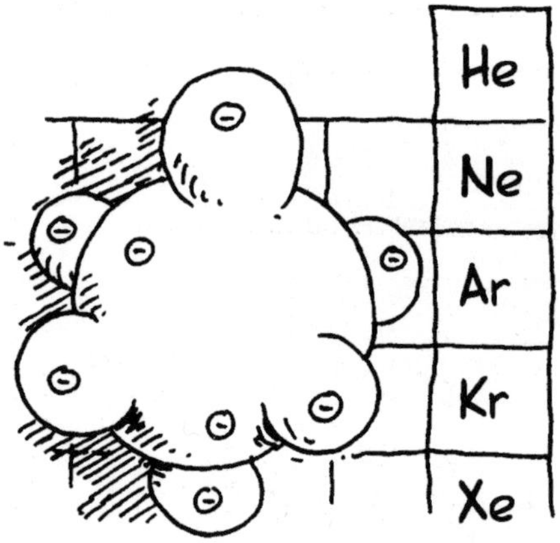

THE LAST COLUMN OF THE PERIODIC TABLE IS UNIQUELY STRANGE. ITS DENIZENS, BECAUSE THEY LIVE FAR TO THE RIGHT, HAVE **HIGH IONIZATION ENERGIES,** SO THEY DON'T EASILY MAKE CATIONS. THEY ALSO HAVE **LOW ELECTRON AFFINITY** BECAUSE THEIR OUTER ORBITALS ARE FULL, SO THEY DON'T MAKE ANIONS EITHER!

元素周期表最后一列的元素特别奇怪。由于它们处于最右边，具有**高电离能**，因此不易变为正离子。同时它们的最外层电子轨道全部满员了，因此又具有**低电子亲合能**，不易变成负离子！

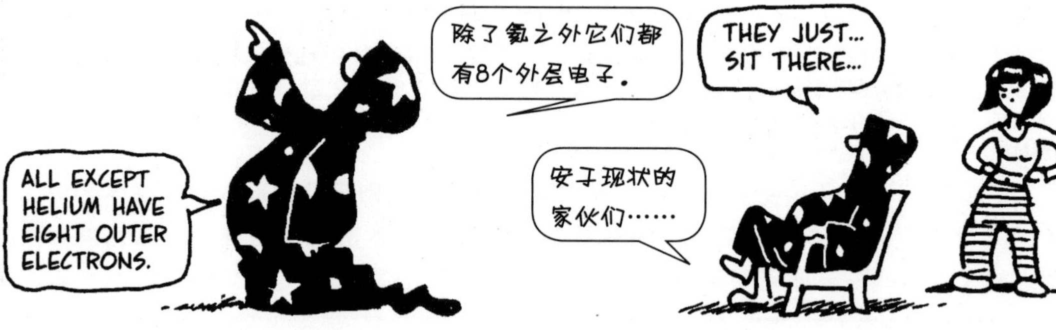

IN FACT, THEY RARELY REACT WITH ANYTHING. THEY JUST FLOAT AROUND IN AN UNCONNECTED, STANDOFFISH, GASEOUS STATE AND SO ARE KNOWN AS **NOBLE GASES.** YOU ALREADY KNOW ABOUT **NEON**, BUT THE MOST COMMON IS **ARGON** (ALMOST 1% OF THE ATMOSPHERE). IT IS USED IN ORDINARY INCANDESCENT LIGHT BULBS, SINCE IT WON'T REACT WITH THE HOT FILAMENT.

事实上，它们极少与其他物质反应，而总是保持着飘浮、游离的气态，因此被称为“**稀有气体**”（Noble Gases）*。也许你听说过氖灯里含有的**氖气**。但是地球上最丰富的稀有气体是氩气（几乎占到大气的1%）。一般的白炽灯泡里就充满了这种气体，这样高温的灯丝就不会被氧化。

* 直译是“贵族气体”。

JUST LIKE REAL NOBILITY, THE NOBLE GASES ARE THE ENVY OF THE COMMON ELEMENTS. EVERYONE WANTS THAT FULL COMPLEMENT OF EIGHT OUTER ELECTRONS.

就像真正的贵族一样，这些“稀有气体”具有所有“普通元素”所嫉妒的特点——8个外层电子轨道全都满员。

WE CALL THIS THE **RULE OF EIGHT:** AN ATOM TENDS TO PICK UP OR GIVE AWAY JUST ENOUGH ELECTRONS TO MAKE EIGHT IN ITS OUTER SHELL—AN **ELECTRON OCTET.**

这叫作**八电子规则**：一个原子倾向于得到或失去一定数量的电子以达到最外层电子数等于8的稳定状态——**电子八隅体**。

METALS TEND TO SHED ELECTRONS...

金属容易失电子

NONMETALS TEND TO ACQUIRE THEM.

非金属容易得电子

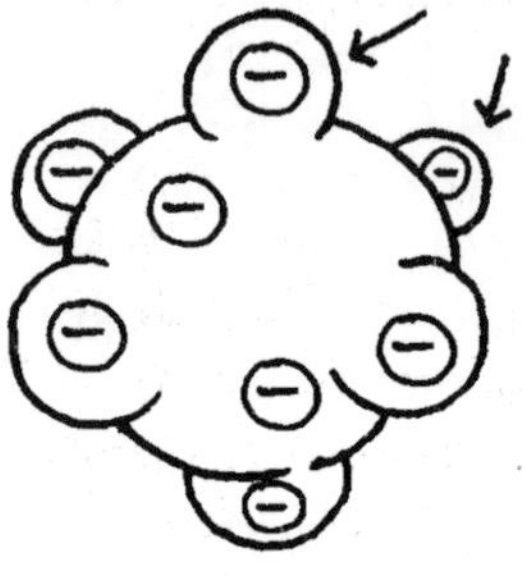

AND THIS BRINGS US TO THE SUBJECT OF OUR NEXT CHAPTER...

这就引出了下一章的话题……

BEFORE GOING ON, PLEASE TAKE A MOMENT TO APPRECIATE HOW AMAZING THIS CHAPTER HAS BEEN. STARTING FROM SOME WEIRD PROPERTIES OF ELEMENTARY ATOMIC PARTICLES, SCIENCE HAS MANAGED TO DESCRIBE THE ATOM, EXPLAIN THE PERIODIC TABLE, AND ACCOUNT FOR MANY CHEMICAL PROPERTIES OF THE ELEMENTS. NO WONDER ATOMIC THEORY HAS BEEN CALLED **"THE SINGLE MOST IMPORTANT IDEA IN SCIENCE."**

在进入下一章之前，让我们体会一下本章是多么有趣。在科学的思维下，从原子内部粒子神奇的结构出发，成功地一层层地描述了整个原子的性质，解释了元素周期表，甚至解释了不同元素的化学性质。怪不得“原子论”被称为**“科学中唯一最重要的思想”**。

Chapter 3 Togetherness

第三章 团结

IF ELEMENTS AND ATOMS WERE ALL THERE WERE, CHEMISTRY WOULD BE A PRETTY DULL SUBJECT. ATOMS WOULD JUST JIGGLE AROUND BY THEMSELVES LIKE A BUNCH OF NOBLE GASES, AND NOTHING WOULD HAPPEN.

如果元素和原子千古不变，那么化学将是一门乏味的学科。原子只是像稀有气体一样自娱自乐，什么都不会发生。

BUT IN REALITY, CHEMISTRY IS A SORT OF FRENZY OF TOGETHERNESS. MOST ATOMS ARE GREGARIOUS LITTLE CRITTERS... AND THAT'S HOW WE'RE GOING TO DRAW THEM, SOMETIMES... AS LITTLE CRITTERS.

但事实上，化学表现为空间中物质的一种紧密团结。大多数原子都是些群居的小家伙，所以下面我们有时候就把它们画成一些可爱的小东西。

THE COMBINATIONS ARE ENDLESS. METALS BOND TO METALS, NONMETALS TO NONMETALS, METALS TO NONMETALS. SOMETIMES ATOMS CLUMP TOGETHER IN LITTLE CLUSTERS AND SOMETIMES IN IMMENSE CRYSTAL ARRAYS. NO WONDER THE SUBJECT IS SO... SEXY!

这种组合是无数金属与金属、非金属与非金属、金属与非金属依靠键引力形成的。原子有时聚成一小簇，有时存在于巨大的晶体中。难怪这个学科是如此地引人入胜！

ATOMS COMBINE WITH EACH OTHER BY EXCHANGING OR SHARING ELECTRONS. THE DETAILS DEPEND ON THE PREFERENCES OF THE PARTICULAR ATOMS INVOLVED. DOES AN ATOM "WANT" TO SHED AN ELECTRON OR TO PICK ONE UP? AND HOW BADLY?

原子通过交换或共享电子彼此结合。具体细节取决于所涉及特定原子的喜好，即一个原子是“希望”摆脱还是“希望”接受一个电子呢？并且这种心情有多么迫切呢？

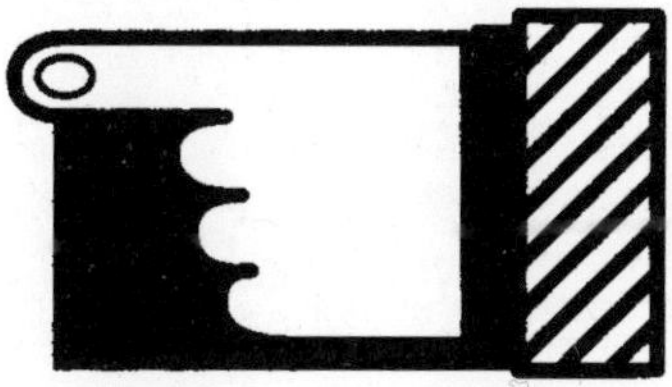

METALS, AS WE'VE SEEN, TEND TO GIVE UP ELECTRONS, THOUGH SOME METALS DO SO MORE ENTHUSIASTICALLY THAN OTHERS. A CHEMIST WOULD SAY THAT METALS ARE MORE OR LESS **ELECTROPOSITIVE.**

正如我们所见，金属倾向于放弃电子，当然，某些金属更热衷于此道。化学家因此说金属或多或少带有**电正性**。

NONMETALS ARE MORE OR LESS **ELECTRONEGATIVE:** THEY TEND TO ACCEPT EXTRA ELECTRONS. SOME NONMETALS, LIKE FLUORINE AND OXYGEN, AVIDLY GRAB ELECTRONS, WHILE OTHERS, SUCH AS CARBON, CAN TAKE THEM OR LEAVE THEM.

非金属则或多或少带有**电负性**：他们倾向于接受额外的电子。一些非金属，如氟和氧，迫切希望得到电子；而其他非金属，如碳，则可以选择接受电子或放弃电子。

IN BETWEEN ARE THE METALLOIDS, WHICH ARE COMPLETELY AMBIVALENT.

在这两类之间的类金属常感到非常纠结。

Ionic Bonds 离子键

WHEN A HIGHLY ELECTROPOSITIVE ATOM MEETS A HIGHLY ELECTRONEGATIVE ONE, THE RESULT IS AN IONIC BOND. THE ELECTROPOSITIVE ATOM EASILY GIVES AWAY ONE OR MORE ELECTRONS AND BECOMES A POSITIVELY CHARGED CATION. THE ELECTRONEGATIVE ATOM LOVES TO ACQUIRE EXTRA ELECTRONS AND IN DOING SO BECOMES AN ANION.

当一个具有高度电正性的原子遇到高度电负性的原子时，就会形成离子键。带有电正性的原子放弃一个或多个电子并变成一个带正电的阳离子。负电性的原子获得额外的电子，并因此成为一个阴离子。

THE TWO IONS THEN EXPERIENCE AN ELECTROSTATIC ATTRACTION.

然后两个离子产生静电引力。

IN FACT, THEY ATTRACT NOT ONLY EACH OTHER, BUT EVERY OTHER CHARGED PARTICLE IN THE NEIGHBORHOOD.

事实上，它们并不仅仅是相互吸引，也会吸引周围环境中的其他带电粒子。

THEIR MUTUAL ATTRACTION PACKS THEM TOGETHER IN A DENSE, REGULAR **IONIC CRYSTAL.** IN THE CASE OF SODIUM AND CHLORIDE,* EACH ION HAS A SINGLE CHARGE SO NEUTRALITY IS ACHIEVED BY THIS SIMPLE CUBIC ARRANGEMENT:

它们的相互吸引使它们堆集成密集并且规整的**离子晶体**。以钠和氯为例*，每个离子都带有一个单电荷，物质的电中性就是由这个简单的立方排布造成的。

IF YOU LOOK CLOSELY AT TABLE SALT, YOU CAN SEE THAT THE CRYSTALS ARE LITTLE CUBES—EACH ONE A MONSTER ARRAY OF SODIUM AND CHLORIDE IONS.

如果你仔细观察食盐，就可以看到立方形的小晶体，每一个都是由钠和氯离子排列而成的。

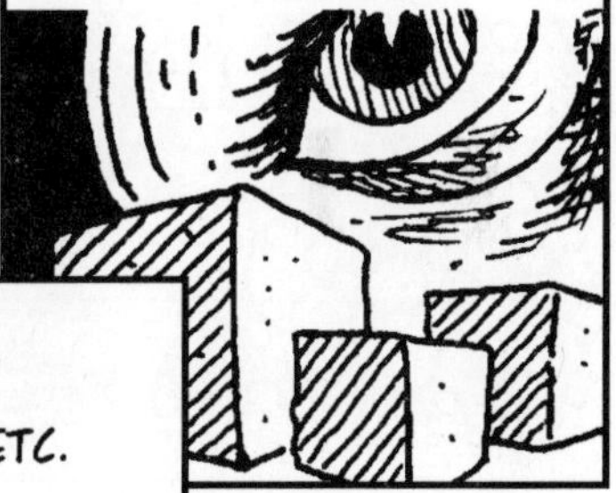

*SINGLE-ATOM ANIONS ARE NAMED BY ADDING "IDE" TO THE ROOT OF THEIR ELEMENTAL NAME: FLUORIDE, OXIDE, ETC.

*元素名称之后加上“化物”即“IDE”，就是单原子阴离子的名称：氟化物、氧化物等。

OTHER IONS MAY FORM DIFFERENT CRYSTALLINE STRUCTURES. WHEN CALCIUM, WHICH GIVES UP TWO ELECTRONS, COMBINES WITH CHLORINE, WHICH ACCEPTS ONLY ONE, TWO CHLORIDE IONS ARE NEEDED TO NEUTRALIZE EACH CALCIUM. WE WRITE AN ION WITH ITS ELEMENT SYMBOL AND CHARGE. SO THE CALCIUM ION IS Ca^{2+}, AND CHLORIDE IS Cl^-.

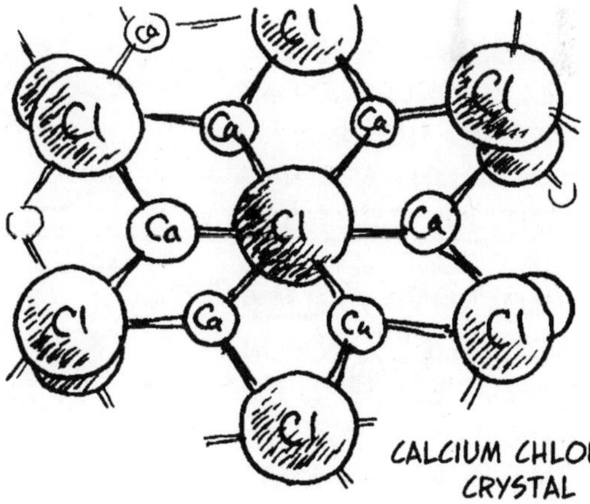

CALCIUM CHLORIDE CRYSTAL

氯化钙晶体

其他离子可形成不同的晶体结构。例如，钙给出两个电子与氯结合，而氯只能接受一个，需要两个氯离子与钙结合才能电中性。我们以元素符号和电荷离子表示离子，如钙离子是Ca^{2+}、氯离子是Cl^-。

THE FORMULA OF THESE IONIC CRYSTALS IS GIVEN "IN LOWEST TERMS." EVEN THOUGH A SODIUM CHLORIDE CRYSTAL MAY CONTAIN TRILLIONS OF ATOMS, WE WRITE ITS **EMPIRICAL FORMULA** AS NaCl. THIS SHOWS THAT THE CRYSTAL HAS ONE SODIUM ION FOR EACH CHLORIDE. IN THE SAME WAY, CALCIUM CHLORIDE IS WRITTEN $CaCl_2$.

这些离子晶体的表达式是“最简式”。尽管氯化钠晶体中可能含有的原子数多得要命，但我们写的**实验式**为NaCl。这表明该晶体中每一个钠离子对应一个氯化物。同样，氯化钙写成$CaCl_2$。

OCCASIONALLY, IONICALLY BONDED ATOMS HAVE NO NATURAL CRYSTALLINE ARRANGEMENT. INSTEAD THEY CLUMP TOGETHER INTO SMALL GROUPS CALLED **MOLECULES.** BORON TRIFLUORIDE, BF_3, IS AN IONIC COMPOUND THAT IS GASEOUS AT ROOM TEMPERATURE.

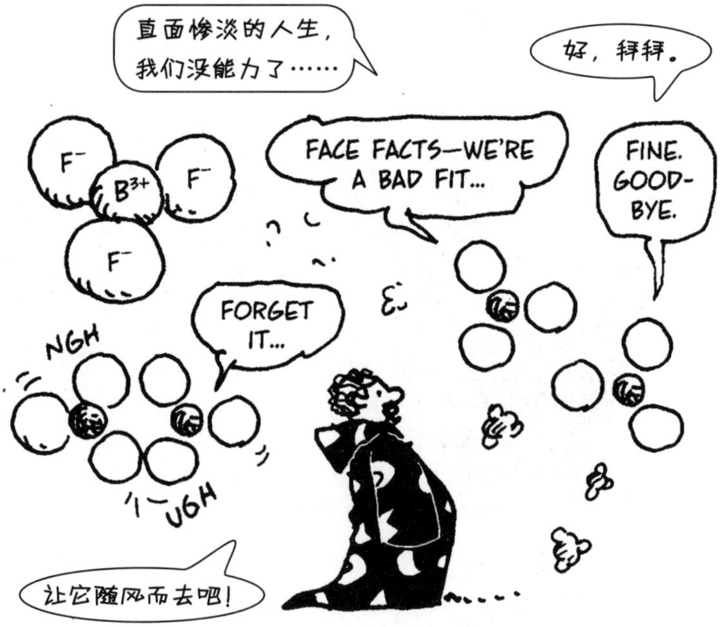

有时形成离子键的原子没有排布成天然晶体。相反，它们簇合在一起成为**分子**。比如三氟化硼（BF_3），在室温下是一种离子化合物的气体。

SOME IONS CONSIST OF MORE THAN ONE ATOM. WE'LL SEE HOW TO BUILD THESE **POLYATOMIC** IONS LATER IN THE CHAPTER. THESE THINGS BEHAVE VERY MUCH LIKE MONOATOMIC IONS, EXCEPT FOR THEIR SHAPE. THE WHOLE STRUCTURE ACTS AS A SINGLE CHARGED UNIT.

某些离子由多个原子组成。我们将在以后的章节中学到这些**多原子**离子如何组成。它们的形状不同于单原子离子，其表现却十分类似，整体结构是一个单电荷单元。

A TYPICAL EXAMPLE IS **SULFATE,** SO_4^{2-}, AN ANION THAT BONDS WITH Ca^{2+} TO MAKE **CALCIUM SULFATE,** $CaSO_4$, AN INGREDIENT OF WALLBOARD.

举一个典型的例子：**硫酸根**（SO_4^{2-}）作为一个阴离子，与Ca^{2+}成键得到**硫酸钙**（$CaSO_4$），这是组成墙灰的一种成分。

NH_4^+	AMMONIUM	铵离子
OH^-	HYDROXIDE	氢氧根离子
NO_2^-	NITRITE	亚硝酸根离子
NO_3^-	NITRATE	硝酸根离子
HCO_3^-	BICARBONATE	碳酸氢根离子
CO_3^{2-}	CARBONATE	碳酸根离子
SO_3^{2-}	SULFITE	亚硫酸根离子
PO_4^{3-}	PHOSPHATE	磷酸根离子

EACH POLYATOMIC ION MUST BE REGARDED AS A SINGLE ION. FOR EXAMPLE, ALUMINUM HYDROXIDE, WHICH COMBINES Al^{3+} AND OH^-, MUST HAVE THREE HYDROXIDES TO BALANCE EACH ALUMINUM. THE FORMULA IS WRITTEN $Al(OH)_3$, AND THE CRYSTAL STRUCTURE LOOKS LIKE THIS:

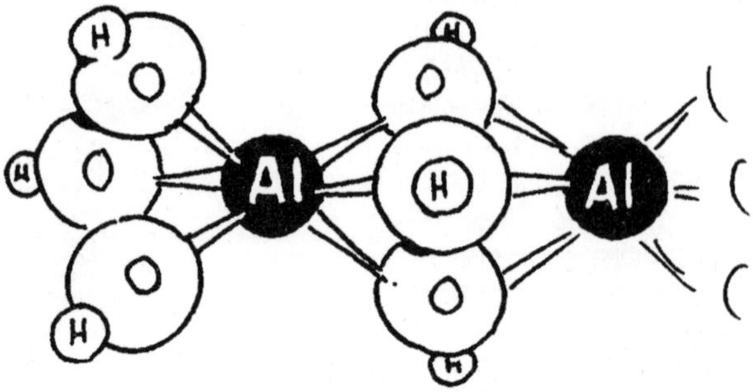

每个多原子离子必须被视为一个单一的离子，例如，氢氧化铝，结合一个Al^{3+}和OH^-，必须有三个氢氧根离子对应每个铝离子。其化学式为$Al(OH)_3$，其晶体结构如上图所示。

IONIC BONDS ARE STRONG. IT TAKES A LOT OF ENERGY TO BREAK THEM. THIS EXPLAINS WHY MOST IONIC CRYSTALS HAVE SUCH HIGH MELTING POINTS: TREMENDOUS HEAT IS NEEDED TO JAR THE IONS LOOSE AND GET THEM SLOSHING AROUND AS A LIQUID.

离子键是很牢固的，需要巨大的能量来打破它们。这就是为什么大多数离子晶体有如此高的熔点的原因，它们需要巨大的热量使离子震动松动，让它们变成液体的形式左右晃动。

食盐的熔点是801℃。

AND YET—HIT A SALT CRYSTAL WITH A HAMMER AND IT CRUMBLES. WHY SHOULD IT BE SO BRITTLE?

问：用一个锤子敲击一块食盐晶体，它就会破碎，为什么它会如此脆弱？

ANSWER: WHEN WHACKED, THE CRYSTAL MAY DEVELOP TINY CRACKS, AND ONE LAYER MAY SHIFT SLIGHTLY ACROSS ANOTHER.

答：当受到重击时，晶体会出现细小裂缝，其层状结构会产生一定的位移。

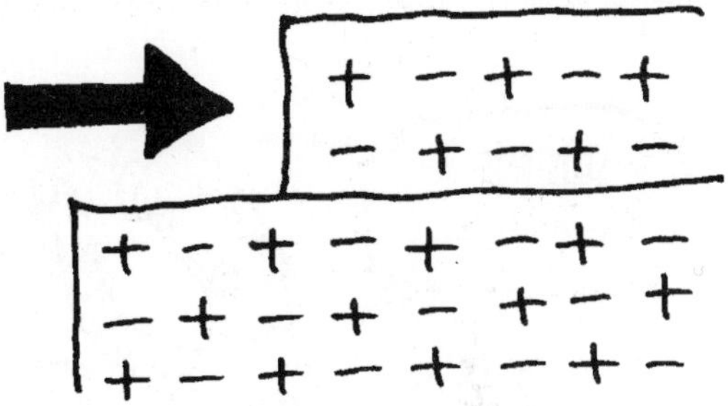

THIS SHIFT CAN ALIGN POSITIVES OPPOSITE POSITIVES AND NEGATIVES OPPOSITE NEGATIVES. NOW THE TWO CHUNKS REPEL EACH OTHER, AND THE CRYSTAL LITERALLY FLIES APART.

这种转变使电正性和电负性的部分各自对齐。于是，这两个晶块之间互相排斥，结晶破裂。

BUT NOT ALL CRYSTALS BEHAVE THIS WAY—METALLIC CRYSTALS, FOR EXAMPLE...

但并非所有晶体都这样，例如金属晶体……

Metallic Bonds

PURE METALS ALSO FORM CRYSTALS, THOUGH YOU PROBABLY DON'T THINK OF THEM THAT WAY. THEY LACK THE TRANSPARENCY AND SPARKLE OF NaCl AND OTHER IONIC CRYSTALS, AND THEY USUALLY AREN'T BRITTLE.

金属键

纯金属也会形成晶体，尽管你可能并不这么认为。它们缺乏透明度，也不像氯化钠等离子晶体那样闪亮，它们通常也是不易碎的。

METALS LIKE TO GIVE UP ELECTRONS. WHEN MANY METALLIC ATOMS GET TOGETHER, THEY SHED AN ENTIRE "ELECTRON SEA" THAT ENGULFS THE METAL IONS.

金属倾向于放弃电子。当许多金属原子聚在一起时，他们释放出一片“电子的海洋”，并淹没了金属离子。

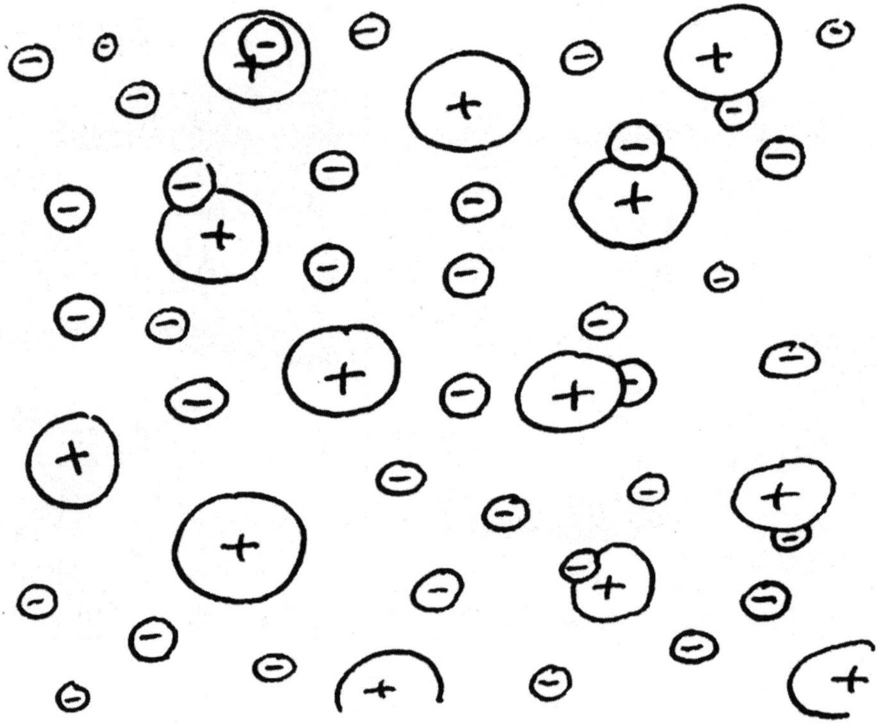

PULLED FROM ALL DIRECTIONS, THE METAL IONS FIND IT HARD TO MOVE, AND THEY PACK TIGHTLY TOGETHER IN CRYSTALLINE STRUCTURES. THERE ARE SEVERAL POSSIBLE PACKING ARRANGEMENTS, ALL OF THEM DENSE. HERE ARE TWO.

来自各个方向的拉力使金属离子难以动弹，它们以晶体结构的形式紧密地团成一团儿。金属离子有多种可能的密集排列形式，这里举两个例子：

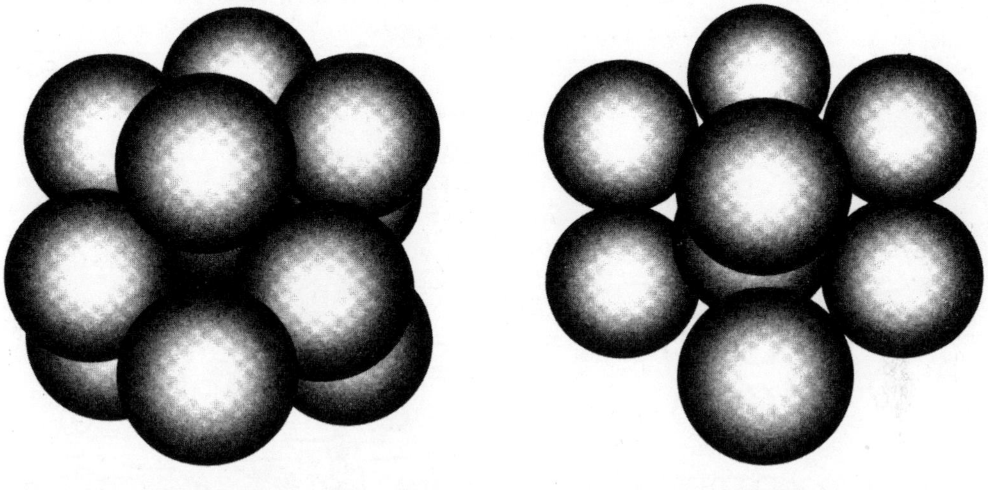

BODY-CENTERED CUBIC: EACH ATOM SURROUNDED BY EIGHT OTHERS

体心立方：每个原子被8个原子包围。

FACE-CENTERED CUBIC: EACH ATOM SURROUNDED BY TWELVE OTHERS

面心立方：每个原子被12个原子包围。

METALS TEND TO BE GOOD CONDUCTORS OF ELECTRICITY. THE LIGHT, FREE ELECTRONS MOVE AROUND EASILY. NEGATIVE CHARGE COMING FROM OUTSIDE CAN PUSH THE "SEA" OF ELECTRONS, MAKING A CURRENT.

金属通常是电的良导体。其中自由电子移动方便，来自外部的负电荷可以推动海量的电子成为电流。

LIKE ANY CRYSTAL, BEING WHACKED BY A HAMMER MAY CAUSE A METAL'S CRYSTALLINE STRUCTURE TO CRACK AND SHIFT.

如同任何晶体，被锤子重击可能会导致金属的晶体结构裂缝和移位。

BUT UNLIKE IONIC CRYSTALS, THE METAL'S IONIC REPULSION IS OVERCOME BY THAT NEGATIVE SEA OF ELECTRONS HOLDING ALL THE ATOMS IN PLACE.

但是，与离子晶体不同，金属的离子排斥力消失在所有原子的电子产生的电负性海洋中。

SO, INSTEAD OF SHATTERING, A METAL TENDS TO BEND OR STRETCH.*

因此，金属往往弯曲或伸展而不是粉碎*。

*DISCLAIMER: THERE ARE EXCEPTIONS, AS WITH NEARLY EVERYTHING IN CHEMISTRY.

* 声明：就像化学中的任何一条规律一样，这一条规律也有例外。

Covalent Bonding and Molecules

共价键和分子

IONIC BONDS FORM WHEN A HIGHLY ELECTRONEGATIVE ATOM MEETS A HIGHLY ELECTROPOSITIVE ONE. ELECTRONS ARE HANDED OFF, AND ONE ATOM GETS SOLE CUSTODY.

一个高度电负性的原子遇到一个高度电正性的原子就会产生一个离子键。两个原子都给予一个电子，而只有一个原子独自获得这对电子。

METALLIC BONDING HAPPENS WHEN A LOT OF ELECTROPOSITIVE ATOMS ARE TRAPPED BY ALL THE ELECTRONS THEY SHARE. IT'S LIKE A COMMUNAL HOUSEHOLD.

当很多电正性的原子被它们塑造的电子所困时，金属键就形成了。这就像一个公共的家庭。

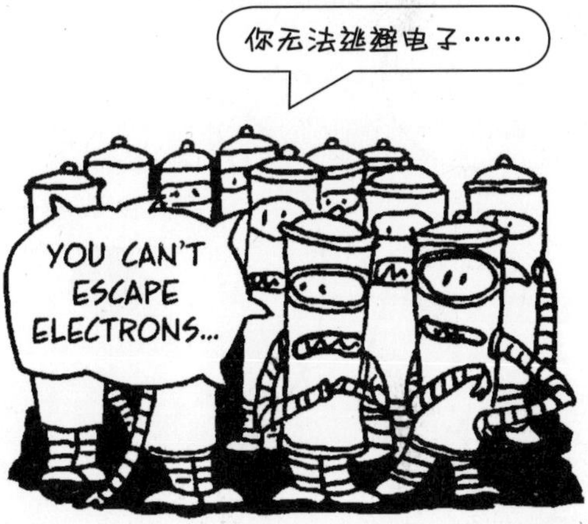

AND THEN THERE'S EVERYTHING ELSE: THE BONDS BETWEEN TWO ELECTRONEGATIVE ATOMS...

还有，在两个电负性的原子之间还会有键……

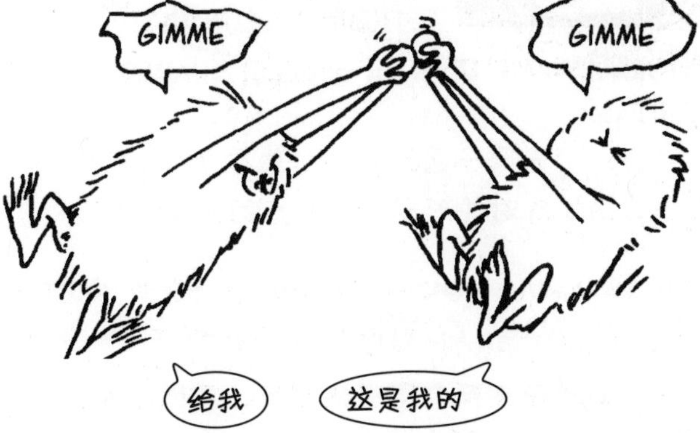

OR BETWEEN ATOMS THAT ARE ONLY SOMEWHAT ELECTRONEGATIVE OR ELECTROPOSITIVE. ONE SHEDS ELECTRONS, BUT RELUCTANTLY... THE OTHER ACCEPTS THEM, BUT HALF-HEARTEDLY... AND THE RESULT IS A SORT OF MARRIAGE, OR JOINT CUSTODY ARRANGEMENT.

还有那些只是略有电负性或电正性的原子之间也能成键。一个原子给出电子，但很勉强……另外一个对接受电子也很漫不经心……结果是即便在一起，也是强扭的瓜。

THE SIMPLEST POSSIBLE EXAMPLE IS HYDROGEN. A LONE HYDROGEN ATOM HAS AN UNPAIRED ELECTRON, WHICH THE ATOM CAN EITHER GIVE UP OR PAIR WITH ANOTHER ELECTRON.

可能最简单的例子是氢。一个单独的氢原子有一个未成对电子，氢原子既可以放弃这个电子也可以使其与其他电子配对。

WHEN ONE HYDROGEN ENCOUNTERS ANOTHER, THEIR ELECTRONS NATURALLY PAIR UP IN A SINGLE, SHARED ORBITAL.

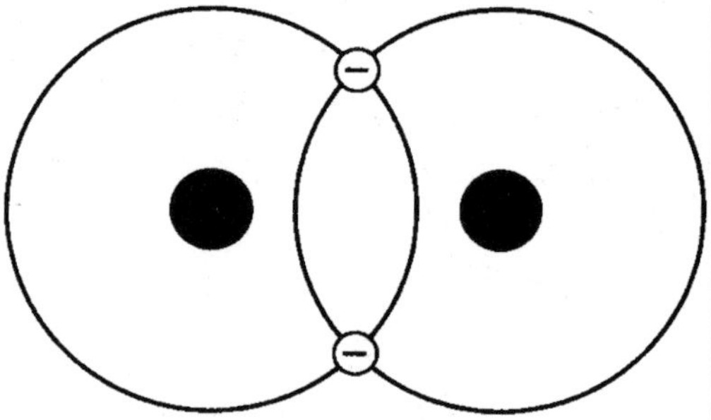

当一个氢原子遇到另一个氢原子时，它们的电子自然地成对并形成一个单一的共享轨道。

THIS PAIR PULLS ON BOTH NUCLEI, SO IT HOLDS THE ATOMS TOGETHER. THE BOND IS CALLED **COVALENT**, BECAUSE BOTH ATOMS CONTRIBUTE EQUALLY.

这对电子在两个原子核之间，并联结两个原子。这种键被称为**共价键**，因为这两个原子的贡献相同。

EACH HYDROGEN ATOM "THINKS" IT HAS A FULL 1s VALENCE SHELL, SO THE RESULTING TWO-SOME, OR HYDROGEN **MOLECULE,** H_2, IS STABLE.

每个氢原子“认为”它有一个全面的价电子层，因此产生的一对**氢原子**，或氢分子H_2，是稳定的。

MORE EXAMPLES: OXYGEN, THE SECOND-MOST ELECTRONEGATIVE ELEMENT (AFTER FLUORINE), HAS SIX VALENCE ELECTRONS. WE INDICATE THIS WITH A "LEWIS DIAGRAM" THAT REPRESENTS EACH OF THESE OUTER ELECTRONS AS A DOT.

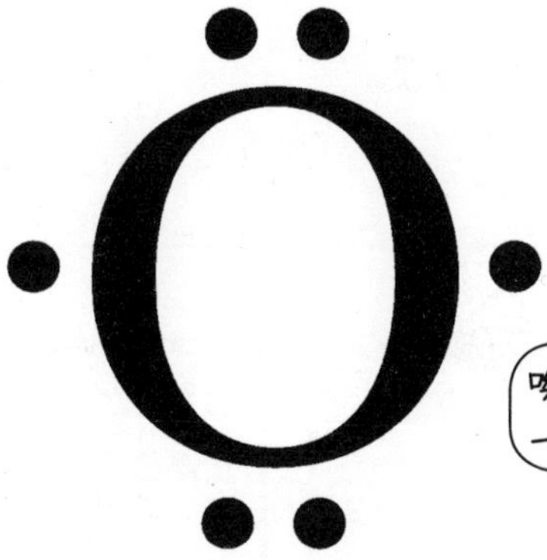

更多的例子：氧，排名第二的电负性元素（氟第一），有6个价电子。我们用“电子式”表达，每个外层电子就是一个点。

WHEN TWO OXYGENS GET TOGETHER, THEY BOND COVALENTLY BY SHARING FOUR ELECTRONS, AS SHOWN IN THIS LEWIS DIAGRAM:

当两个氧在一起，它们共享4个电子共价成键，如电子式所示：

HERE, TOO, BOTH ATOMS NOW HAVE A FULL OUTER OCTET. (COUNT THE ELECTRONS!) WHEN FOUR ELECTRONS ARE SHARED IN THIS WAY, WE CALL IT A **DOUBLE BOND** AND SOMETIMES WRITE IT AS O=O.

这里，现在两个原子有完整的最外层电子数（数一下电子数量）。当4个电子以这种方式共享时，我们称之为一个**双键**，有时写成O = O。

NITROGEN, WITH FIVE VALENCE ELECTRONS, FORMS TRIPLE COVALENT BONDS TO MAKE N_2 OR N≡N.

氮有5个价电子，以3个共价键的形式，形成氮气或N≡N。

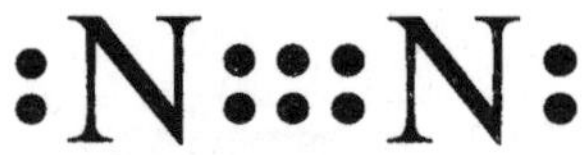

MANY OTHER NON-METALS, INCLUDING THE HALOGENS, FORM DIATOMIC (TWO-ATOM) MOLECULES IN THIS WAY.

许多其他非金属包括卤素，以这种方式形成双原子（两个原子）分子。

COVALENT BONDING INVOLVES ELECTRON SHARING BETWEEN A SPECIFIC PAIR OF ATOMS. IT'S LIKE A HANDSHAKE.

共价键的形成涉及配对的原子之间电子的共享，就像握手一样。

SINCE ATOMS HAVE ONLY A LIMITED NUMBER OF "HANDS," COVALENT COMPOUNDS ARE USUALLY FOUND IN THE FORM OF **MOLECULES,** OR SMALL, DISCRETE GROUPS OF ATOMS.

由于原子只有有限数量的“手”，因此，共价化合物通常是以**分子**或原子的小规模离散群的形式出现。

EVERY MOLECULE IN A PURE SUBSTANCE HAS THE SAME COMPOSITION. WE WRITE ITS FORMULA ACCORDING TO THE NUMBER OF EACH KIND OF ATOM PRESENT.

纯物质中的每个分子都具有相同的成分。我们可以根据分子中各类原子的数量写出它的分子式。

H_2O, WATER
水

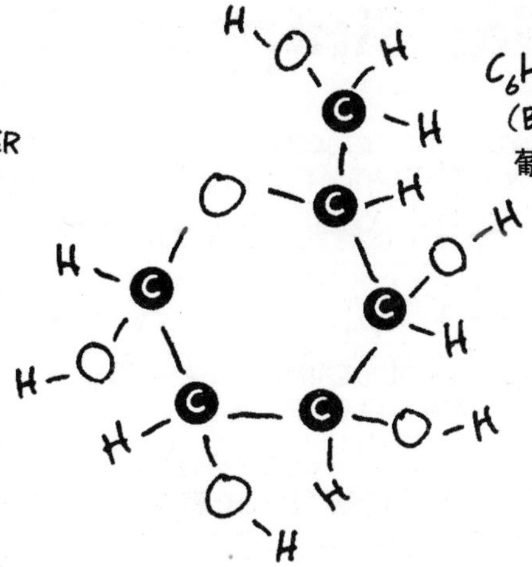

$C_6H_{12}O_6$, GLUCOSE (BLOOD SUGAR)
葡萄糖（血糖）

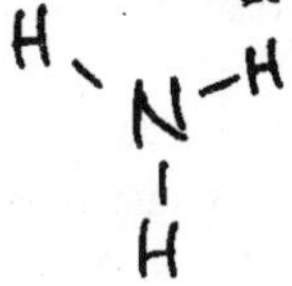

NH_3, AMMONIA
氨

OCCASIONALLY WE DO SEE COVALENTLY BONDED CRYSTALS. DIAMOND, FOR EXAMPLE, CONSISTS OF A SO-CALLED **COVALENT NETWORK** OF CARBON ATOMS.

有时，我们也会见到共价键晶体。比如钻石，就是由碳原子形成的**共价网络**所构成的。

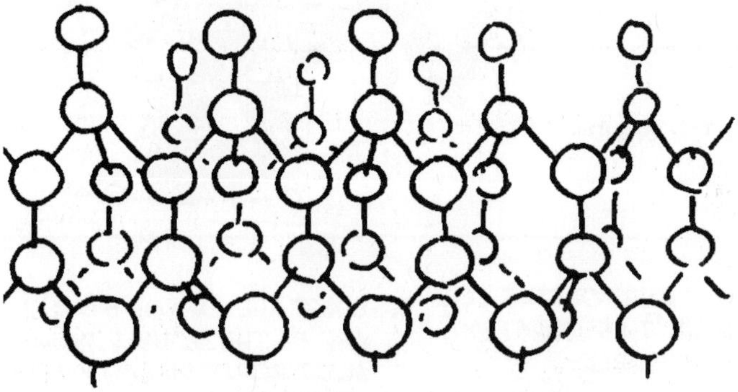

DIAMOND
钻石

* 英语“洋葱”一词onion也以ion（离子）结尾。

Molecular Shapes 分子的形状

SO FAR, WE'VE LOOKED ONLY AT COVALENT BONDS BETWEEN TWO IDENTICAL ATOMS. NOW LET'S SEE HOW DIFFERENT ATOMS CAN SHARE ELECTRONS.

到目前为止，我们只看到两个相同的原子之间的共价键。现在让我们来看看不同的原子是如何共享电子的。

CARBON DIOXIDE, FAMOUS EXHAUST GAS, CO_2: CARBON HAS FOUR VALENCE ELECTRONS AND OXYGEN HAS SIX, SO WE WRITE:

二氧化碳CO_2，著名的废气：碳有4个价电子，氧有6个，所以我们写成：

THESE CAN COMBINE LIKE SO:

它们可以像这样结合：

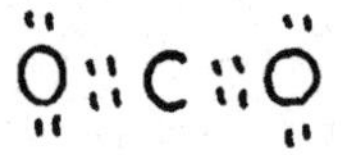

AND CO_2 HAS TWO DOUBLE BONDS.

CO_2有两个双键。

WHAT IS THE ACTUAL SHAPE OF THE CO_2 MOLECULE? TO ANSWER THIS QUESTION, USE THIS BRILLIANT PRINCIPLE:

什么是CO_2分子的真实形状？要回答这个问题，请使用这个重要的原则：

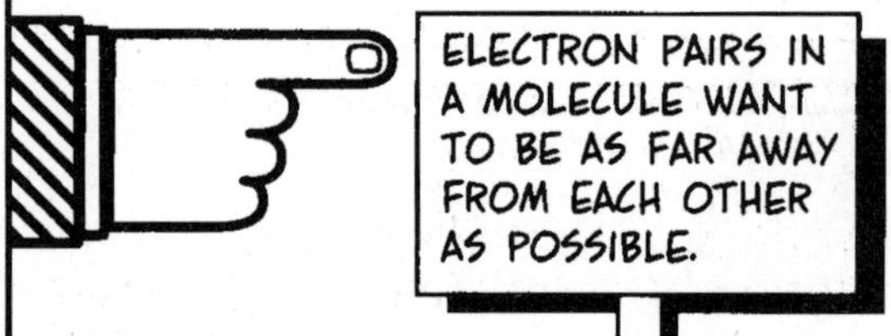

在分子中，电子对会尽可能地远离对方。

SINCE ALL CARBON'S VALENCE ELECTRONS ARE IN THE DOUBLE BONDS, THE BONDS MUST POINT DIRECTLY AWAY FROM EACH OTHER.

由于碳所有的外层电子在双键上，因此双键必须直接指向远离对方的方向。

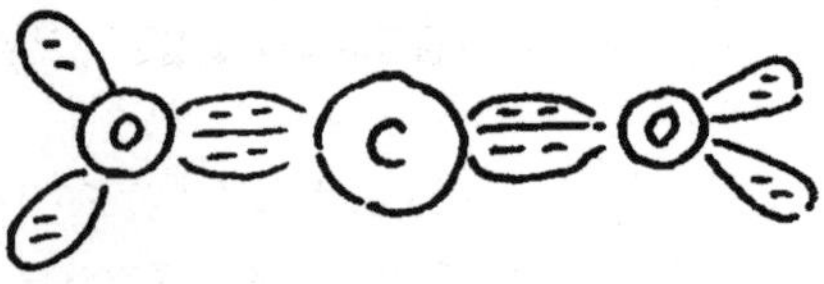

THE THREE ATOMS LIE IN A STRAIGHT LINE.

这三个原子位于一条直线上。

IN **SULFUR TRIOXIDE,** SO_3, SULFUR AND OXYGEN EACH HAVE SIX VALENCE ELECTRONS.

在**三氧化硫**（SO_3）中硫和氧各有6个价电子。

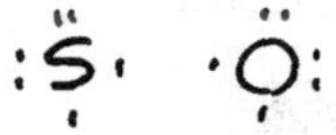

THREE OXYGENS CAN BOND TO SULFUR.

3个氧原子可以与硫成键。

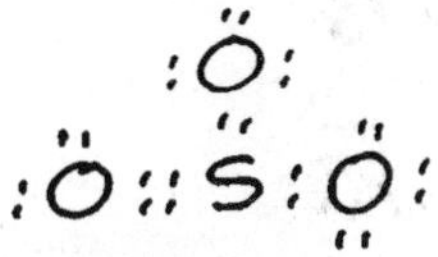

(THE DOUBLE BOND COULD GO ON ANY ONE OF THE OXYGENS.)

（双键可以挨着其中任何一个氧原子。）

USING THE PRINCIPLE THAT ELECTRON PAIRS MUST AVOID EACH OTHER (EXCEPT FOR THE ONES IN THE DOUBLE BOND—THEY'RE STUCK), WE CONCLUDE THAT SO_3 IS TRIANGULAR AND LIES IN A PLANE.

根据电子对必须远离对方的原则（在双键上被固定的两个电子对除外），我们得出这样的结论：SO_3是一个平面三角形。

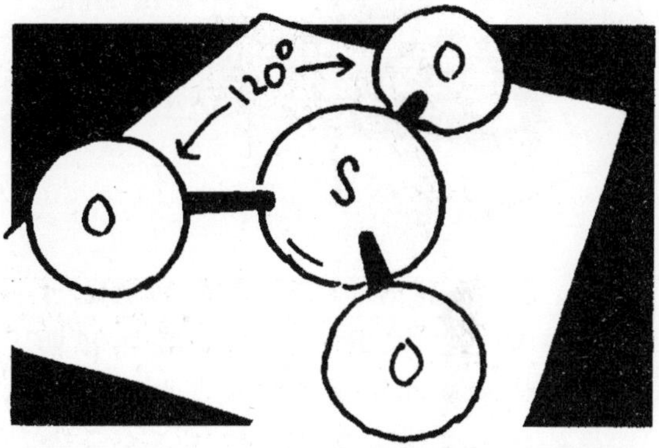

CARBON TETRACHLORIDE, CCl_4, AN INDUSTRIAL SOLVENT, COMBINES

四氯化碳（CCl_4），一种工业用溶剂，

WITH FOUR SINGLE BONDS.

以4个单键结合而成。

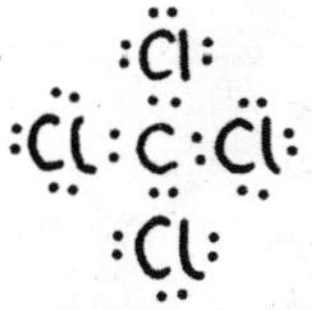

FOR MAXIMUM BOND SEPARATION, THIS MOLECULE HAS A TETRAHEDRAL SHAPE, WITH THE OUTER ATOMS AT THE POINTS OF A TRIANGULAR PYRAMID.

为了使键最大限度地分离，这种分子形成四面体，外层原子位于一个三角形金字塔的顶点。

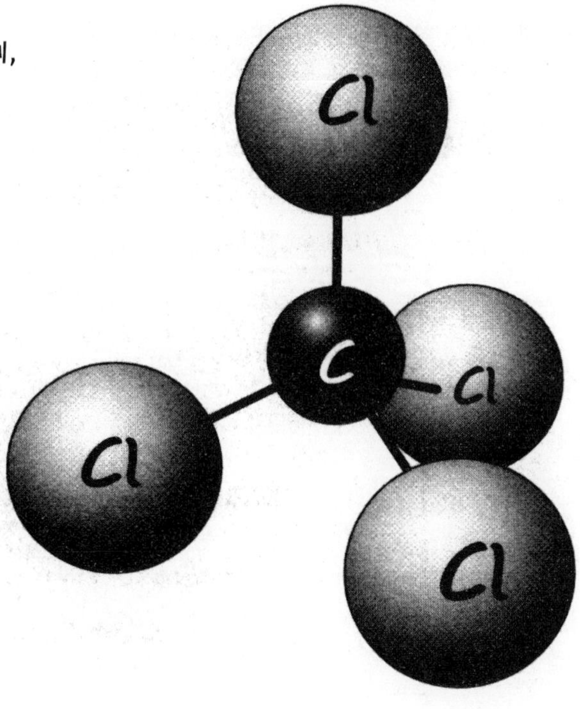

AMMONIA, NH_3 YOU MIGHT EXPECT THIS TO BE A TRIANGLE, BUT THE LEWIS DIAGRAM SAYS OTHERWISE. THE FOURTH ELECTRON PAIR REPELS THE OTHERS, AND WE GET A TETRAHEDRON WITH H AT THREE OF THE VERTICES.

氨（NH_3），你可能以为这是一个三角形。根据电子式事实并非如此。因为第四个电子对排斥其他键，我们得到氢原子在三个顶点上的一个四面体。

WATER, H_2O, IS SIMILAR. IT HAS TWO ELECTRON PAIRS WITH NOTHING ATTACHED TO THEM. THEY, TOO, MUST BE TAKEN INTO ACCOUNT.

水（H_2O）也类似。它有两个额外的电子对没有与其他原子相连，但它们也必须被考虑在内。

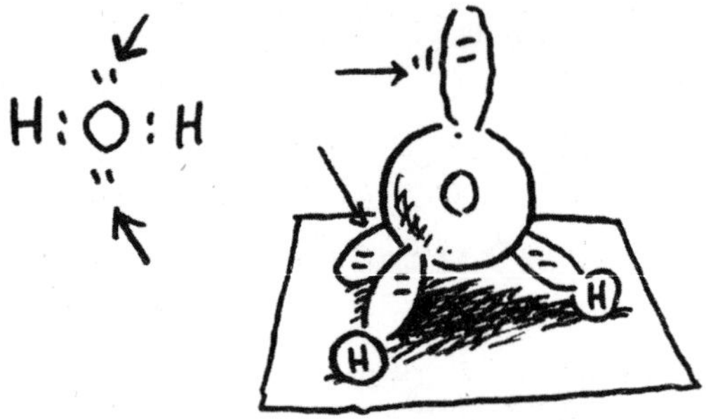

MOLECULES LIKE NH_3 AND H_2O ARE CALLED **BENT.**

如氨和水分子被称为**非直线型**分子。

THIS COVERS THE SHAPES OF THE MOST COMMON MOLECULES, ALTHOUGH THERE ARE SOME ODDITIES LIKE SF_6, WHERE THE SULFUR HAS SIX ELECTRON PAIRS.

前面所述涵盖了最常见的分子形状，但像SF_6气体，其中的硫有6个电子对，就会显得很诡异。

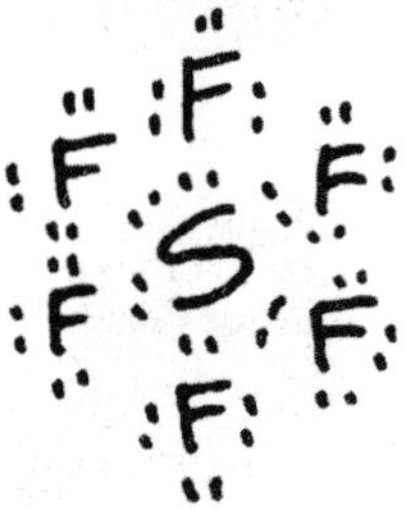

SF_6 IS OCTAHEDRAL.

SF_6是八面体。

Shape and Orbital Bond Theory (advanced)

形状和轨道键理论（高级）

ON THE PREVIOUS TWO PAGES, WE USED THE PRINCIPLE THAT ELECTRON PAIRS IN MOLECULES STAY AWAY FROM EACH OTHER. WE CAN ACCOUNT FOR THIS FACT IN TERMS OF ELECTRON ORBITALS.

前面，我们使用了分子中电子对互相远离的原则，我们可以从电子轨道上解释这一点。

WHEN H BONDS WITH H, TWO s ORBITALS MERGE. THIS IS CALLED A σ (SIGMA) BOND.

当H原子与H原子成键，两个S轨道便合并。这就是所谓的σ（西格玛）键。

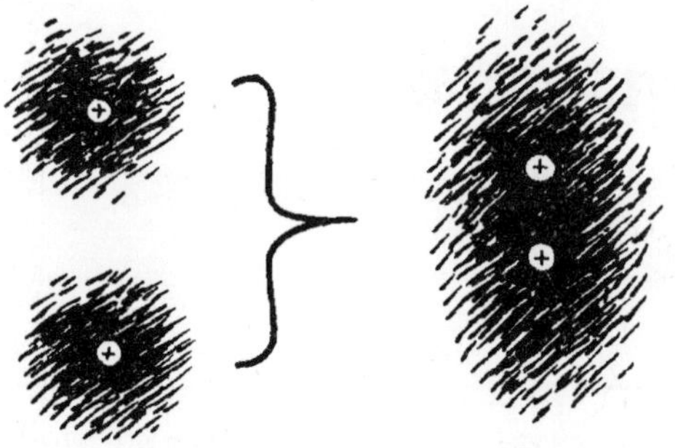

IN O_2, TWO ELECTRONS IN p ORBITALS ARE SHARED IN A π (PI) BOND.

在O_2分子中，每个原子提供两个p轨道电子共享一个π（pi）键。

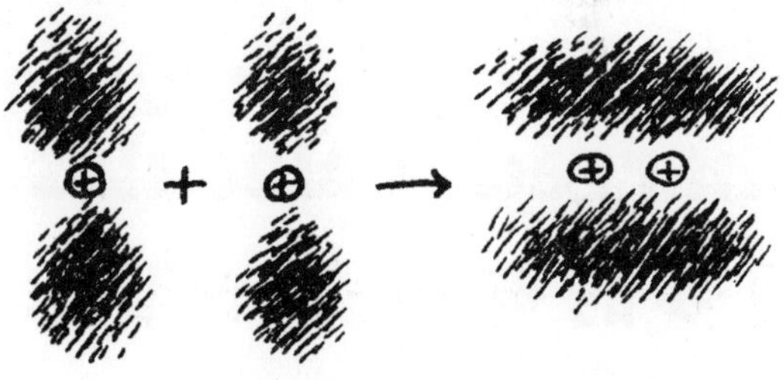

(WE HAVE OMITTED THE NONBONDING ORBITALS.)

（我们省略了非键轨道。）

BUT IN GENERAL, WE GET SOMETHING CALLED **HYBRID ORBITALS.** FOR EXAMPLE:

总体来说，我们得到了称为**杂化轨道的东西**，例如：

CARBON, WITH $2s^2 2p^2$, HAS TWO PAIRED s ELECTRONS AND TWO UNPAIRED p ELECTRONS.

碳，$2s^2 2p^2$，有两个成对的s电子和两个未成对的p电子。

WHEN A HYDROGEN ATOM APPROACHES, ITS NUCLEUS PULLS ON C'S ELECTRONS, RAISING THEIR ENERGY.

当一个氢原子靠近时，其原子核拉向碳的S电子，增强了它们的能量。

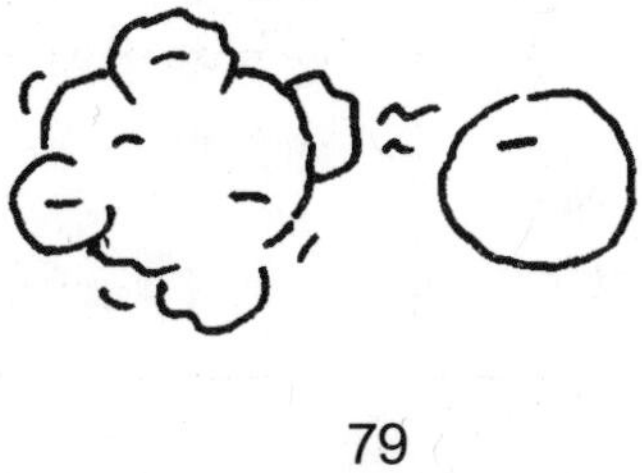

ONE s ELECTRON IS "PROMOTED" TO A p ORBITAL, AND NOW ALL ARE UNPAIRED.

一个s电子被“提升”到p轨道，现在一切外层电子都未成对。

THE UNPAIRED ORBITALS "HYBRIDIZE" AND BECOME LOPSIDED. SUCH AN ORBITAL IS CALLED AN sp **HYBRID.** ONE OF THEM LOOKS LIKE THIS.

未配对轨道进行“杂化”，成为一端大一端小的结构。这种轨道称为sp**杂化**轨道。其中一个看起来是这样的。

AND FOUR OF THEM LOOK LIKE THIS. (HERE EACH ONE IS BONDED TO A HYDROGEN ATOM.)

四个则是这样的。（这里是每一个都与一个氢原子键合。）

THE LOPSIDED LOBES MUST REPEL EACH OTHER, SO THE CH_4 MOLECULE MUST BE A TETRAHEDRON. **THE MOLECULE'S GEOMETRY IS CAUSED BY THE SHAPE OF HYBRID ORBITALS.**

杂化轨道之间必须互相排斥。因此，甲烷分子必然是一个四面体。**分子的几何形状就是根据这种杂化轨道的形状而形成的。**

More on Lewis Diagrams and Charged Molecules

更多的电子式和带电分子

IN A LEWIS DIAGRAM, EACH ATOM ENDS UP WITH A COMPLETE OCTET (USUALLY—SEE BELOW). THIS CAN OFTEN HAPPEN IN MORE THAN ONE WAY. FOR INSTANCE, WE JUST SAW SO_3, BUT SO_2 ALSO EXISTS, AND IS ACTUALLY THE MORE COMMON OXIDE OF SULFUR.

在路易斯式中，每个原子都是一个完整的电子结构（下文常见）。这种结构经常以不止一种方式出现。例如，我们仅仅看到三氧化硫，但二氧化硫也是存在的，而且实际上它是更常见的硫氧化物。

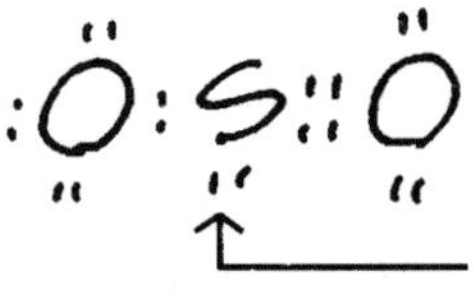

NOTE UNBONDED PAIR

注意未成键电子对

SULFUR'S EXTRA ELECTRON PAIR IMPLIES THAT THE MOLECULE IS BENT.

硫的额外电子对意味着分子会弯曲。

INCIDENTALLY, THE DOUBLE BOND ISN'T REALLY ON ONE OXYGEN OR THE OTHER, BUT SOMEHOW HALFWAY ON BOTH AT THE SAME TIME, A QUANTUM-MECHANICAL MYSTERY KNOWN AS **RESONANCE.**

顺便说一句，双键不是真的连在一个氧或其他原子上，但在某种程度上几乎同时存在于两个原子上，量子力学称为神秘的**共振**。

$$O{=}S{-}O \rightleftharpoons O{-}S{=}O$$

WE CAN ALSO WRITE A LEWIS DIAGRAM FOR **SULFATE,** SO_4^{2-}, WITH NO DOUBLE BONDS AT ALL. THIS LOOKS NICE AND NATURAL, EXCEPT THAT TWO EXTRA ELECTRONS ARE REQUIRED TO COMPLETE ALL THE BONDS. SO SO_4^{2-} IS REALLY A COVALENTLY BONDED POLYATOMIC ION WITH A CHARGE OF −2.

我们也可以将完全没有双键的**硫酸根**电子式记为SO_4^{2-}。这当然看起来不错且自然除了这两个额外的电子，必须完成所有的键。因此，SO_4^{2-}是真正的-2价共价键多原子离子。

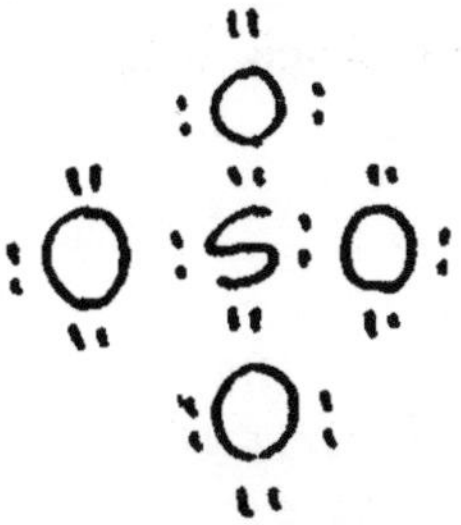

MORE POLYATOMIC IONS:

更多的多原子离子：

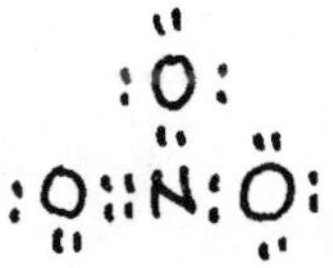

NITRATE, NO_3^-, HAS ONE EXTRA ELECTRON AND RESONANCE BETWEEN THREE DIFFERENT FORMS.

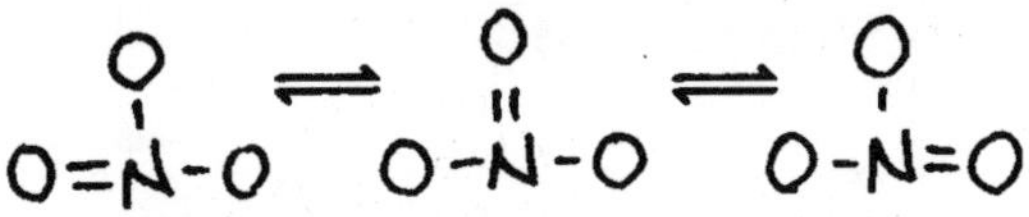

硝酸根，（NO_3^-），有一个额外的电子以及3种不同形式的共振。

HYDROXIDE, OH^-, HAS ONE EXTRA ELECTRON.

氢氧根（OH^-），有一个额外的电子。

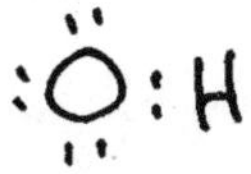

USUALLY, ALL ELECTRONS ARE PAIRED AND EVERY ATOM GETS A FULL OCTET—BUT THERE ARE EXCEPTIONS. IN NITROGEN DIOXIDE, NO_2, NITROGEN HAS AN UNPAIRED ELECTRON.

通常，所有的电子都是成对的，每一个原子都是一个完整的八隅体结构，但是也有例外。在二氧化氮（NO_2）中，氮有一个未成对电子。

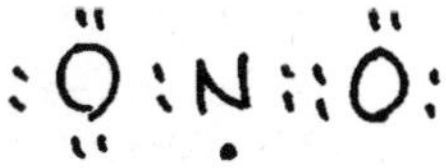

AND IN BERYLLIUM FLUORIDE, BeF_2, Be GETS ONLY HALF AN OCTET.

而氟化铍（BeF_2），铍只有半个八电子结构。

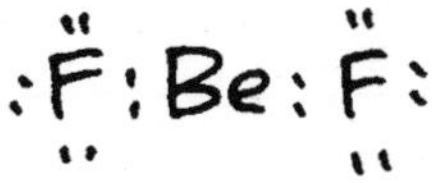

"MOSTLY" IONIC? WHAT IS **THAT** SUPPOSED TO MEAN?

“主要”是离子？那是什么意思？

Polarity 极性

MANY BONDS ARE NOT PURELY COVALENT OR IONIC, BUT SOMEWHERE IN BETWEEN.

许多键并非单纯的共价键或离子键，而是介于两者之间。

CONSIDER WATER, H_2O. OXYGEN, WITH AN ELECTRONEGATIVITY VALUE (EN) OF 3.5, IS MORE ELECTRONEGATIVE THAN HYDROGEN (EN = 2.1).* THIS MEANS THAT THE ELECTRONS IN THE O—H BOND ARE NOT EQUALLY SHARED, BUT TEND TO HOVER CLOSER TO THE OXYGEN ATOM.

以水（H_2O）为例，氧的电负值（En）为3.5，高于氢（En=2.1）。*这意味着，在O-H键中的电子并没有平均共享，而是更接近氧原子。

THE EFFECT OF THIS NOT-PURELY-COVALENT BOND IS THAT THIS MOLECULE HAS POSITIVELY AND NEGATIVELY CHARGED **POLES.** THE HYDROGEN END HAS A FRACTIONAL POSITIVE CHARGE, WHILE THE OXYGEN END HAS A FRACTIONAL NEGATIVE CHARGE, BECAUSE THE ELECTRONS ARE CLOSER TO ONE END.

这个不太纯粹的共价键作用，使这种分子有正**极**和负**极**。氢一端有部分正电荷，而氧一端有部分负电荷，这是因为电子更接近某一端。

*ON AN ARTIFICIAL SCALE RANGING FROM *0.7* FOR CESIUM, THE MOST ELECTROPOSITIVE ELEMENT, TO *4.0* FOR FLUORINE, THE MOST ELECTRONEGATIVE.

* 可以用一种人为的指标对电负性进行排序，从电正性最强的元素铯0.7开始，直到电负性最强的氟4.0。

A BOND LIKE O–H, IN WHICH THE ELECTRONS ARE CLOSER TO ONE END, IS CALLED **POLAR.** POLAR BONDS ARE INTERMEDIATE BETWEEN COVALENT BONDS (EQUAL SHARING) AND IONIC BONDS (COMPLETE TRANSFER OF ELECTRONS).

在O-H键中，电子更接近O一端，这种现象被称为**极性**，极性键介于共价键（平均共享）和离子键（电子完全转移）之间。

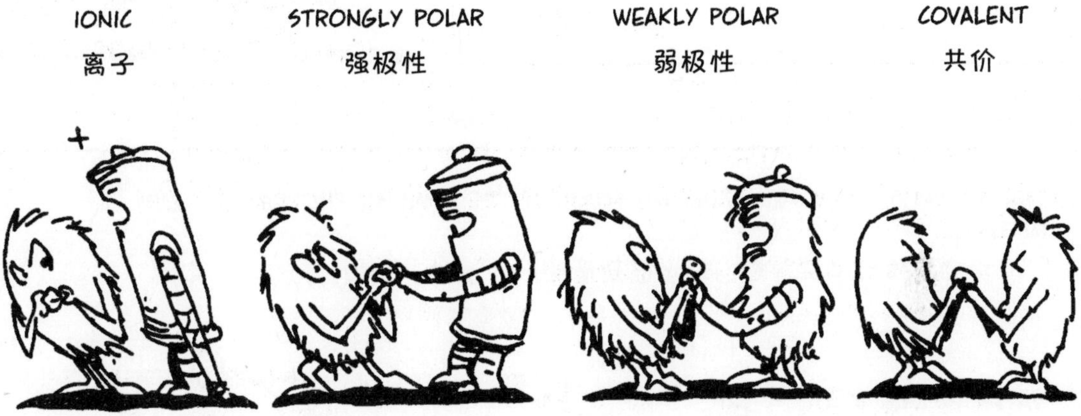

THE POLARITY OF BONDS AFFECTS THE WAY CHARGE IS DISTRIBUTED OVER A MOLECULE.

键的极性影响着电荷在一个分子中的分布。

A BOND'S POLARITY DEPENDS ON THE DIFFERENCE IN ELECTRONEGATIVITY BETWEEN TWO ATOMS. BIGGER DIFFERENCES MEAN MORE POLARITY, WITH A DIFFERENCE OF **2.0** OR MORE BEING CONSIDERED IONIC.

BOND 键	EN DIFF. 电负差	BOND TYPE 键种类
N≡N	0	COVALENT 共价键
C–H	0.4	ESSENTIALLY COVALENT 本质上的共价键
O–H	1.4	MODERATELY POLAR 中等极性
H–F	1.9	STRONGLY POLAR 强极性
Li–F	3.0	IONIC 离子键

键的极性取决于两个原子之间的电负性差异。更大的差异意味着更强的极性，电负差在2.0或以上可认为是离子键。

SAMPLE ELECTRONEGATIVITIES 部分元素的电负性

H	2.1	Na	0.9
Li	1.0	Mg	1.2
C	2.5	S	2.5
N	3.0	Cl	3.0
O	3.5	K	0.8
F	4.0	Ca	1.0

THE POLARITY OF WATER EXPLAINS SOME OF ITS FAMILIAR PROPERTIES. FOR INSTANCE:

水的极性可以解释我们对水很熟悉的一些属性。例如：

WATER IS LIQUID AT ROOM TEMPERATURE. THE PARTIAL CHARGES AT EACH END OF A WATER MOLECULE MAKE THE MOLECULES ATTRACT EACH OTHER, END TO END. WATER BONDS WEAKLY TO ITSELF. THIS INTERNAL COHESION HOLDS WATER TOGETHER IN LIQUID FORM.

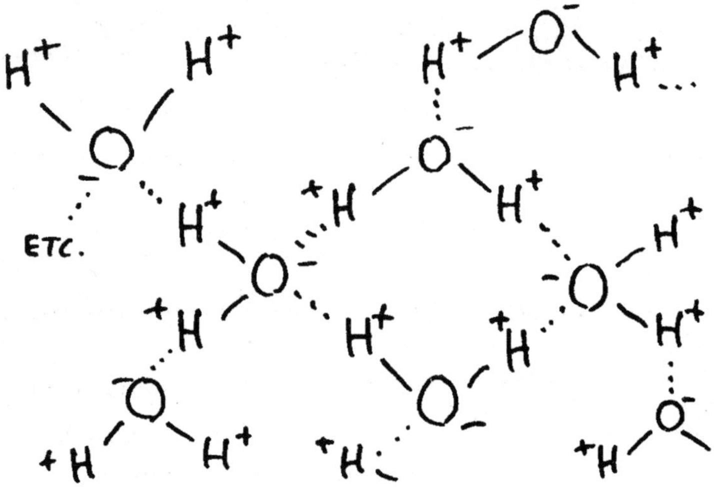

水在室温下是液体。在单个水分子各末端的部分电荷使水分子相互吸引，首尾相接。水分子自身产生微弱的氢键。这种内部凝聚力使水凝聚成液态。

BY CONTRAST, THE MUCH HEAVIER BUT LESS POLAR SO_2 HAS LITTLE MUTUAL ATTRACTION, SO IT FORMS A GAS AT ROOM TEMPERATURE.

相比之下，二氧化硫分子重得多但极性小，所以不相互吸引，在室温下形成气体。

POLARITY ALSO EXPLAINS WHY WATER IS SO GOOD AT DISSOLVING IONIC COMPOUNDS SUCH AS TABLE SALT. THE CRYSTAL'S IONIC BONDS SLOWLY GIVE WAY TO THE PULL OF WATER'S POLES, AS IONS BREAK OFF THE CRYSTAL AND ATTACH THEMSELVES TO WATER MOLECULES.

极性也解释了为什么水对食盐这种离子化合物的溶解度好。水的极性拉动并慢慢破坏晶体的离子键，离子离开晶体并依附于水分子。

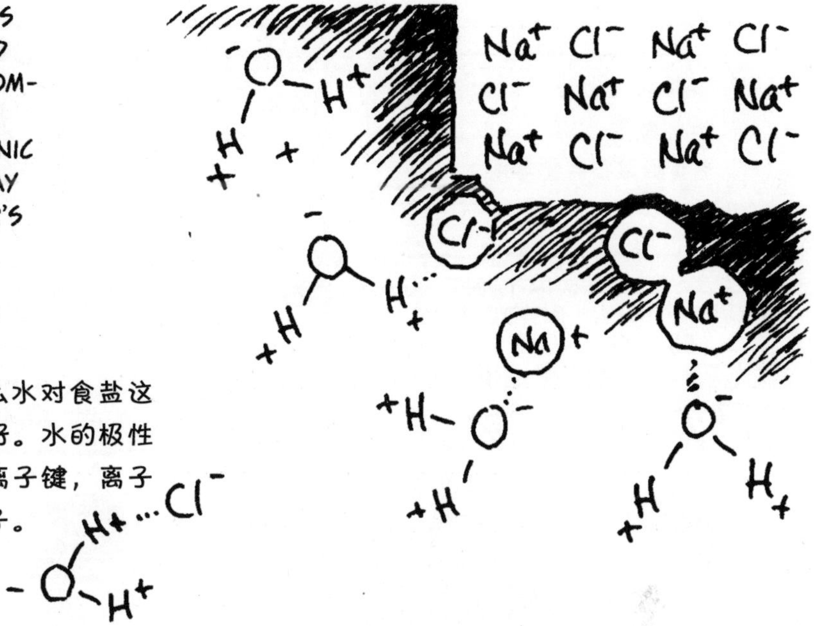

SIMILARLY, THE WEAK ATTRACTION OF A POLAR H TO ANOTHER MOLECULE IS CALLED **HYDROGEN BONDING.** IT HAPPENS TO BE A KEY FEATURE OF THE CHEMISTRY OF LIFE (SEE PAGE 306).

同样，H极与另一个分子之间极弱的吸引力被称为**氢键**。这正好是生命化学的主要特点（见第306页）。

IONIC, COVALENT, METALLIC: THESE ARE THE MAIN TYPES OF CHEMICAL BONDS. WE'VE SEEN HOW THESE INTERATOMIC INTERACTIONS ARISE FROM THE ELECTRICAL PROPERTIES OF ATOMS, AND HOW THEY AFFECT THE STRUCTURES OF SUBSTANCES. NOW WE WANT TO FIND OUT WHAT THEY HAVE TO DO WITH THE CHEM...

离子键、共价键、金属键，这些都是化学键的主要类型。我们已经从原子的带电性了解了这些原子之间是如何相互作用的以及它们是如何影响物质结构的，现在我们要找出它们与化学之间有什么关系……

你看这一切看不见的亚原子物质都很有趣……但这本书不是应该在讲**化学**吗？
LOOK—ALL THIS INVISIBLE SUB-ATOMIC STUFF IS INTERESTING AND ALL... BUT ISN'T THIS BOOK SUPPOSED TO BE ABOUT **CHEMISTRY?**
UH... YES...?
嗯……是……？
那么，爆炸在哪呢？冒泡的毒液呢？我什么时候能闻到恶臭？快点！！！
SO WHERE ARE THE **EXPLOSIONS?** THE BUBBLING, NOXIOUS **GOO?** WHEN CAN I SMELL SOMETHING **VILE?** COME **ON!!!**
呃……咳……这个领域的秘密难道还不能满足你吗？
ER... COUGH... THE SECRETS OF THE UNIVERSE AREN'T ENOUGH FOR YOU?
WELL, I GUESS NOT!
哼！当然不能！
AH... WELL THEN... I THINK YOU MAY LIKE THE NEXT CHAPTER...
啊……好……那么，我想你可能会喜欢下一章……

Chapter 4 第四章
Chemical Reactions 化学反应

OOPS! SOMEHOW WE FIND OURSELVES MAROONED ON A DESERT ISLAND. HOW ARE WE GOING TO SURVIVE? MAYBE WE CAN MAKE SOMETHING USEFUL OUT OF THE MATERIALS AT HAND...

哎呀！不知怎的，我们发现自己被困在一个荒岛上了。我们该如何生存？也许我们可以用手头有的材料做些有用的东西……

Combustion, Combination, Decomposition
燃烧，化合，分解

LET'S WRITE A **REACTION EQUATION** FOR FIRE. WOOD CONTAINS MANY DIFFERENT MATERIALS, BUT IT'S MAINLY MADE OF C, H, AND O IN THE RATIO 1:2:1. WE CAN WRITE THE EMPIRICAL FORMULA FOR WOOD AS CH_2O, AND THEN FIRE LOOKS LIKE THIS:*

让我们写出火的**反应方程式**。木材含有多种不同的材料，但它的主要成分是比例为1:2:1的C、H和O。我们可以以CH_2O作为木材的实验式，燃烧的过程可以这样表示*：

$$CH_2O\ (s) + O_2\ (g) \longrightarrow CO_2\ (g)\uparrow + H_2O\ (g)\uparrow$$

*WE'RE LEAVING OUT PARTIALLY OR WHOLLY NONCOMBUSTED PRODUCTS SUCH AS SOOT, SMOKE, CO, ETC.

*我们不考虑部分或全部未完全燃烧的产物，如烟尘、烟雾、一氧化碳等。

THE NOTATION EXPLAINED: THE SUBSTANCES ON THE LEFT OF THE HORIZONTAL ARROW $\longrightarrow$ ARE CALLED **REACTANTS.** ON THE RIGHT ARE THE **REACTION PRODUCTS.** $\xrightarrow{\Delta}$ WILL MEAN THAT HEAT WAS ADDED. THE SMALL LETTERS IN PARENTHESES SHOW THE PHYSICAL STATE OF THE CHEMICALS: g = GAS; s = SOLID; l = LIQUID; aq = DISSOLVED IN WATER. ↑ MEANS AN ESCAPING GAS, AND ↓ WILL MEAN A SOLID SETTLING OUT OF SOLUTION, OR **PRECIPITATING.**

符号解释：水平箭头（→）左侧的物质称为**反应物**，右边的是**反应产物**。$\xrightarrow{\Delta}$意味着反应需要热量。括号中的小写字母表示化学物质的物理状态：g=气体，s=固体，l=液体，aq=水溶液，↑指释放气体，↓指溶液中出现固体而产生**沉淀**。

SO OUR EQUATION READS: SOLID WOOD PLUS GASEOUS OXYGEN AND HEAT MAKES GASEOUS CARBON DIOXIDE PLUS WATER VAPOR. THIS IS A TYPICAL **COMBUSTION REACTION.** (YOU CAN TEST FOR THE WATER BY HOLDING A COOL GLASS OVER THE FLAME; DROPLETS WILL CONDENSE ON IT.)

因此，我们的方程式如下：木材加氧气和热量，产生二氧化碳气体和水蒸气。这是一个典型的**燃烧反应**。（你可以做一个有关水的测试：在火焰上放置一块冷却的玻璃，玻璃上会凝结出水滴。）

NOW THAT WE HAVE FIRE, WE'LL MAKE A BETTER FUEL: **CHARCOAL.** WE PUT DRY WOOD AND COCONUT SHELLS IN A PIT (TO LIMIT AVAILABLE OXYGEN) AND FIRE IT UP. THE REACTION IS*

既然我们已经有火了，我们就可以制作更好的燃料——**木炭**。我们把干燥的木材和椰子壳放进一个坑里（为了限制可以使用的氧气），然后点燃它。反应是*：

$$CH_2O \xrightarrow{\triangle} C(s) + H_2O(g)\uparrow$$

THIS IS A **DECOMPOSITION** REACTION (OF THE FORM AB ⟶ A + B). IT MAKES ELEMENTAL CARBON, OR CHARCOAL.

这是一个**分解**反应（公式是 AB→A + B）。它产生了单质碳，即木炭。

WE BUILD A STONE STOVE AND FUEL IT WITH CHARCOAL. CHARCOAL'S COMBUSTION IS A **COMBINATION** REACTION (A+B ⟶ AB):

我们建造一个石炉并用木炭做燃料，木炭燃烧是一种**化合**反应（A + B→AB）：

$$C(s) + O_2(g) \longrightarrow CO_2(g)\uparrow$$

*MORE OR LESS. AGAIN WE IGNORE TRACE REACTANTS AND PRODUCTS.
* 我们再次忽略部分微量的反应物和产物。

IN THIS OVEN WE CAN MAKE **POTTERY.** WE SCOOP A FINE-GRAINED MINERAL, KAOLINITE, FROM THE LAKE BOTTOM AND GRIND IT WITH A LITTLE WATER TO MAKE A SMOOTH KAOLIN CLAY, $Al_2Si_2O_5(OH)_4$. WE SHAPE THIS INTO VESSELS AND FIRE THEM IN A HOT OVEN:

我们可利用这个炉子来制作**陶器**。我们从湖底搜集到一种细粒的矿物——高岭石，然后加入少量的水来研磨它，使其变成光滑的高岭土，化学式为$Al_2Si_2O_5(OH)_4$。我们将高岭土塑造成容器并在高温下烧制：

$$3Al_2Si_2O_5(OH)_4\,(s) \xrightarrow{\Delta} Al_6Si_2O_{13}\,(s) + 4SiO_2\,(s) + 6H_2O\,(g)\uparrow$$

THE FIRST PRODUCT IS CALLED **MULLITE.** THE SECOND, SiO_2, IS **SILICA,** OR SAND—AND MELTED, IT'S **GLASS.** WHEN THE CLAY IS FIRED, MULLITE FUSES WITH THE GLASSY SILICA TO FORM A VERY HARD, WATERPROOF POT.

第一个产物叫作**莫来石**；第二个产物二氧化硅是**硅石**，即沙子，而且熔融后可形成**玻璃**。当黏土烧制后，莫来石与玻璃质的硅石结合，就形成非常坚硬的、防水的陶器。

Balancing Equations
配平方程

NOTE THAT SOME OF THE SUBSTANCES IN THE POTTERY REACTION HAVE NUMERICAL COEFFICIENTS IN FRONT OF THEM. THE EQUATION MEANS THREE MOLECULES OF KAOLIN CLAY YIELD ONE MOLECULE OF MULLITE, FOUR OF SILICA, AND SIX OF WATER.

请注意在制备陶器反应中的一些物质前面的系数。该方程意味着三分子高岭石产生一分子莫来石、四分子的硅石和六分子的水。

$$3Al_2Si_2O_5(OH)_4(s) \xrightarrow{\Delta} Al_6Si_2O_{13}(s) + 4SiO_2(s) + 6H_2O(g)\uparrow$$

THE COEFFICIENTS **BALANCE** THE EQUATION. THE SAME NUMBER OF EACH KIND OF ATOM APPEARS ON BOTH SIDES: 6 Al, 6 Si, 27 O, and 12 H. HOW DO WE FIND THESE COEFFICIENTS?

这些系数**配平**了反应方程。每个原子以相同的数目出现在方程式两侧：6Al、6Si、27O和12H。我们该如何找到这些系数呢？

START WITH AN UNBALANCED EQUATION

以未配平的方程开始：

$$Al_2Si_2O_5(OH)_4(s) \xrightarrow{\Delta} Al_6Si_2O_{13}(s) + SiO_2(s) + H_2O(g)\uparrow$$

	L	R
Al	2	6
Si	2	3
O	9	16
H	4	2

WRITE DOWN THE NUMBER OF ATOMS ON EACH SIDE.

记下每边的原子数。

BALANCE ONE ELEMENT. WE START WITH Al. MULTIPLY BY 3 ON THE LEFT TO GET:

配平一个元素。我们从铝开始。把方程左边乘以3：

$$3Al_2Si_2O_5(OH)_4(s) \xrightarrow{\Delta} Al_6Si_2O_{13}(s) + SiO_2(s) + H_2O(g)\uparrow$$

	L	R
Al	6	6
Si	6	3
O	27	16
H	12	2

AGAIN COUNT ATOMS ON EACH SIDE.

再次计算每边的原子数。

BALANCE ANOTHER ELEMENT. WE CAN BALANCE Si BY PUTTING A 4 IN FRONT OF SiO_2:

配平另一个元素。我们可以把4放在SiO_2的前面配平Si:

$$3Al_2Si_2O_5(OH)_4(s) \xrightarrow{\Delta} Al_6Si_2O_{13}(s) + 4SiO_2(s) + H_2O(g)\uparrow$$

	L	R
Al	6	6
Si	6	6
O	27	22
H	12	2

AGAIN COUNT ATOMS ON EACH SIDE.

再次计算每边的原子数。

FINALLY, A 6 IN FRONT OF H_2O BALANCES BOTH H AND O.

最后，在H_2O前面写下6配平H和O。

$$3Al_2Si_2O_5(OH)_4(s) \xrightarrow{\Delta} Al_6Si_2O_{13}(s) + 4SiO_2(s) + 6H_2O(g)\uparrow$$

	L	R
Al	6	6
Si	6	6
O	27	27
H	12	12

- WRITE THE EQUATION WITHOUT COEFFICIENTS.
- LIST THE ELEMENTS IN THE EQUATION.
- CHECK THE NUMBER OF EACH KIND OF ATOM ON BOTH SIDES.
- BALANCE ATOMS ONE ELEMENT AT A TIME BY ADJUSTING COEFFICIENTS.
- REDUCE TO LOWEST TERMS IF NECESSARY.

· 写下没有系数的方程。
· 列出方程中的元素。
· 检查方程两边每一种原子的数量。
· 通过调整系数，一次配平一个元素的原子。
· 尽可能使系数降到最低。

THE ACT, OR ART, OF BALANCING EQUATIONS IS CALLED **REACTION STOICHIOMETRY.**

配平方程的过程，或者说艺术，被称为**反应的化学计量学**。

HERE ARE SOME PRACTICE EXAMPLES. SUPPLY COEFFICIENTS IN EACH EQUATION.

这里有一些练习的例子。给以下每个方程配平系数。

$$Al(s) + Fe_2O_3(s) \xrightarrow{\Delta} Al_2O_3(s) + Fe(s)$$

$$KClO_3(s) \xrightarrow{\Delta} KCl(s) + O_2(g)$$

$$C_4H_{10}(g) + O_2(g) \longrightarrow CO_2(g) + H_2O(g)$$

$$N_2(g) + H_2(g) \longrightarrow NH_3(g)$$

$$P_4(s) + F_2(g) \longrightarrow PF_5(g)$$

$$Zn(NO_3)_2(s) \xrightarrow{\Delta} ZnO(s) + NO_2(g) + O_2(g)$$

$$H_3PO_4(l) \xrightarrow{\Delta} H_2O(l) + P_4O_{10}(s)$$

$$Cu(s) + AgNO_3(aq) \longrightarrow Cu(NO_3)_2(aq) + Ag\downarrow$$

$$Fe(s) + O_2(g) \longrightarrow Fe_2O_3(s)$$

$$FeCl_3(s) + H_2O(l) \longrightarrow HCl(aq) + Fe(OH)_3\downarrow$$

The Mole 摩尔

THE EQUATION'S COEFFICIENTS LET US FIND THE RELATIVE **MASSES** OF PRODUCTS AND REACTANTS. THE CALCULATION USES A UNIT CALLED THE **MOLE.** ONE MOLE OF A SUBSTANCE IS THE AMOUNT WHOSE MASS EQUALS THE MOLECULAR OR ATOMIC WEIGHT OF THE SUBSTANCE **EXPRESSED IN GRAMS.**

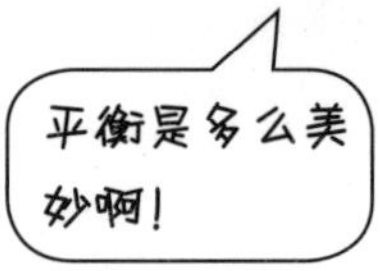

该方程的系数让我们找到产物和反应物相对**质量**之间的联系。我们使用一个称为**摩尔**的单位来计算。1摩尔物质的质量等于**以克表示**的物质分子或原子的重量。

THAT'S KIND OF A MOUTHFUL FOR A SIMPLE IDEA. LET'S ILLUSTRATE BY EXAMPLE:

这是种简化问题的想法。让我们通过实例说明：

	"MOLECULAR" WEIGHT 分子质量	MOLAR WEIGHT 摩尔质量
O_2	32 AMU	32 GRAMS
SiO_2	60 AMU	60 GRAMS
$Al_2Si_2O_5(OH)_4$	258 AMU	258 GRAMS
Fe	56 AMU	56 GRAMS
PROTON	1 AMU	1 GRAM
NaCl	58.5 AMU	58.5 GRAMS

NOTE: HERE MOLECULAR WEIGHT REALLY MEANS THE MASS OF A BASIC PARTICLE OF THE SUBSTANCE EXPRESSED IN AMU. IN AN IONIC CRYSTAL LIKE NaCl, WE MEAN A BASIC COMPONENT OF THE CRYSTAL.

注意：这里的分子质量的真正意义指一个基本的物质粒子的质量，用AMU表示。离子晶体，比如盐，我们认为是晶体的基本组成部分。

THE MOLE IS USED TO SCALE UP FROM ATOMIC DIMENSIONS TO METRIC WEIGHTS. TO BE PRECISE, A GRAM IS ABOUT 602,200,000,000,000,000,000,000 BIGGER THAN AN AMU. THAT IS, 1 g = 6.022×10^{23} AMU.

摩尔这个单位用于把原子尺度转换到公制重量单位，准确地说，1克约有602,200,000,000,000,000,000,000AMU。也就是说，1g = 6.022×10^{23}AMU

译注：摩尔表示物质的量（物质的量表示一定数目粒子的集体的物理量，属于专有名词，研究对象只能是粒子）。

THIS THEN, IS THE **NUMBER OF PARTICLES IN A MOLE.** A MOLE OF **ANYTHING** HAS THIS MANY PARTICLES! 6.022×10^{23} IS CALLED **AVOGADRO'S NUMBER,** AFTER AMEDEO AVOGADRO, WHO FIRST SUGGESTED THAT EQUAL VOLUMES OF GAS HAVE EQUAL NUMBERS OF MOLECULES.

这就是1**摩尔所含的粒子数。**1摩尔的**任何东西**会有这么多的**粒子！**6.022×10^{23}被称为**阿伏伽德罗常数。**阿美德奥·阿伏伽德罗*首先提出，相同体积的气体具有同等数目的分子。

*编者注：阿美德奥·阿伏伽德罗，意大利科学家，他提出了著名的阿伏伽德罗定律。

NOW SUPPOSE I START WITH 100 kg OF CLAY. HOW MANY KILOGRAMS OF POTTERY WILL I GET? WE START WITH THE BALANCED EQUATION:

现在假设我用了100千克黏土，我会得到多少陶器呢？我们开始配平方程：

$$3Al_2Si_2O_5(OH)_4(s) \xrightarrow{\Delta} \underbrace{Al_6Si_2O_{13}(s) + 4SiO_2(s)} + 6H_2O(g)\uparrow$$

THE CLAY 黏土 — THE POTTERY 陶器

THEN WRITE A **MASS-BALANCE TABLE**, SHOWING THE NUMBER OF GRAMS OF EACH REACTANT AND PRODUCT:

然后写出一个**质量平衡表**，表示出各反应物和产物的克数：

REACTANTS 反应物	MOLAR WEIGHT 摩尔质量	PRODUCTS 产物	MOLAR WEIGHT 摩尔质量
3 MOL $Al_2Si_2O_5(OH)_4$	3 X 258 = 774 g	1 MOL $Al_6Si_2O_{13}$	426 g
		4 MOL SiO_2	4 X 60 = 240 g
		6 MOL H_2O	6 X 18 = 108 g
TOTAL 总计	774 g	TOTAL 总计	774 g

THIS SAYS 774 g OF KAOLIN CLAY MAKES 426 + 240 = 666 g OF POTTERY.

SO 1 g KAOLIN MAKES (666/774) g = 0.86 g OF POTTERY

AND 100 kg MAKES (0.86)(100kg)(1000 g/kg) = 86,000 g = 86 kg.

这表示774克高岭土能够得到426 + 240 = 666克的陶器。1克高岭土可以得到（666/774）克= 0.86克陶器，那么，100千克高岭土就可以得到0.86×100千克×1000克/千克= 86000克= 86千克陶器。

WE CAN EQUALLY WELL WORK BACKWARD. IF WE WANT 100 kg OF POTTERY, HOW MUCH WET CLAY SHOULD WE MIX UP? (ANS: (100)(774/666) kg.)

我们同样也可以倒推。如果我们想得到100千克的陶器，需要混合多少湿黏土？答：(100)(774/666) 千克。

More Reactions
更多的反应

WE'VE MADE VESSELS AND A STOVE. NOW LET'S COOK UP SOME **BUILDING MATERIALS.** WE HEAT LIMESTONE, CHALK, AND/OR SEASHELLS, WHICH ARE ALL MADE OF CALCIUM CARBONATE, $CaCO_3$. THE PRODUCT IS **QUICKLIME,** CaO.

我们已经做好了容器和炉灶。现在，让我们生产一些**建筑材料**。我们加热石灰石、白垩，也可加入贝壳，它们的主要成分都是碳酸钙（$CaCO_3$）。生成的产品是**生石灰**（CaO）。

$$CaCO_3(s) \xrightarrow{\Delta} CaO(s) + CO_2(g)\uparrow$$

BAKING CaO TOGETHER WITH POWDERED VOLCANIC ROCK MAKES **CEMENT.** ADD WATER, SAND, AND PEBBLES—**CONCRETE!** LET'S BUILD!

烘焙粉末状的火山岩和生石灰可以生产出**水泥**。加入水、沙子和小石子可以得到**混凝土**！让我们开始行动吧！

WE CAN EVEN PAINT OUR HOUSE. **WHITEWASH,** OR **SLAKED LIME,** $Ca(OH)_2$, COMBINES CaO AND H_2O:

我们甚至可以用**石灰水**或**熟石灰**来粉刷我们的房子，熟石灰也就是石灰水——$Ca(OH)_2$，是由CaO和H_2O化合而成的：

$$CaO(s) + H_2O(l) \longrightarrow Ca(OH)_2(aq)$$

SLAKED LIME ALSO MAKES A GOOD PUTTY AND MORTAR... AND OVER TIME, WHITEWASH SLOWLY COMBINES WITH CO_2 FROM THE AIR AND HARDENS INTO A WHITE, STUCCO-LIKE MATERIAL:

熟石灰虽不起眼，但却是很好的涂料……随着时间的推移，石灰水慢慢与空气中的二氧化碳结合变硬形成白色灰泥状物质：

$$Ca(OH)_2(s) + CO_2(g) \longrightarrow CaCO_3(s) + H_2O(g)\uparrow$$

LIMESTONE AGAIN!

又变成石灰石了！

NOW LET'S MAKE **SOAP,** SO WE CAN WASH UP.

现在，让我们做块**肥皂**，然后我们就可以洗洗了。

FIRST BURN SEAWEED TO GET A WHITE, POWDERY MIXURE OF Na_2CO_3 (SODA ASH) AND K_2CO_3 (POTASH). SEPARATE OUT THE SODA ASH (NEVER MIND HOW).

首先，燃烧海藻以得到白色Na_2CO_3（纯碱）和K_2CO_3（碳酸钾）粉末状混合物。分离出纯碱（忽略这个分离过程）。

COMBINE SODA ASH WITH WHITEWASH TO MAKE THIS REACTION:

混合纯碱和生石灰，就会发生如下反应：

$$Ca(OH)_2(aq) + Na_2CO_3(aq) \longrightarrow 2NaOH(aq) + CaCO_3(s)\downarrow$$

A WHITE CLOUD OF $CaCO_3$ SETTLES TO THE BOTTOM. DECANT—CAREFULLY!—THE CLEAR NaOH SOLUTION. THIS IS **CAUSTIC LYE,** STRONG STUFF!

白色絮状的碳酸钙沉淀到水底。倒出澄清的氢氧化钠溶液。当心！这是腐蚀性的碱液，具有强碱性。

WE BOIL SOME **WILD BOAR FAT** WITH THE CAUSTIC LYE. THE FAT WILL NOT DISSOLVE IN WATER, BUT THE SODIUM IONS PUT A POLAR "TAIL" ON THE FAT MOLECULE, ALLOWING IT TO INTERACT WITH WATER IN A SOAPY WAY. WHAT'S THE REACTION?
我们把**野猪的脂肪**和腐蚀性的碱液共煮。脂肪不会溶于水，但钠离子使脂肪分子产生极性的“尾巴”，以皂化的方式使它与水相互作用。这算什么反应呢？
嗯，**野猪**很生气，后果很严重！
WELL, THE **BOAR** IS **NOT** HAPPY!

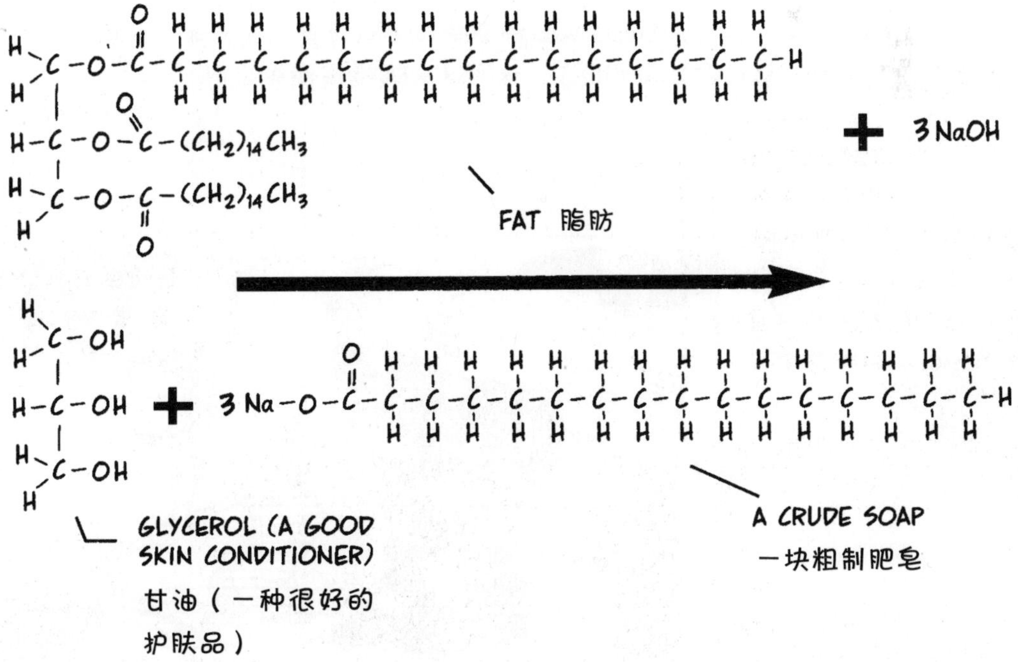
$+ 3NaOH$
FAT 脂肪
GLYCEROL (A GOOD SKIN CONDITIONER)
甘油（一种很好的护肤品）
A CRUDE SOAP
一块粗制肥皂

Redox Reactions
氧化还原反应

NOW LET'S MAKE SOME FLARES, SO WE CAN SIGNAL PASSING SHIPS. THIS WILL REQUIRE **EXPLOSIVE POWDER.** ITS INGREDIENTS ARE **CHARCOAL, SULFUR,** AND POTASSIUM NITRATE OR **SALTPETER,** KNO_3.

现在，让我们做一些照明弹，然后我们可以用作信号指示过往船只。这需要**炸药**。其成分是**木炭**、**硫黄**和硝酸钾或**硝石**（KNO_3）。

WE ALREADY HAVE CHARCOAL... SULFUR WE SCRAPE UP IN ELEMENTAL FORM FROM THE NEARBY VOLCANO (IT'S THE YELLOW STUFF)... K IS IN POTASH, AND NITRATE WILL COME FROM $Ca(NO_3)_2$, WHICH WE FIND IN **BAT GUANO.**

我们已经有木炭……我们从附近的火山刮到单质形式硫黄（它是黄色的东西）……钾存在于碳酸钾中，硝酸盐来自于$Ca(NO_3)_2$，这可以从**蝙蝠粪便**中找到。

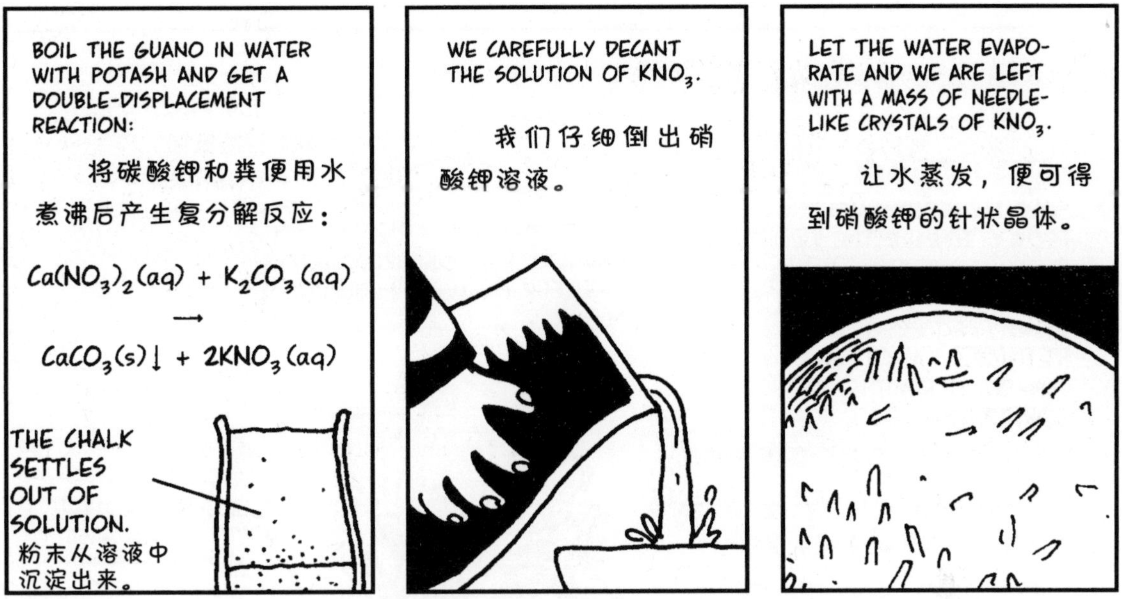

WHAT WILL THE REACTION PRODUCTS BE WHEN WE SET THIS STUFF OFF?

当我们点燃它，会产生什么反应产物呢？

$$C + S + KNO_3 \longrightarrow ??$$

IT TURNS OUT THAT WE CAN MAKE A GOOD GUESS AT THE PRODUCTS BY FOLLOWING THE **ELECTRONS.**

事实证明，我们可以通过追踪**电子**对产物作一个很好的猜测。

EXPLOSIONS BELONG TO AN IMPORTANT CLASS OF REACTIONS INVOLVING THE **TRANSFER OF ELECTRONS** FROM ONE ATOM TO ANOTHER. SUCH REACTIONS ARE CALLED **OXIDATION-REDUCTION** REACTIONS, OR **REDOX** FOR SHORT.

爆炸属于一类涉及电子从一个原子转移到另一个原子的重要反应，这种反应被称为氧化还原反应，简称为REDOX（氧化还原反应）。

EXAMPLE: IN COMBUSTION,

例如：燃烧，

$C + O_2 \rightarrow CO_2$,

FOUR ELECTRONS MOVE FROM C TOWARD THE TWO O ATOMS. WE SAY C IS **OXIDIZED.** O, WHICH GAINS ELECTRONS, IS **REDUCED.** ANOTHER EXAMPLE IS RUSTING, OR **CORROSION:**

4个电子从C向两个氧原子移动。我们说C**被氧化**。O得到电子，**被还原**。另一个例子是生锈或腐蚀：

$4Fe + 3O_2 \rightarrow 2Fe_2O_3$

Fe SHEDS ELECTRONS AND IS OXIDIZED; O GAINS THEM AND IS REDUCED.

Fe失去电子被氧化，O得到电子被还原。

NOTE: OXYGEN ITSELF NEED NOT BE INVOLVED! OXIDATION MEANS THE TRANSFER OF ELECTRONS TO **ANY** ATOM!

注意：氧本身不必参与氧化反应！氧化是指电子转移到**任何**原子的过程！

Oxidation Numbers 氧化数

HOW MANY ELECTRONS DOES EACH ATOM GAIN OR LOSE?

每个原子得到或失去多少个电子？

THE **OXIDATION STATE** OR **OXIDATION NUMBER** OF AN ELEMENT IN A COMPOUND SHOWS ITS SURPLUS OR DEFICIT OF ELECTRONS. THAT IS, THE OXIDATION NUMBER IS THE **NET CHARGE ON THE ATOM.*** FOR INSTANCE, IN CaO, Ca HAS THE OXIDATION NUMBER +2—IT GIVES AWAY TWO ELECTRONS—AND O'S OXIDATION NUMBER IS −2, BECAUSE IT ACCEPTS TWO.

一个元素在化合物中的**氧化态**或**氧化数**显示其得到或失去的电子数。也就是说，氧化数是**原子的净电荷***。例如，在氧化钙中，Ca具有氧化数+2， 它给出两个电子，而O的氧化数为-2，因为它接受两个电子。

*OR WHAT IT WOULD BE, IF THE BOND WERE FULLY IONIC. IN ASSIGNING OXIDATION NUMBERS, WE PRETEND THAT THE ELECTRONS ARE COMPLETELY TRANSFERRED FROM ONE ATOM TO ANOTHER, EVEN THOUGH IN REALITY THEY MAY BE ONLY UNEQUALLY SHARED.

* 即如果化学键完全是离子键，将会是怎样的情况？分配氧化数时，我们就假定电子完全从一个原子转移到另一个原子，虽然在现实中，它们可能只是分配得不平衡而已。

1) THE OXIDATION NUMBER OF AN ELEMENT IN ELEMENTAL FORM IS ZERO.

2) SOME ELEMENTS HAVE THE SAME OXIDATION NUMBER IN ALMOST ALL THEIR COMPOUNDS:

- H: +1 (EXCEPT IN METAL HYDRIDES LIKE NaH, WHERE IT'S -1)
- ALKALI METALS Li, Na, K, ETC.: +1
- GROUP 2 METALS Be, Mg, ETC.: +2
- FLUORINE: −1
- OXYGEN: ALMOST ALWAYS -2

3) IN A NEUTRAL COMPOUND, THE OXIDATION NUMBERS ADD UP TO ZERO.

4) IN A POLYATOMIC ION, THE OXIDATION NUMBERS ADD UP TO THE CHARGE ON THE ION.

1）一个元素单质中的元素氧化数为零。

2）有些元素在几乎所有化合物中都具有相同的氧化数：

- 氢：+1（金属氢化物NaH除外，为-1）
- 碱金属锂、钠、钾等：+1
- 第二族金属铍、镁等：+2
- 氟：-1
- 氧：几乎总是-2

3）在一个中性的化合物中，氧化数之和为零。

4）多原子离子中，各原子的氧化数加起来等于离子所带电荷数。

AN ATOM'S OXIDATION NUMBER DEPENDS ON THE OTHER ATOMS AROUND IT. FOR INSTANCE, IN HCl, CHLORINE ACQUIRES ONE ELECTRON (FOR AN OXIDATION STATE OF −1) BECAUSE Cl IS MORE ELECTRONEGATIVE (EN = 3.0) THAN HYDROGEN (EN = 2.1).

一个原子的氧化数取决于它周围的其他原子。例如，在氯化氢中，氯获得一个电子（氧化态为-1），因为氯的负电性（EN=3.0）比氢（EN=2.1）的更强。

BUT IN THE **PERCHLORATE** ION, ClO_4^-, CHLORINE HAS AN OXIDATION NUMBER OF +7. ALL ITS VALENCE ELECTRONS GO TO OXYGEN, WHICH IS EVEN MORE ELECTRONEGATIVE (EN = 3.5) THAN CHLORINE.

但在**高氯酸根**离子中（ClO_4^-），氯有一个正七价氧化数，它的所有价电子都转移给氧，因为氧的电负性比氯强（EN=3.5）。

HERE ARE SOME ELEMENTS AND THEIR COMMON OXIDATION NUMBERS. THE BIGGER THE PLUS, THE MORE OXIDIZED.

这里有一些元素与其常见的氧化数。氧化数越大，氧化程度越高。

	MOST REDUCED 最低还原态	INTERMEDIATE 中间态	MOST OXIDIZED 最高氧化态
H	NiH_2 (-1)	H_2 (0)	H_2O, OH^- (+1)
C	CH_4 (-4)	C (0)	CO_2, CO_3^{2-} (+4)
O	H_2O, CO_2, CaO, ETC. (-2)	H_2O_2 (-1) (过氧化氢) (HYDROGEN PEROXIDE)	O_2 (0)
N	NH_3 (-3)	N_2 (0), N_2O (+1), NO (+2)	NO_3^- (+5)
S	H_2S, K_2S (-2)	S (0), SO_2 (+4)	SO_3, SO_4^{2-} (+6)
Fe	Fe (0)	FeO (+2)	Fe_2O_3 (+3)
Cl	HCl (-1)	Cl_2 (0)	ClO_4^- (+7)

→ OXIDATION 氧化

还原 REDUCTION ←

IN REDOX REACTIONS, SOME SUBSTANCES—**REDUCING AGENTS** OR **REDUCTANTS**—DONATE ELECTRONS, AND OTHERS—**OXIDIZING AGENTS**, OR **OXIDANTS**—GAIN THEM.

在氧化还原反应中，一些物质比如**还原剂**给出电子，其他物质如**氧化剂**则得到电子。

GOING BACK TO THE EXPLOSIVE BLACK POWDER, WHAT ARE THE MOST LIKELY OXIDIZING AND REDUCING AGENTS? LET'S IGNORE THE SULFUR FOR THE TIME BEING AND CONCENTRATE ON THE CARBON AND SALTPETER:

再回到黑火药爆炸的问题，什么是最有可能的氧化剂和还原剂？让我们忽略硫黄，集中考虑碳和硝石：

$$C + KNO_3 \longrightarrow ?$$

OF THOSE FOUR ELEMENTS, WE CAN ELIMINATE K AND O, BECAUSE THEY ARE ALREADY FULLY OXIDIZED (K AT +1) AND REDUCED (O AT -2) RESPECTIVELY. IT IS VERY HARD TO OXIDIZE O^{2-} OR REDUCE K^{+}! BUT C (0) CAN BE OXIDIZED TO +4 AS EITHER CO_2 OR CO_3^{2-}, AND N (+5) CAN BE REDUCED TO 0 AS N_2. SO WE EXPECT SOMETHING LIKE THIS BEFORE BALANCING:

在这4个因素中，我们可以排除K和O，因为它们已经分别地完全被氧化（K为+1价）或被还原（O为-2价）。通常氧化O^{2-}或还原K^{+}是个艰难的活儿！但C（0）可被氧化为+4价成为CO_2或CO_3^{2-}，而且N（+5）可被还原为0价成为N_2。因此，我们预期在配平之前得到如下反应式：

$$C(s) + KNO_3(s) \longrightarrow CO_2(g)\uparrow + N_2(g)\uparrow + K_2CO_3(s)$$

WE CAN BALANCE THIS BY FOLLOWING THE ELECTRONS: EACH MOL OF C GIVES UP 4 MOL ELECTRONS, AND EACH MOL OF N ACCEPTS 5. THIS BALANCES IF **20** MOL ELECTRONS MOVE FROM 5C TO 4N. (WE GET THE OTHER COEFFICIENTS BY BALANCING K AND O.)

我们可以通过电子转移配平以下反应式：每摩尔C放弃4摩尔电子，每摩尔N接受5摩尔电子。如果20摩尔电子从5摩尔C移动至4摩尔N就配平了。（我们通过配平K和O，得到其他系数。）

$$5C(s) + 4KNO_3(s) \longrightarrow 3CO_2(g)\uparrow + 2N_2(g)\uparrow + 2K_2CO_3(s)$$

THIS REACTION WILL ACTUALLY PRODUCE A PRETTY GOOD FIZZ, BUT CENTURIES OF EXPERIMENT HAVE SHOWN THAT ADDING SULFUR MAKES A MUCH BIGGER POP.

这种反应实际上会发出一种相当不错的嘶嘶声，但过去数百年的实验已经表明，加入硫黄会有更大的威力。

硫黄就是**金色**的！或者是黄色的……

ELEMENTAL S (0), REDUCES EASILY TO -2 IN K_2S. IN FACT, CHEMISTS NOW KNOW THAT FORMING K_2S IS "EASIER" THAN FORMING K_2CO_3. DOING SO CONSUMES LESS ENERGY—AND LEAVES MORE ENERGY TO POWER THE BANG. SO WE EXPECT SOMETHING LIKE:

单质S（0），在K_2S中容易被还原成-2。事实上，化学家现在已经知道，形成K_2S比形成K_2CO_3"容易"。这样做既可以消耗较少的能量，又能留下更多的能量产生爆炸。

$$C(s) + KNO_3(s) + S(s) \longrightarrow CO_2(g)\uparrow + N_2(g)\uparrow + K_2S(s)$$

EACH C LOSES 4 ELECTRONS

每个C失去4个电子

EACH N GAINS 5 ELECTRONS

每个N得到5个电子

EACH S GAINS 2 ELECTRONS

每个S获得两个电子

THIS BALANCES WHEN 3 MOLS C GIVE UP 12 MOLS ELECTRONS, OF WHICH 10 MOLS ELECTRONS GO TO 2 MOLS N AND 2 MOLS ELECTRONS GO TO ONE MOL S:

当3摩尔C放弃12摩尔电子时，其中10摩尔电子转移给2 摩尔N，2摩尔电子转移给1摩尔S：

$$3C(s) + 2KNO_3(s) + S(s) \longrightarrow 3CO_2(g)\uparrow + N_2(g)\uparrow + K_2S(s) + \textbf{BANG!}\ 砰!$$

NOW WE CAN MAKE A FORMULA FOR BLACK POWDER. WE START WITH THE MASS-BALANCE TABLE:

现在，我们可以做一个黑火药的化学反应式。我们从物料平衡表开始：

REACTANTS 反应物	MOLAR WEIGHT 摩尔质量	PRODUCTS 产品	MOLAR WEIGHT 摩尔质量
3 mol C	3 x 12 = 36 g	3 mol CO_2	3 x 44 = 132 g
2 mol KNO_3	2 x 101 = 202 g	1 mol N_2	28 g
1 mol S	32 g	1 mol K_2S	110 g
TOTAL 总计	270 g	TOTAL 总计	270 g

FOR ONE GRAM OF POWDER, WE NEED (36/270) g = 0.13 g C, (202/270) g = 0.75 g KNO_3, AND (32/270) g = 0.12 g S. MULTIPLY BY 100 TO SEE WHAT WE NEED TO MAKE 100 g OF POWDER:

对于1克的火药，我们需要(36/270)克=0.13克 C, (202/270)克=0.75克KNO_3和(32/270)克=0.12克S，再乘以100，来看看要得到100克火药需要些什么：

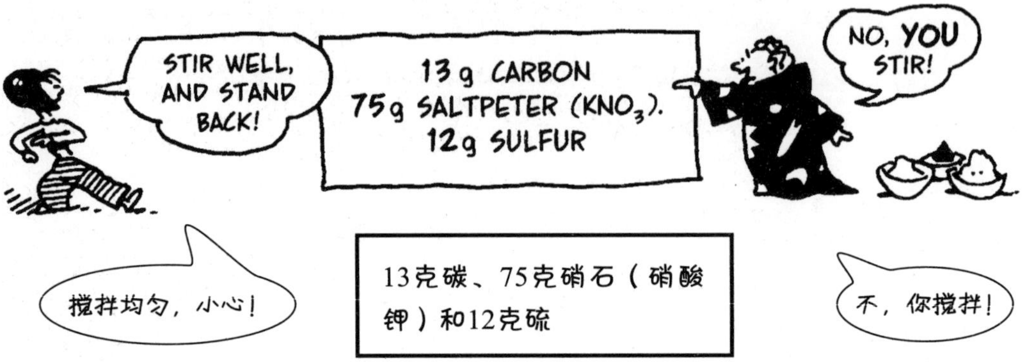

NOT BAD! A CLASSIC GUNPOWDER RECIPE CALLS FOR 10 g SULFUR, 15 g CARBON, AND 75 g SALTPETER. THE DIFFERENCE FROM OUR RESULT IS DUE TO TRACES OF OTHER REACTION PRODUCTS THAT WE NEGLECTED. THE REAL RECIPE IS A PRODUCT OF TRIAL AND ERROR.

不错！一个经典的火药配方需要10克硫、15克碳和75克硝石。由于我们忽略了其他反应物，导致我们的结果与真实火药配方有所差异。真正的配方是一个反复试验与纠错的产物。

IF YOU TRY THIS AT HOME (**NOT RECOMMENDED** IN THE FIRST PLACE), ALWAYS BE SURE TO GRIND THE INGREDIENTS **SEPARATELY**— UNLESS YOU WANT TO BLOW OFF YOUR FINGERS, OR EVEN YOUR WHOLE HAND.

如果你在家里尝试（不推荐作为首选），一定要把这些成分分开研磨——除非你想炸飞你的手指，甚至你的整个手。

CONGRATULATIONS! YOU'RE RESCUED!
恭喜你们，你们获救了！
WHO WANTS TO BE RESCUED? WE WERE JUST SHOWING OFF!
谁愿意被拯救？我们只是在做秀。
COULD YOU BRING US SOME LAB GEAR AND MAGAZINES?
你能给我们带来一些实验室装置和杂志吗？

Chapter 5 第五章
Heat of Reaction 反应热

IN THE LAST CHAPTER, WE LOOKED AT CHEMICAL REACTIONS AS TRANSFERS OF MATTER. WE KEPT A CAREFUL ACCOUNTING OF ATOMS AS THEY REARRANGED THEMSELVES.

NOW WE LOOK AT REACTIONS ANOTHER WAY: AS TRANSFERS OF **ENERGY.**

在上一章，我们把化学反应看作是物质的转移。当它们重新排列时，我们一直仔细地计算原子数目。

现在让我们来看看化学反应的另一面：**能量转移**。

PHYSICISTS DEFINE ENERGY MECHANICALLY, AS THE ABILITY TO DO **WORK.*** WORK IS WHAT HAPPENS WHEN A FORCE OPERATES ON AN OBJECT OVER A DISTANCE: WORK = FORCE X DISTANCE. THE METRIC UNIT OF ENERGY IS THE NEWTON-METER, OR **JOULE.**

物理学家将机械能定义为**做功的能力***。功是当一个力远距离作用于物体上时产生的现象：功=力×距离。其单位是牛顿·米或**焦耳**。

1 JOULE = WORK DONE BY A FORCE OF ONE NEWTON OPERATING OVER A DISTANCE OF ONE METER.

1J=1N·m，即1牛顿力的作用使物体在力的方向上移动1米距离所做的功为1焦耳。

*NOT TO BE CONFUSED WITH USEFUL WORK.

*不要与有用功混淆。

CHEMISTS CARE ABOUT WORK, TOO (AN EXPLOSION DOES WORK), BUT WE ALSO CARE ABOUT OTHER FORMS OF ENERGY: **CHEMICAL** ENERGY, **RADIANT** ENERGY, AND **HEAT.** EACH OF THESE HAS THE ABILITY TO DO WORK.

化学家不仅关心做功（爆炸产生的功），也对其他形式的能量感兴趣：**化学能**、**辐射能**和**热量**。这些都有做功的能力。

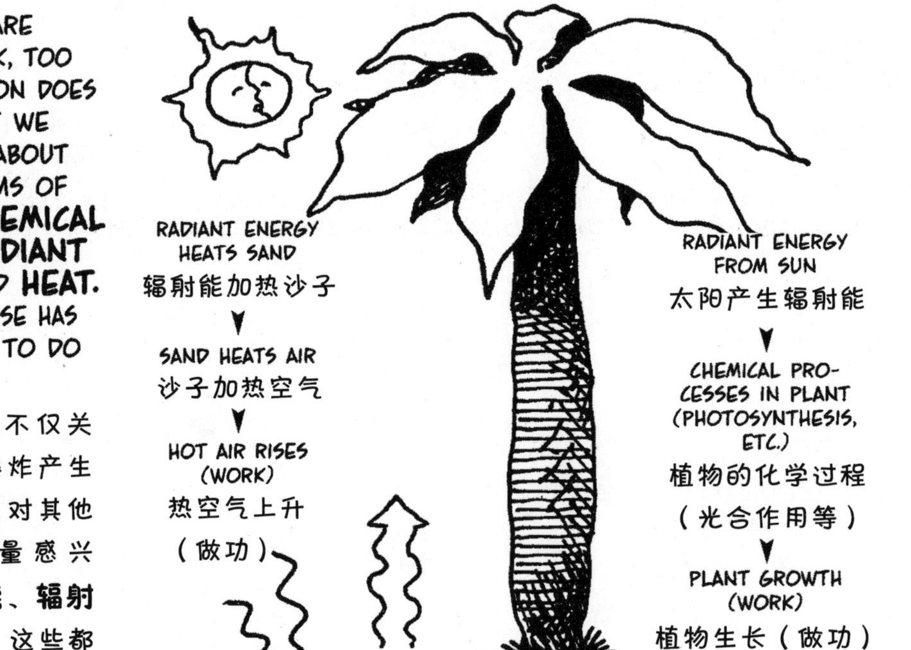

ONE KIND OF ENERGY CAN BE CONVERTED INTO ANOTHER KIND, BUT ENERGY IS NEVER CREATED OR DESTROYED. THAT'S A LAW—THE LAW OF **CONSERVATION OF ENERGY.**

能量既不会凭空产生，也不会凭空消失，只能从一种形式转化为另一种形式，这是一个定律，即**能量守恒定律。**

LET'S EXAMINE MECHANICAL ENERGY MORE CLOSELY. IF I PUSH THIS COCONUT, IT MOVES... AND THE LONGER AND/OR HARDER I PUSH, THE FASTER IT GOES. (THIS IS CLEARER IN OUTER SPACE, AWAY FROM FRICTION AND GRAVITY.) BY DOING WORK ON THE COCONUT, I ADD ENERGY TO IT: **KINETIC ENERGY (K.E.),** THE ENERGY OF MOTION.

让我们更加深入地研究一下机械能。如果我推动这个椰子，它就会移动……它移动的距离越长，或者我越用力推动，它就滚得越快。（这毫无疑问需要在没有摩擦和重力的太空中进行。）通过对椰子做功，就给它增加了一个能量——动能，即运动所具有的能量。

$$K.E. = \frac{1}{2}mv^2$$

BACK ON EARTH, I PUSH THE COCONUT AGAIN, BUT IN AN UPWARD DIRECTION. THE COCONUT FLIES UP, BUT IT SLOWS UNDER THE PULL OF GRAVITY. EVENTUALLY IT STOPS AND BEGINS TO FALL. WHAT BECAME OF THE ENERGY I ADDED??

在地球上，我把椰子抛向空中。椰子飞起来，但在引力作用下向上运动的速度减缓，最终停止并开始下落。我施加的力变成了什么样的能量？？

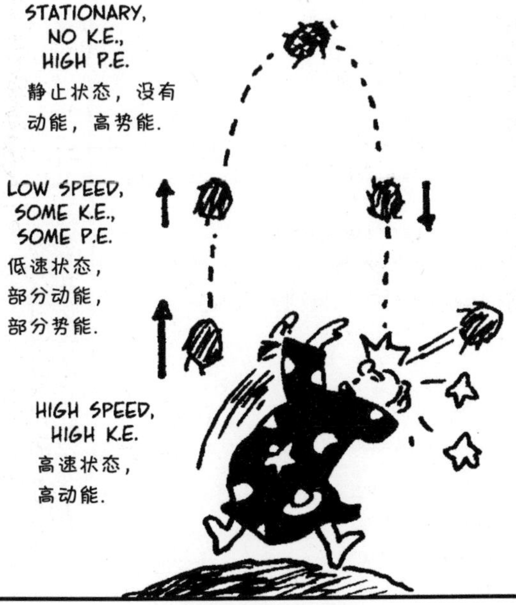

AS THE COCONUT SLOWS AND LOSES K.E., IT GAINS **POTENTIAL ENERGY (P.E.).** THIS IS ENERGY THAT DEPENDS ON THE BODY'S POSITION IN THE EARTH'S GRAVITATIONAL FIELD. K.E. + P.E. IS CONSTANT.

随着椰子减速，并失去**动能**，它获得**势能**，这种能量取决于物体在地球引力场中的位置。动能+势能的和为常数。

IT TURNS OUT THAT **ALL** FORMS OF ENERGY CAN BE UNDERSTOOD IN TERMS OF KINETIC AND POTENTIAL ENERGY. RADIANT ENERGY, FOR INSTANCE, IS THE K.E. OF MOVING PHOTONS, OR LIGHT PARTICLES.* THERE IS POTENTIAL ENERGY STORED IN CHEMICAL BONDS. AND HEAT IS... HEAT IS... WHAT **IS** HEAT, ANYWAY?

事实证明，**所有**形式的能都可以用动能和势能来说明。例如辐射能，是移动的光子或光粒子的动能*。还有潜在的能量储存在化学键中。

热是……什么**是**热啊？

*THE "LIGHT" NEED NOT BE VISIBLE. MOVING PHOTONS CONVEY THE ENERGY OF ALL ELECTROMAGNETIC RADIATION, FROM X-RAYS TO RADIO WAVES.

*“光”不一定是可见的。移动的光子传递了从X射线到电磁波所有形式的电磁辐射能量。

HEAT, WE KNOW, HAS SOMETHING TO DO WITH **TEMPERATURE,** AND TEMPERATURE IS FAMILIAR ENOUGH. WE EVEN KNOW HOW TO MEASURE IT, WITH A THERMOMETER.

我们知道热与**温度**有关，而我们太熟悉温度了。我们甚至知道如何用温度计来测量温度。

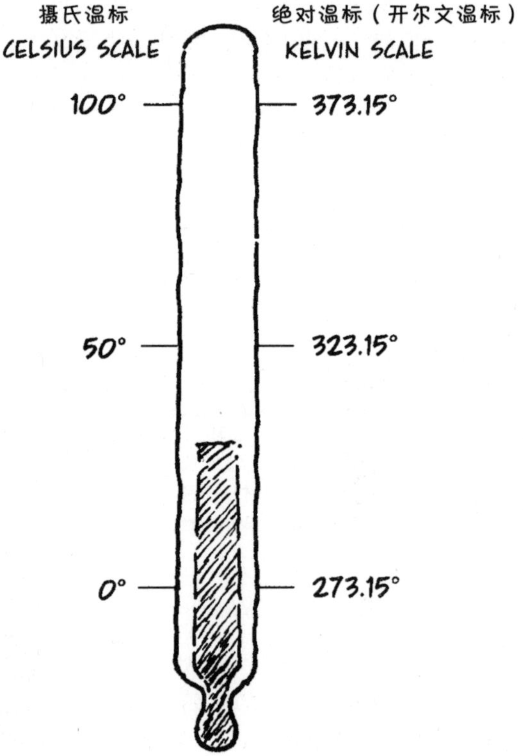

THE UNITS ARE **DEGREES CELSIUS** (°C). THE CELSIUS SCALE SETS:

0°C = MELTING POINT OF WATER
100°C = BOILING POINT OF WATER

THE **KELVIN** SCALE HAS DEGREES THE SAME SIZE AS CELSIUS, BUT STARTS LOWER:

0°K = ABSOLUTE ZERO, WHERE ALL MOLECULAR AND ATOMIC MOTION STOPS = −273.15°C.

°C = °K − 273.15

其单位是**摄氏度**（°C）。摄氏温标规定：

0°C=水的熔点
100°C=水的沸点

开尔文温标的单位与摄氏单位大小相同，但零点降低。

0°K=绝对零度，所有分子和原子的运动停止 = -273.15°C

°C=°K-273.15

COLLOQUIALLY, WE SAY SOMETHING IS HOT WHEN WE REALLY MEAN IT HAS A HIGH TEMPERATURE. A CHEMIST WOULD NEVER SAY THIS! **HEAT AND TEMPERATURE ARE NOT THE SAME.**

通俗地说，我们说什么是热的，意味着它具有较高的温度。但化学家永远不会这么说！因为**热和温度是两码事。**

SAME TEMPERATURE CHANGE
DOUBLE THE HEAT CHANGE
同样的温度变化，双倍的热交换

TO ILLUSTRATE THE DIFFERENCE, SUPPOSE WE COOK TWO COCONUTS, RAISING THEIR TEMPERATURE BY 75°C (FROM 25° TO 100°, SAY). THEN THE TWO COCONUTS TOGETHER HAVE THE **SAME TEMPERATURE CHANGE** AS ONE COCONUT, BUT THEY ABSORB **TWICE AS MUCH HEAT,** BECAUSE THEY CONTAIN TWICE AS MUCH MATTER TO HEAT UP.

为了说明差异，假设我们煮俩椰子，将它们的温度提高75°C（由25°C上升至100°C）。虽然两个椰子一起和一个椰子有**相同的温度变化**，但它们吸收的热量是一个椰子的两倍，因为它们有着两倍物质需要升温。

WHAT, THEN, IS THE RELATIONSHIP BETWEEN TEMPERATURE AND HEAT?
那么，热与温度之间有什么关系？

TO BEGIN WITH, WHEREVER WE LOOK, HEAT TRANSFERS ARE ASSOCIATED WITH **TEMPERATURE DIFFERENCES.** WE KNOW FROM EXPERIENCE THAT HEAT FLOWS FROM HOT TO COLD.

我们看到，热量传递与**温度的差异**相关，我们由经验得知，热量热的一端传向冷的一端。

THAT IS, WHEN A HIGHER-TEMPERATURE OBJECT MEETS A LOWER-TEMPERATURE OBJECT, ENERGY FLOWS FROM THE WARMER ONE TO THE COOLER ONE UNTIL THEIR TEMPERATURES EQUALIZE. AN EXAMPLE IS WHEN WE IMMERSE SOMETHING COOL IN HOT WATER. (ASSUME THAT THE "SOMETHING" DOESN'T MELT.)

也就是说，当一个温度较高的物体遇到一个温度较低的物体时，能量从较热的一个传到较冷的一个，直到两处温度相同。有个例子是把凉的东西浸在热水里（假定凉的东西不会融化）。

INITIAL STATE $T_2 < T_1$	HEAT FLOW TAKES PLACE	FINAL STATE $T_2 < T_{FINAL} < T_1$
初始状态 $T_2 < T_1$	热流产生	最终状态 $T_2 < T_{最终} < T_1$

(FINAL TEMPERATURES ARE EQUAL, AND BETWEEN THE ORIGINAL EXTREMES)

（最终温度相同，并且位于两者初始温度之间）

THE AMOUNT OF ENERGY TRANSFERRED IS THE HEAT: **HEAT IS THE ENERGY CHANGE ASSOCIATED WITH A DIFFERENCE IN TEMPERATURE.**

热量是能量的传递：**热量是与温度变化有关的能量变化。**

Internal Energy 内能

WHERE DOES HEAT ENERGY GO? TO ANSWER THIS QUESTION, CONSIDER THIS COCONUT, WHICH REALLY STANDS FOR ANY CHEMICAL SYSTEM WITH A DEFINITE BOUNDARY BETWEEN ITSELF AND ITS SURROUNDINGS.

热能到哪里去了？要回答这个问题，先看一下这个椰子，它表示任何一个在其自身与环境之间有明确边界的化学系统。

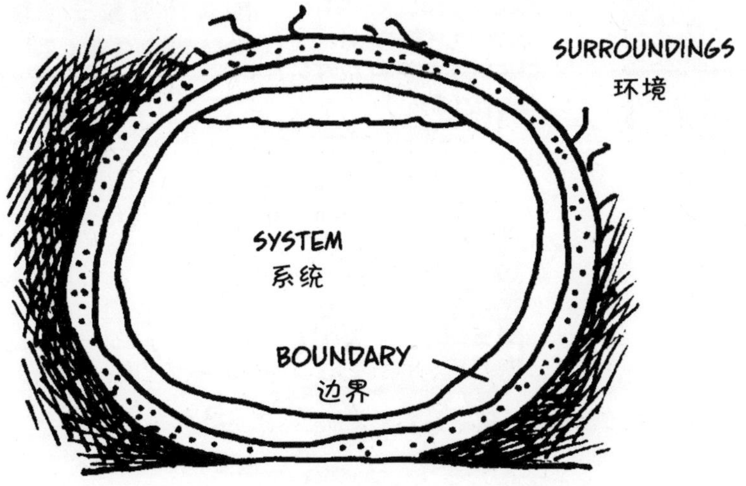

AT CLOSE RANGE, THE COCONUT SEETHES WITH ENERGY. ALL ITS MOLECULES ARE JIGGLING RANDOMLY, SO THEY HAVE KINETIC ENERGY. THEY ALSO HAVE POTENTIAL ENERGY: ELECTRIC ATTRACTIONS AND REPULSIONS ACCELERATE AND DECELERATE PARTICLES, ANALOGOUS TO THE WAY GRAVITY WORKS ON A THROWN OBJECT.

近距离观察，椰子充满了能量。它的所有分子随机振动，所以它们有动能。它们也有势能：电荷相互吸引和排斥使粒子加速和减速，类似重力作用于被抛出对象的方式。

A SYSTEM'S **INTERNAL ENERGY** IS THE TOTAL KINETIC AND POTENTIAL ENERGY OF ALL ITS PARTICLES.

一个系统的**内能**是所有粒子动能和势能的总和。

A SYSTEM'S **TEMPERATURE** IS A MEASURE OF THE AVERAGE TRANSLATIONAL KINETIC ENERGY* OF ALL ITS PARTICLES, I.E., HOW FAST THEY FLY OR WIGGLE.

一个系统的**温度**是其中所有粒子平均平动动能的量度*，例如，它们飞行或摆动的速度。

编者注：温度是大量分子热运动的集体表现，含有统计意义。对于个别分子来说，温度是没有意义的。

THIS MAKES SENSE, GIVEN WHAT WE KNOW ABOUT TEMPERATURE. A HIGHER-T SYSTEM RAISES THE TEMPERATURE OF A LOWER-T SYSTEM BECAUSE HIGHER-ENERGY PARTICLES TRANSFER ENERGY TO LOWER-ENERGY ONES.

从我们所知的温度知识来说，这是很有道理的。较高的温度系统能够提高一个低温系统的温度，因为高能量粒子能够把能量转移到低能量的粒子上。

*TRANSLATIONAL ENERGY IS ENERGY ASSOCIATED WITH PARTICLES MOVING THROUGH SPACE. THE ENERGY OF SPINNING AND INTERNAL VIBRATION IS NOT INCLUDED.

*平动动能是与粒子在空间运动相关的能量，自旋能和内振动不包括在内。

THIS IS A BIT MORE COMPLICATED THAN IT SOUNDS. IN GASES, T MEASURES HOW ENERGETICALLY MOLECULES FLY AROUND, BUT IN METALS, T ALSO INCLUDES THE ENERGY OF MOVING ELECTRONS... IN CRYSTALS, WIGGLING IONS HAVE P.E. AS WELL AS K.E., BECAUSE PARTICLES PULL AGAINST EACH OTHER... AND MOLECULES (OR PARTS OF MOLECULES) CAN ROTATE OR VIBRATE INTERNALLY. EVERY SUBSTANCE IS DIFFERENT!

以下内容听起来会更复杂一点。在气体中，温度衡量分子飞行的能量，但在金属中，温度还包括移动电子的能量……在晶体中，摆动的离子有势能和动能，因为粒子互相拉动……分子（或分子的一部分）可以自主地旋转或振动。每一种物质的能量都是不同的！

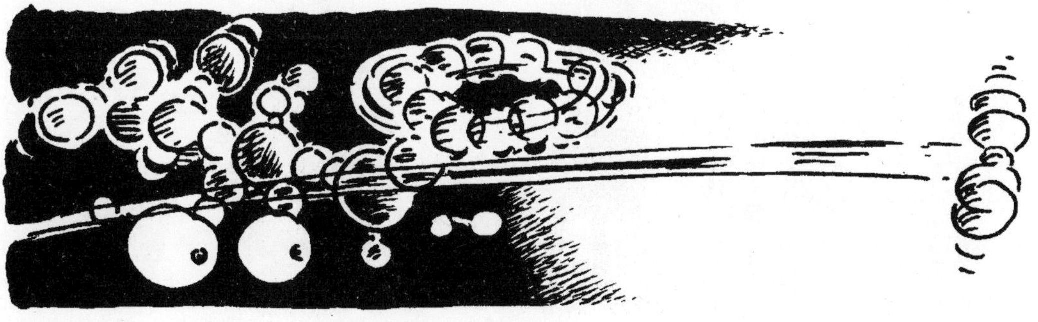

WHEN HEAT IS ADDED AND INTERNAL ENERGY RISES, SOME OF THE ADDED ENERGY DOES NOT CONTRIBUTE TO A RISE IN TEMPERATURE, BUT RATHER IS ABSORBED AS P.E., ROTATION, OR INTERNAL VIBRATION.

随着能量增加和内能上升，一些增加的能量没有使温度上升，而是被吸收成为势能，用于旋转或内部振动。

Different chemicals have different temperature responses to heat.

不同的化学物质对热量有不同的温度反应。

Heat Capacity
热容量

THE **HEAT CAPACITY** OF A SUBSTANCE IS THE ENERGY INPUT REQUIRED TO RAISE ITS TEMPERATURE BY 1°C. WE CAN SPEAK OF HEAT CAPACITY PER GRAM ("SPECIFIC HEAT") OR PER MOLE ("MOLAR HEAT CAPACITY").

某物质的**热容量**是温度提高1°C所需的能量，我们可以将其称作每克**热容量**（“比热容”）或每摩尔热容量（“摩尔热容量”）。

JAMES PRESCOTT **JOULE** (1818–1889) MEASURED THE HEAT CAPACITY OF WATER. HE ATTACHED A FALLING WEIGHT TO A PADDLE WHEEL IMMERSED IN WATER. BY MEASURING THE SLIGHT RISE IN TEMPERATURE OF THE WATER,* JOULE FOUND THE WORK EQUIVALENT OF A TEMPERATURE CHANGE. RESULT:

詹姆斯·普雷斯科特·焦耳（1818—1889年）测量了水的热容量。他将一个落锤固定在浸泡于水中的桨轮上。通过测量，水的温度略有上升*，焦耳发现了做功对应于温度的变化，结果相当于：

WATER'S HEAT CAPACITY PER GRAM OR **SPECIFIC HEAT** IS

4.184 Joules/g°C

水的每克热容量或**比热容**为4.184焦耳/克℃

EXAMPLE: TO RAISE THE TEMPERATURE OF 5g OF WATER BY 7°C REQUIRES AN ADDED ENERGY OF

5 X 7 X 4.184
= 146 JOULES.

例如：要把5克水的温度提高7°C需要补充能量5×7×4.184=146焦耳。

HERE, AT LAST, IS THE PRECISE RELATIONSHIP BETWEEN TEMPERATURE AND HEAT:

最后，这里给出精确的温度与热量之间的关系：

Heat change =
Mass x ΔT x Specific heat

热量变化=质量×ΔT×比热容

*YOU CAN RAISE TEMPERATURE BY DOING WORK ON AN OBJECT. FOR INSTANCE, WHEN YOU HAMMER A NAIL, THE NAIL HEAD WARMS UP.

*你可以击打一个物体，使其升温。例如，当你钉钉子时，钉子头是热的。

FROM THAT SINGLE FORMULA AND WATER'S SPECIFIC HEAT, WE CAN FIND ALL OTHER SPECIFIC HEATS! LET'S START WITH COPPER. IMMERSE 2 kg COPPER AT 25°C IN 5 kg WATER AT 30°C. LET THE TEMPERATURE STABILIZE. CHECK THE THERMOMETER. IT READS 29.83°C. THE WATER BARELY CHANGED TEMPERATURE, BUT THE COPPER REALLY HEATED UP!

通过简单的公式和水的比热容，我们可以得到所有其他比热容！让我们先从铜开始。在25°C的水中浸泡2千克铜，在30°C的水中浸泡5千克铜，让温度稳定下来。温度计显示为29.83°C，而水的温度几乎没有变化，但铜很明显地被加热了！

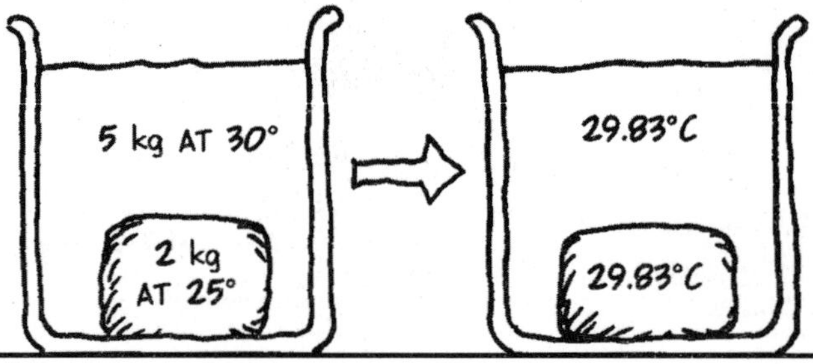

THE TEMPERATURE CHANGES (ΔT) ARE

$\Delta T_{WATER} = -0.17°$
$\Delta T_{COPPER} = 4.83°$

温度变化（ΔT）是

$\Delta T_{水}$=-0.17℃
$\Delta T_{铜}$=4.83℃

WE CAN IMMEDIATELY CALCULATE WATER'S HEAT LOSS. (HEAT CHANGES ARE DENOTED BY THE LETTER q):

$q_{WATER} = (5000g)(-0.17°C)(4.18\ J/g°C)$
$= -3553$ Joules

我们可以立即计算出水的热量损失（热量变化用字母q表示）：

$q_{水}$=（5000克）（-0.17°C）（4.18焦耳/克°C）
=-3553焦耳

BUT THE WATER'S LOSS IS PRECISELY COPPER'S GAIN (ASSUMING NO HEAT LEAKS OUT OF THE VESSEL). THAT IS,

q_{COPPER} = 3553 Joules.

SINCE THERE WERE 2000g OF COPPER, THE FORMULA SAYS:

$3553\ J = (2000g)(4.83°)C_{Cu}$

(C_{Cu} = COPPER'S SPECIFIC HEAT)

SOLVING FOR C_{Cu},

$$C_{Cu} = \frac{3553\ J}{(2000\ g)(4.83°)} = \mathbf{0.37\ J/g°C}$$

但水损失的热量恰恰是铜得到的热量（假设无热量泄漏出容器）。也就是说，

$q_{铜}$=3553焦耳。

由于有2000克铜，公式表示为：

3553焦耳=（2000克）（4.83°C）$C_{铜}$

（$C_{铜}$=铜的比热容）

求解铜的比热容，

$$C_{铜}= \frac{3553\ 焦耳}{(2000克)(4.83°C)}=0.37\ 焦耳/克°C$$

AMAZINGLY, COPPER'S SPECIFIC HEAT IS LESS THAN **ONE-TENTH** THAT OF WATER. WATER CAN SOAK UP HEAT WITH LITTLE RISE IN TEMPERATURE, WHILE COPPER'S TEMPERATURE RISES ALMOST EFFORTLESSLY.

令人惊讶的是，铜的比热容比水的**十分之一**还少。水能吸收热量但是温度几乎没有上升，然而铜的温度上升却毫不费力。

LIQUID WATER HAS MANY HYDROGEN BONDS BETWEEN ITS MOLECULES (SEE CHAPTER 3). THESE BONDS MAKE IT HARD TO GET A WATER MOLECULE MOVING! ADDED HEAT LARGELY GOES INTO THE P.E. ASSOCIATED WITH THESE ATTRACTIONS.

液态水分子之间有许多的氢键（见第三章）。这些键导致水分子的移动非常困难！加热在很大程度上增加了与这些作用相关的势能。

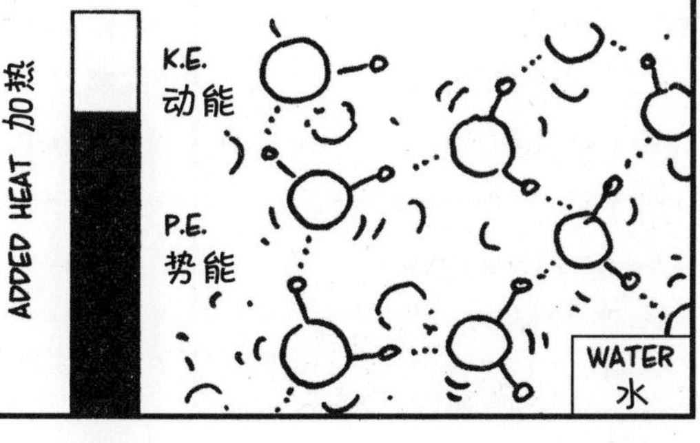

COPPER, ON THE OTHER HAND, HAS A "SEA" OF HIGHLY MOBILE ELECTRONS. ADDED ENERGY SIMPLY MAKES THEM FLY AROUND FASTER. THAT IS, HEAT ALMOST ALL GOES INTO K.E., AND TEMPERATURE RISES ACCORDINGLY.

另一方面，铜拥有高度流动的电子"海"。加热反而会让它们飞得更快。也就是说，所有热量几乎都成为动能，而且，温度也会相应升高。

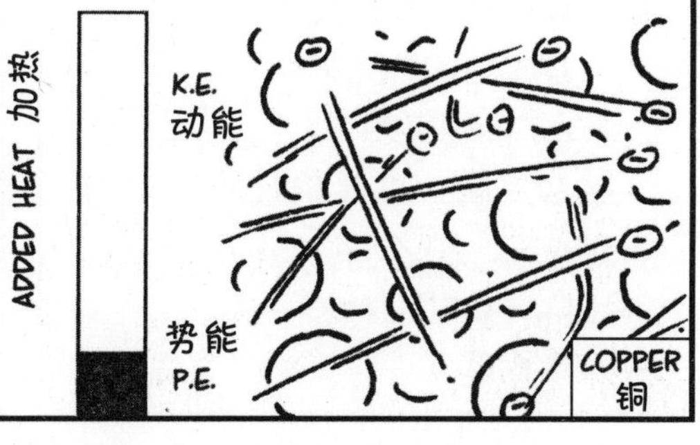

THIS EXPLAINS WHY WATER IS USED AS A COOLANT IN MACHINERY, FROM CAR ENGINES TO NUCLEAR REACTORS. THE HEAT TRANSFER FROM HOT METAL TO COOL WATER DROPS THE METAL'S TEMPERATURE DRAMATICALLY, WHILE RAISING WATER'S RELATIVELY LITTLE.

这解释了为什么水会作为从汽车发动机到核反应堆的机械冷却剂，热从热金属传导到冷水，导致金属的温度显著降低，而水升温不明显。

MANY OTHER SPECIFIC HEATS CAN BE FOUND THE SAME WAY. IF WE REPLACE COPPER WITH IRON IN THE EXPERIMENT (SAME TEMPERATURES, SAME MASSES), WE FIND

$$\Delta T_{WATER} = -0.206°$$

$$\Delta T_{IRON} = 4.794°$$

FROM THE EXACT SAME COMPUTATION AS BEFORE, WE FIND

$$C_{IRON} = 0.45\ J/g°C$$

ALSO VERY LOW.

许多其他形式的比热容可以以同样的方式算出来。如果我们在实验中用铁（相同的温度和质量）取代铜，我们发现：

$$\Delta T_{水} = -0.206°C$$

$$\Delta T_{铁} = 4.794°C$$

和以前一样精确计算，我们发现：

$$C_{铁} = 0.45\ J/g°C$$

也很低。

NOW MEASURE IRON AGAINST ETHANOL, OR GRAIN ALCOHOL. ASSUME THE SAME MASSES AND A 5° TEMPERATURE DIFFERENCE AT THE START.

$$\Delta T_{ETHANOL} = -0.36°$$

$$\Delta T_{IRON} = 4.65°$$

AND WE CALCULATE AS BEFORE:

$$C_{ETHANOL} = 2.4\ J/g°C$$

CLOSER TO WATER.

现在测量铁对酒精或乙醇。假定同样的质量，以5℃的温差开始。

$$\Delta T_{乙醇} = -0.36°C$$

$$\Delta T_{铁} = 4.65°C$$

经计算，我们得到：

$$C_{乙醇} = 2.4\ J/g°C$$

与水相似。

WE CAN CONTINUE MEASURING ONE THING AGAINST ANOTHER UNTIL WE "BOOTSTRAP" A WHOLE TABLE OF SPECIFIC HEATS.

我们可以继续用两种物质对应测量，直到填满整个比热容表为止。

SUBSTANCE	SPECIFIC HEAT (J/g°C)
MERCURY, Hg	0.14
COPPER, Cu	0.37
IRON, Fe	0.45
C (GRAPHITE)	0.68
SIMPLE MOLECULES	
ICE, H_2O (s)	2.0
WATER VAPOR, H_2O (g)	2.1
ANTIFREEZE, (CH_2OHCH_2OH)	2.4
ETHANOL, (CH_3CH_2OH)	2.4
LIQUID WATER, H_2O(l)	4.2
AMMONIA, NH_3(l)	4.7
COMPLEX MATERIALS	
BRASS	0.38
GRANITE	0.79
GLASS	0.8
CONCRETE	0.9
WOOD	1.8

物质	比热容 (J/g°C)
汞	0.14
铜	0.37
铁	0.45
碳（石墨）	0.68
简单分子	
冰、水（液态s）	2.0
水蒸气、水（气态g）	2.1
乙二醇（CH_2OHCH_2OH）	2.4
乙醇（CH_3CH_2OH）	2.4
液态水、水（液态l）	4.2
氨气、液氨（液态l）	4.7
复合材料	
黄铜	0.38
花岗岩	0.79
玻璃	0.8
混凝土	0.9
木头	1.8

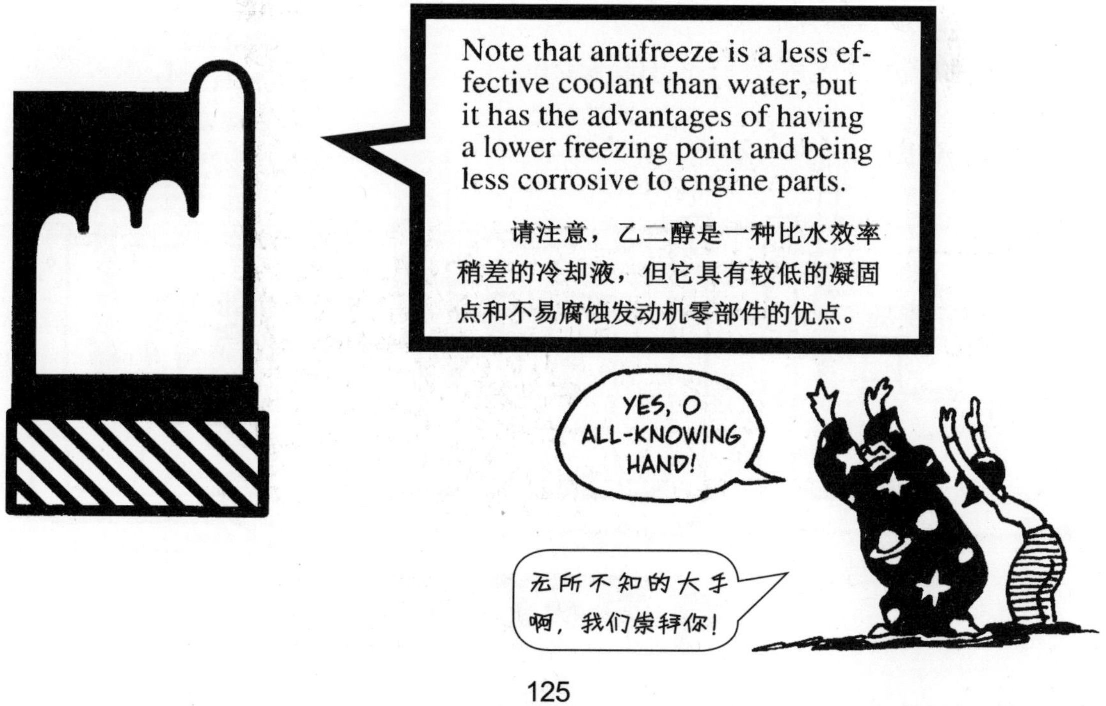

Calorimetry
量热测定

THE POINT OF ALL THESE PRELIMINARIES IS TO FIND THE **HEAT CHANGES OF CHEMICAL REACTIONS:** HOW MUCH ENERGY IS RELEASED OR ABSORBED AS HEAT WHEN A REACTION TAKES PLACE. WE ARE NOW IN A POSITION TO MEASURE THIS.

所有这些预备知识的重点都是为了找到**化学反应的热量变化**：当反应发生时，有多少能量作为热量被释放或吸收。我们现在可以来测量这个了。

THE METHOD IS SIMILAR TO THE WAY WE FOUND SPECIFIC HEATS: RUN THE REACTION IN A VESSEL OF KNOWN HEAT CAPACITY C AND MEASURE THE CHANGE IN TEMPERATURE. SINCE THE VESSEL ABSORBS WHAT THE REACTION GIVES OFF—OR VICE VERSA—THE HEAT CHANGE q OF THE REACTION IS $-q_{VESSEL} = -C\Delta T$.

方法类似于我们之前得到比热容的方式：在已知热容量C的容器中启动反应，并测量温度变化。由于容器会吸收反应放出的能量，反之亦然，反应的热量交换q是：$-q_{容器} = -C\Delta T$。

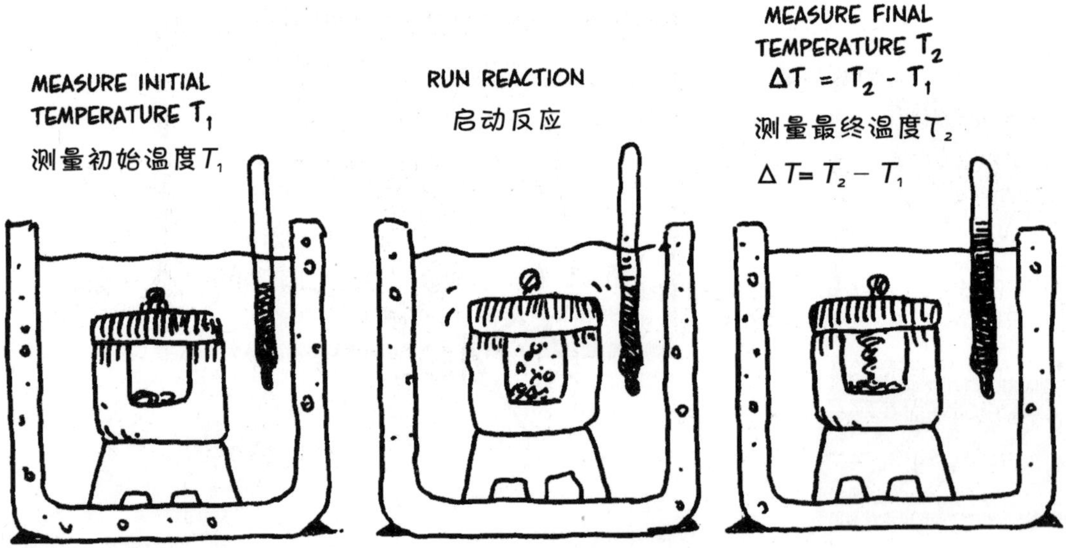

$$q = -C\Delta T$$

THE REACTION VESSEL AND ITS SURROUNDING PARAPHERNALIA TOGETHER ARE CALLED A **BOMB CALORIMETER.** THE REACTION CHAMBER, OR "BOMB," IS USUALLY IMMERSED IN WATER, WHICH CAN BE STIRRED TO DISTRIBUTE THE HEAT. A THERMOMETER COMPLETES THE APPARATUS.

反应容器及其周边工具一起被称为**弹式量热计**。反应室或“弹仓”，通常浸于水中，可通过振动散发热量。再加上一个温度计就构成了这个完整的装置。

Example 范例

COMBUSTION OF **OCTANE** C_8H_{18}, A COMPONENT OF GASOLINE:

辛烷C_8H_{18}是汽油的一种成分，其燃烧过程如下：

$$2C_8H_{18}(l) + 25O_2(g) \longrightarrow 16CO_2(g) + 18H_2O(g)$$

TO MEASURE THE HEAT GIVEN OFF, WE NEED A STRONG, HEAVY BOMB TO WITHSTAND THE HIGH TEMPERATURE AND PRESSURE GENERATED. A THICK-WALLED STEEL CONTAINER OUGHT TO DO... LET'S SUPPOSE ITS HEAT CAPACITY IS 15,000 J/°C. WE IMMERSE IT IN **2.5** L OF WATER, WHICH HAS A MASS OF **2500** g.

为了测量释放出的热量，我们需要一个结实的重型弹仓来承受产生的高温和高压。这是厚壁钢容器应该做的事……让我们假设它的热容量为15000J/°C，将其浸入2.5升水中，水的质量是2500克。

THE WATER'S HEAT CAPACITY IS

$(2500g)(4.184 J/g°C) = 10,460 J/°C.$

SO THE CALORIMETER'S TOTAL HEAT CAPACITY IS

$10,460 + 15,000 = 25,460 J/°C.$

SUPPOSE T_1, THE INITIAL TEMPERATURE OF THE CALORIMETER, IS 25°.

水的热容量为

(2500g) (4.184J/g°C)= 10460 J/°C

因此，量热计的总热容量是

10460+15000 = 25460 J/°C

假设T_1时，量热计的初始温度是25°C。

WE DROP ONE GRAM OF OCTANE INTO THE BOMB... IGNITE IT WITH A SPARK... IT BURNS... THE HEAT SPREADS THROUGHOUT THE CALORIMETER... WE AGAIN CONSULT THE THERMOMETER, AND FIND $T_2 = 26.88°$. THEN

$$\Delta T = T_2 - T_1 = 1.88°$$

THE MAGIC FORMULA IS

$$q = -C_{CALORIMETER}(\Delta T)$$

WE PLUG IN AND FIND

$$q = -(25{,}460\ J/°C)(1.88°C) = -47{,}800\ J$$
$$= -47.8\ kJ$$

AND WE CONCLUDE THAT OCTANE RELEASES 47.8 kJ/g OF HEAT WHEN BURNED.

我们将1克的辛烷放入弹仓……用火花点燃它……它开始燃烧……热量蔓延整个量热计……我们再查看一下温度计，显示$T_2 = 26.88℃$。然后

$$\Delta T = T_2 - T_1 = 1.88°C$$

这个神奇的公式为

$$q = -C_{量热计}(\Delta T)$$

我们代入ΔT得到

$$q = -(25460\ J/°C)(1.88°C) = -47{,}800\ J = -47.8\ kJ$$

我们可以得出这样的结论：辛烷燃烧时释放47.8千焦耳/克的热量。

Enthalpy
焓（热函）

THE BOMB CALORIMETER IS GREAT, WONDERFUL, FANTASTIC, BUT A BIT UNREALISTIC, BECAUSE THE REACTION VESSEL IS SEALED. SOME REACTIONS IN THE BOMB MAY PRODUCE HIGH PRESSURES, WHICH CAN AFFECT TEMPERATURE.

FOR EXAMPLE, AN EXPLOSION IN THE OPEN AIR GIVES OFF GASES THAT EXPAND RAPIDLY AND PUSH THE SURROUNDING AIR OUTWARD. IN OTHER WORDS, THE GASES DO **WORK** ON THE SURROUNDINGS.

例如，在露天爆炸释放出的气体能够迅速扩散，推动周围的空气向外移动。换言之，这些气体会对周围环境**做功**。

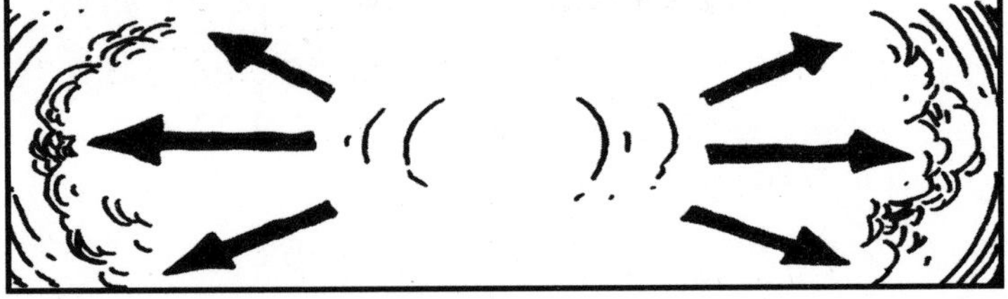

弹式量热计是伟大的、美妙的、神奇的，但有点不切实际，因为反应容器是密封的，弹仓可能会产生高压，从而影响温度。

IN THAT CASE, THE ENERGY CHANGE **ΔE** OF THE REACTION HAS TWO COMPONENTS, **WORK** AND **HEAT**:

$\Delta E = \Delta H + \text{WORK}$

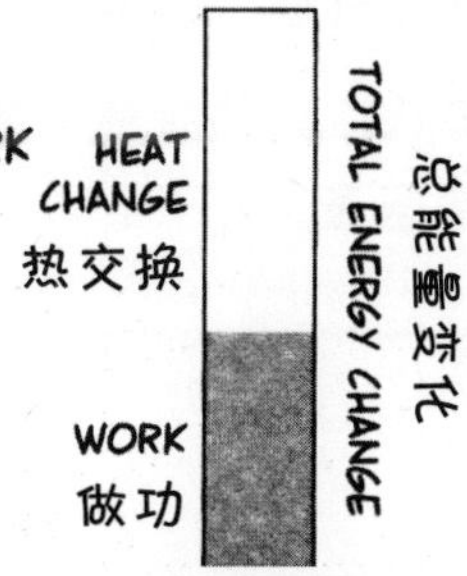

PUSHING AIR OUT OF THE WAY COOLS THE REACTION PRODUCTS!

在这种情况下，反应的能量变化ΔE有两部分，即**做功**和**热量**：

$$\Delta E = \Delta H + 做功$$

通过推动空气做功，冷却了反应产物！

ΔH HERE MEANS THE HEAT CHANGE WHEN THE REACTION IS RUN OUTDOORS.

这里的ΔH指反应在室外发生时的热量交换。

IN THE BOMB CALORIMETER, THE GASES DO NO WORK, BECAUSE THE EXPLOSION IS CONFINED IN A FIXED VOLUME. **ALL** THE ENERGY IS RELEASED AS HEAT.

$$\Delta E = q$$

THEREFORE

$$q = \Delta H + \text{WORK}$$

SO

$$q > \Delta H$$

THE HEAT CHANGE IN THE BOMB IS GREATER THAN THAT IN THE OUTSIDE WORLD.

在弹式量热计中，气体不会做功，因为爆炸被限制在一个固定体积内。**所有的**能量都以热量的形式被释放。

$$\Delta E = q$$

因此，

$$q = \Delta H + 做功$$

所以，

$$q > \Delta H$$

弹仓中的热量变化比在弹仓外的更大。

FROM NOW ON, WE TREAT REACTIONS AS IF THEY TAKE PLACE "OUTDOORS"—MEANING AT **CONSTANT PRESSURE.** IN THAT CASE, THE HEAT RELEASED OR ABSORBED IS CALLED THE **ENTHALPY CHANGE**, AND WRITTEN **ΔH**.

从现在开始，我们认为反应就像发生在"户外"，指在**恒定压强**下。在这种情况下，释放或吸收的热量称为焓变，记为ΔH。

TO MEASURE ENTHALPY CHANGE, WE USE A CALORIMETER THAT MAINTAINS CONSTANT PRESSURE. THEN THE PROCEDURE IS THE SAME AS WITH A BOMB CALORIMETER: MEASURE INITIAL AND FINAL TEMPERATURES T_1 AND T_2, THEN MULTIPLY T_2-T_1 TIMES THE HEAT CAPACITY OF THE CALORIMETER.

为了测量焓变，我们使用了保持恒定压力的量热计。步骤与弹式量热计相同：测量初始温度和最终温度T_1和T_2，再用T_2-T_1乘以量热计的热容量。

Example 范例

EXPLOSION OF BLACK POWDER (HERE WE GIVE A MORE REALISTIC EQUATION THAN PREVIOUSLY):

黑火药爆炸（在这里，我们给出一个比之前更现实的方程）：

$$4KNO_3(s) + 7C(s) + S(s) \longrightarrow 3CO_2\uparrow + 3CO\uparrow + 2N_2\uparrow + K_2CO_3(s) + K_2S(s)$$

SUPPOSE OUR CALORIMETER HAS A KNOWN HEAT CAPACITY OF 337.6 kJ/°C. WE START WITH 500g OF POWDER. THE TEMPERATURE CHANGE ΔT IS FOUND TO BE 4.78°C, AND WE COMPUTE

假设我们的量热计有一个已知的热容量为337.6kJ/°C。我们先加入500克粉末。温度变化ΔT是4.78°C，我们计算如下：

$$\Delta H = -(337.6\ \text{kJ/°C})(4.78\text{°C}) = -1614\ \text{kJ}$$

FROM THIS WE CAN FIND THE ENTHALPY CHANGE PER GRAM, ΔH/g.

由此我们可以发现每克的焓变，ΔH/g。

$$\Delta H/\text{gram} = \frac{-1614}{500} = -3.23\ \text{kJ/g}$$

Example 范例

HERE IS A REACTION THAT ABSORBS HEAT:

这里是一个吸收热量的反应：

$$CaCO_3(s) \xrightarrow{\Delta} CaO(s) + CO_2\uparrow$$

WE START WITH THE CALORIMETER HOT ENOUGH TO DRIVE THE REACTION. AT THE END, THE CALORIMETER IS **COOLER** THAN AT THE BEGINNING. IF WE START WITH ONE MOLE OF $CaCO_3$, WE FIND THAT

我们用一个热到足以驱动反应的量热计，最后，量热计会比最初温度**低**。如果我们加入1摩尔的碳酸钙，我们发现：

$$\Delta T = -0.53°C$$

SO 因此，

$$\Delta H = -(337.6\ kJ/°C)(-0.53°C)$$
$$= 179\ kJ/mol$$

REACTIONS THAT RELEASE HEAT ($\Delta H < 0$) ARE CALLED **EXOTHERMIC.** REACTIONS THAT ABSORB HEAT FROM THE SURROUNDINGS ($\Delta H > 0$) ARE CALLED **ENDOTHERMIC.**

释放热量（$\Delta H < 0$）的反应被称为**放热反应**。从周围环境吸收热量（$\Delta H > 0$）的反应被称为**吸热反应**。

Heats of Formation
生成热

GREAT! NOW WE CAN MEASURE ΔH FOR JUST ABOUT ANY REACTION! TOO BAD THERE ARE SO MANY REACTIONS... THIS COULD TAKE A WHILE... LUCKILY, INGENIOUS (OR LAZY) CHEMISTS HAVE THOUGHT UP A **SHORT CUT:** INSTEAD OF MEASURING ENTHALPY CHANGES, WE CAN **CALCULATE** THEM.

太好了！现在，我们可以测量有关任何反应的焓！可糟糕的是，有这么多的反应……这可能需要花费一段时间……幸运的是，聪明（或懒惰）的化学家们想出了一条**捷径**：不需要测量焓的变化，我们就可以**计算**出它们。

THE BASIC CONCEPT IS CALLED **ENTHALPY OF FORMATION,** WRITTEN ΔH_f: THE ENTHALPY CHANGE THAT OCCURS WHEN A MOLE OF SUBSTANCE IS FORMED FROM ITS CONSTITUENT ELEMENTS. FOR INSTANCE, WHEN A MOLE OF LIQUID WATER IS FORMED FROM HYDROGEN AND OXYGEN, OUR CALORIMETER MEASURES

这个基本概念被称为生成焓，写做ΔH_f：

当1摩尔物质由它的组成元素的单质形成时焓变就发生了。例如，当氢气和氧气形成1摩尔水分子时，用量热计测量到的热量为：

$$H_2(g) + \frac{1}{2}O_2(g) \longrightarrow H_2O(l) \quad \Delta H_f = \Delta H = -285.8 \text{ kJ/mole}$$

EACH SUBSTANCE HAS A HEAT OF FORMATION, WHICH CAN EITHER BE MEASURED OR INFERRED. EVERY ELEMENT IN ITS MOST STABLE FORM (SUCH AS C, O_2 OR S) HAS $\Delta H_f = 0$.

每种物质都有生成热，这可以被测量或推断得到。每一种元素的最稳定的单质（如C，O_2或S）的生成焓为零，即$\Delta H_f = 0$

SUBSTANCE 物质	ΔH_f, kJ/mol
$CO(g)$	−110.5
$CO_2(g)$	−393.8
$CaCO_3(s)$	−1207.6
$CaO(s)$	−635.0
$H_2O(l)$	−285.8
$H_2O(g)$	−241.8
$S(s)$	0
$KNO_3(s)$	−494.0
$K_2CO_3(s)$	−1151.0
$C_3H_5(NO_3)_3(l)$	−364.0
$N_2(g)$	0
$O_2(g)$	0

HOW DO WE USE HEATS OF FORMATION? HERE'S THE IDEA. IMAGINE ANY REACTION: REACTANTS ⟶ PRODUCTS. LET'S IMAGINE IT AS **TWO SUCCESSIVE REACTIONS:** REACTANTS ⟶ CONSTITUENT ELEMENTS ⟶ PRODUCTS.

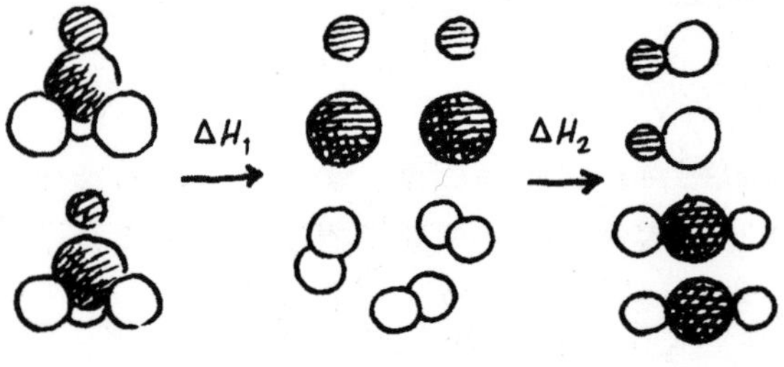

我们如何利用生成热？有个好主意，想象任何一个反应：反应物→产品，假设它为**两个连续的反应**：反应物→组成的元素→产品。

BREAKING THE REACTANTS INTO ELEMENTS HAS A HEAT CHANGE OF **MINUS** THE REACTANTS' TOTAL ENTHALPHY OF FORMATION:

ΔH_1 = −TOTAL ΔH_f OF ALL REACTANTS.

BUILDING THE PRODUCTS HAS A HEAT CHANGE EQUAL TO THE PRODUCTS' COMBINED ENTHALPHY OF FORMATION.

ΔH_2 = TOTAL ΔH_f OF ALL PRODUCTS.

反应物分子分解成单一元素单质的热量与反应物总焓变的**负值**相同：

ΔH_1=-所有反应物的ΔH_f。
生成物需要的热量变化等于生产物的生成焓。
ΔH_2=所有生成物的总ΔH_f。

THE ENTHALPY CHANGE OF THE ENTIRE REACTION, THEN, IS THE TOTAL ENTHALPY CHANGE OF THE TWO INTERMEDIATE REACTIONS:

整个反应的焓变，是两个中间体反应的焓变之差：

$$\Delta H = \Delta H_1 + \Delta H_2$$

$$= \Delta H_f(\text{PRODUCTS}) - \Delta H_f(\text{REACTANTS})$$

生成物 反应物

THAT IS, IN **ANY** REACTION, ΔH IS SIMPLY THE DIFFERENCE BETWEEN THE ENTHALPIES OF FORMATION OF THE PRODUCTS AND THE REACTANTS.

也就是说，在**任何**反应中，ΔH都不过是生成物和反应物之间的生成焓之差。

THIS, BY THE WAY, IS AN EXAMPLE OF A PRINCIPLE CALLED **HESS'S LAW**: ENTHALPY CHANGE DEPENDS ONLY ON THE BEGINNING AND END STATES, NOT ON ANYTHING IN BETWEEN. IF A REACTION HAS INTERMEDIATE STAGES, THEN ΔH IS THE SUM OF THE INTERMEDIATE ENTHALPY CHANGES.

顺便说一句，这是**盖斯定律**的一个实例：

焓变只取决于开始与结尾状态，与其反应步骤无关。如果反应有中间各步骤的话，那么，ΔH就是中间各步骤焓变之和。

Examples 范例

LIMESTONE COOKS TO QUICKLIME:

石灰石制成生石灰：

$CaCO_3(s) \xrightarrow{\Delta} CaO(s) + CO_2\uparrow \quad \Delta H = ?$

WE MAKE AN **ENERGY-BALANCE TABLE,** SIMILAR TO THE MASS-BALANCE TABLES OF THE LAST CHAPTER. WE READ THE HEATS OF FORMATION FROM THE TABLE ON P. 132

我们做一个**能源平衡表**，类似前一章的质量平衡表。我们从第132页的表上得到生成热数据：

反应物 REACTANT	n=产品的摩尔数 n = no. of moles	ΔH_f	$n\Delta H_f$	生成产物 PRODUCT	n	ΔH_f	$n\Delta H_f$
$CaCO_3$	1	-1207.6	-1207.6	CaO	1	-635	-635
				CO_2	1	-393.8	-393.8
TOTAL 总计			-1,207.6				-1,028.8

THEN $\Delta H = \Delta H_f$(PRODUCTS) − ΔH_f(REACTANTS)

$= -1028.8 - (-1207.6) = 1207.6 - 1028.8$

$= 178.8$ kJ FOR EACH MOLE OF CaO PRODUCED.

那么，焓 $\Delta H = \Delta H_f$（生成）$-\Delta H_f$（反应物）

$= -1028.8 - (-1207.6) = 1207.6 - 1028.8$

$= 178.8$千焦耳，为生成每摩尔氧所需要的。

THE REACTION IS **ENDOTHERMIC,** AS WE HAVE SEEN.

该反应是正如我们所看到的**吸热反应**。

EXPLOSION OF NITROGLYCERINE:

硝化甘油爆炸反应：

$4C_3H_5(NO_3)_3(l) \longrightarrow 6N_2\uparrow + O_2\uparrow + 12CO_2\uparrow + 10H_2O\uparrow$

REACTANT 反应物	n	ΔH_f	$n\Delta H_f$	PRODUCT 生成物	n	ΔH_f	$n\Delta H_f$
$C_3H_5(NO_3)_3$	4	-364	-1456	N_2	6	0	0
				O_2	1	0	0
				$H_2O(g)$	10	-241.8	-2418.0
				$CO_2(g)$	12	-393.8	-4725.6
TOTAL 总计			-1456				-7143.6

$\Delta H = -7143.6 - (-1456) = -5687.6$ kJ FOR FOUR MOLES OF NITROGLYCERINE.

ONE MOLE OF NITRO RELEASES ONE-FOURTH AS MUCH:

ΔH/mole = (-5687.6)/4 = -1421.9 kJ/mol.

ONE MOLE OF NITROGLYCERINE WEIGHS 227 g, SO WE CAN ALSO CALCULATE ΔH/gram:

ΔH/g = (-1421.9)/227 = -6.26 kJ/g.

4摩尔硝化甘油产生的焓为ΔH = -7143.6 -(-1456) = -5687.6kJ，1摩尔硝化物产生四分之一的焓变：

ΔH/摩尔= (-5687.6) / 4 = -1421.9kJ/mol。

1摩尔的硝化酸甘油重达227克，所以我们也可以用ΔH/克来计算：

ΔH/g = (-1421.9)/ 227 = - 6.26kJ /g

COMBUSTION OF NATURAL GAS (METHANE, CH_4)

天然气（甲烷，CH_4）的燃烧

$$CH_4(g) + 2O_2(g) \longrightarrow CO_2(g) + 2H_2O(g)$$

反应物 REACTANT	n	ΔH_f	$n\Delta H_f$	生成物 PRODUCT	n	ΔH_f	$n\Delta H_f$
CH_4	1	-74.9	-74.9	$CO_2(g)$	1	-393.8	-393.8
				$H_2O(g)$	2	-241.8	-483.6
TOTAL 总计			-74.9				-877.4

$\Delta H = -877.4 - (-74.9) = -802.5$ kJ/mol, OR ABOUT 或大约 -50.2 kJ/g

WHEN O_2 IS THE OXIDANT IN A REDOX REACTION (AS ABOVE), THE ENTHALPY CHANGE IS CALLED THE **HEAT OF COMBUSTION.** COMBUSTION REACTIONS ARE HIGHLY EXOTHERMIC. BURNING HYDROGEN, FOR INSTANCE, RELEASES **286** kJ/mol OR **143** kJ/g. (= THE HEAT OF FORMATION OF WATER. SEE P. 132) SOME OTHER HEATS OF COMBUSTION, IN kJ PER GRAM OF FUEL:

当氧气作为氧化还原反应（同上）中的氧化剂时，焓的变化被称为**燃烧热**。燃烧反应是强放热反应。氢气燃烧生成水，产生286kJ/mol或143kJ/g的热量（等于水的生成热，参见第132页）。另外一些燃料的产热数据（用kJ/g表示）如下：

HYDROGEN 氢气	143
NATURAL GAS (CH_4) 天然气（甲烷）	50
GASOLINE 汽油	48
CRUDE OIL 原油	43
COAL 煤	29
PAPER 纸	20
DRIED BIOMASS 干的生物质	16
AIR-DRIED WOOD 风干木材	15

IN THIS CHAPTER WE'VE SEEN HEAT CHANGES IN TWO DIFFERENT CONTEXTS: FIRST, ASSOCIATED WITH TEMPERATURE CHANGES, AND SECOND, ASSOCIATED WITH REACTIONS. IN THE NEXT CHAPTER, WE FIND HEAT IN ANOTHER, SURPRISING PLACE: CHANGES OF **STATE.**

在这一章中，我们已经看到了两种不同情况下热量变化：第一与温度的变化有关，第二与反应有关，而在下一章，我们会发现热量在另一个领域的惊人表现：**物态变化**。

* 英语物态（state）和州是同一个词。

THAT IS, WHEN A SUBSTANCE CHANGES FROM A SOLID STATE TO LIQUID (OR LIQUID TO GAS, OR GAS TO SOLID, ETC.), HEAT IS ADDED OR TAKEN AWAY—AND THIS HAPPENS WITH NO CHANGE IN TEMPERATURE. AT TIMES, IN OTHER WORDS, HEAT CAN CHANGE **STRUCTURE** RATHER THAN TEMPERATURE.

也就是说，当一种物质从固体变成液态（或者从液体变成气体，从气体变成固体等）放热或带走热量，并不伴随着温度的变化。换句话说，有时热量可以改变**结构**，而不是温度。

TO UNDERSTAND THIS PUZZLE, WE NEED TO GO A BIT DEEPER INTO THE WORLD OF SOLIDS, LIQUIDS, AND GASES...

要理解这个难题，我们需要更深地进入固体、液体和气体的世界……

Chapter 6 第六章

Matter in a State 物质的状态

UNDER ORDINARY CONDITIONS—OUTSIDE OF STARS, SAY—MATTER COMES IN THREE STATES: SOLID, LIQUID, AND GAS.

在通常情况下，除恒星之外，物质有三种物态：固态、液态和气态。

IN SOLIDS, PARTICLES ARE LOCKED TOGETHER IN A RIGID STRUCTURE. A SOLID HAS BOTH A DEFINITE SHAPE AND VOLUME.

在固体状态下，粒子被锁在刚性结构中。固体有一定的形状和体积。

IN LIQUIDS, PARTICLES CLING TOGETHER, BUT OVERALL STRUCTURE IS LACKING. A LIQUID HAS A DEFINITE VOLUME, BUT ITS SHAPE CONFORMS TO ITS CONTAINER.

在液体状态下，粒子紧密地聚集在一起，但是没有固定的结构。液体有固定的体积，但是其形状由盛它的容器所决定。

IN GASES, STRUCTURE IS ABSENT. PARTICLES FLY AROUND ALMOST TOTALLY INDEPENDENTLY. A GAS HAS NEITHER A FIXED SHAPE NOR VOLUME, BUT WILL EXPAND TO FILL ANY CONTAINER.

在气体状态下，不存在固定的结构。粒子都是独立地飞行。气体既没有固定的形状也没有固定的体积，但它会扩散并充满整个容器。

WHAT HOLDS SOLIDS AND LIQUIDS TOGETHER? THE ANSWER LIES WITH **INTERMOLECULAR FORCES** (IMFs) WITHIN THE SUBSTANCE. THESE ARE ATTRACTIONS BETWEEN MOLECULES (AS OPPOSED TO THE BONDS WITHIN A MOLECULE).

什么使固体和液体聚在一起呢？答案是物质的**分子间作用力**（IMFs）。这是分子间的吸引力（与一个分子内的键力是不同的）。

难道你不喜欢看到周围的空间布满固体和液体吗？

我不知道……可能会很有趣！

ONE IMF WE HAVE ALREADY ENCOUNTERED IS THE **HYDROGEN BOND.** IN WATER MOLECULES, ELECTRONS STAY CLOSER TO THE OXYGEN ATOM, SO THE HYDROGEN ATOMS EFFECTIVELY CARRY A POSITIVE CHARGE. THIS ATTRACTS THEM TO THE NEGATIVE POLE OF ANOTHER WATER MOLECULE.

氢键是一种我们已知的**分子间作用力。**在水分子中，电子更靠近氧原子，所以，氢原子仅带一个正电荷，这使它们会被吸引到另一个水分子的负极。

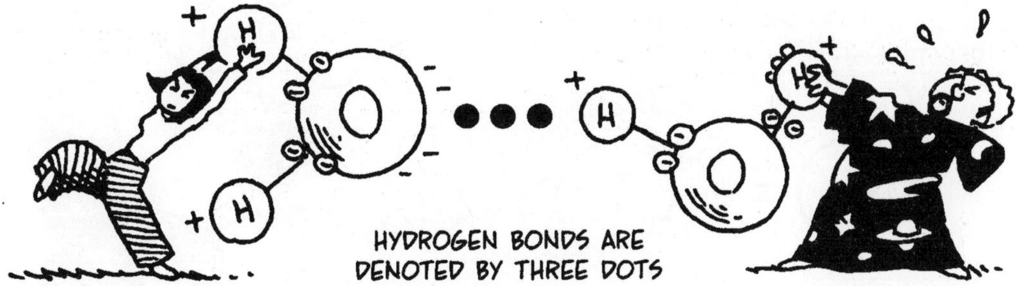

HYDROGEN BONDS ARE DENOTED BY THREE DOTS

氢键用三个点表示

BECAUSE OF ITS TWO ELECTRIC POLES, A WATER MOLECULE IS CALLED A **DIPOLE.** MANY OTHER MOLECULES ARE DIPOLES, TOO, AND THEY ATTRACT EACH OTHER END TO CHARGED END. DIPOLES MAY ALSO ATTRACT IONS.

水分子有两个电极，它被称为偶极子。许多其他分子也是偶极子，它们依靠各自带电的末端相互吸引。偶极子也可能吸引离子。

NONPOLAR MOLECULES CAN BECOME DIPOLES. FOR EXAMPLE, WHEN AN ION NEARS A MOLECULE, THE ION'S CHARGE CAN PUSH OR PULL THE MOLECULE'S ELECTRONS TOWARD ONE END. THE MOLECULE BECOMES AN **INDUCED DIPOLE,** AND ONE END IS ATTRACTED TO THE ION. A DIPOLE CAN INDUCE ANOTHER DIPOLE, TOO.

非极性分子可以成为偶极子。例如，当一个离子接近一个分子时，离子的电荷可以推或拉分子的电子到分子的另一端去。这种一端被离子吸引的分子成为**诱导偶极子**。一个偶极子也可以诱导产生另一个偶极子。

* 英文“吸引力”（attraction）一词在字面上可拆成“吸引”（attract）和“离子”（ion）两词。

EVEN THE GHOSTLY FLIGHT OF ELECTRONS **WITHIN AN ATOM** OR MOLECULE CAN MAKE IT AN "INSTANTANEOUS" DIPOLE—WHICH CAN THEN INDUCE A NEARBY ATOM OR MOLECULE TO BECOME A DIPOLE, ETC. THE RESULTING RIPPLING ATTRACTION IS CALLED THE **LONDON DISPERSION FORCE.**

即使是在一个**原子**或分子中的电子诡异的行动，也会导致分子成为“瞬间”偶极子，这样就可以引起附近的原子或分子成为偶极子，由此产生的影响被称为色散力。

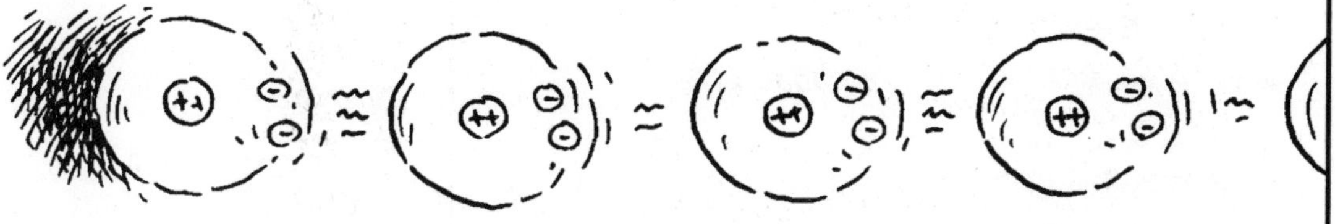

A TEMPORARY CHARGE IMBALANCE SETS OFF A RIPPLE OF DIPOLE-DIPOLE ATTRACTIONS.

分子中瞬间的电荷不平衡引发了偶极-偶极作用。

ALTHOUGH THEY ARE CALLED INTER-**MOLECULAR** FORCES, THESE ATTRACTIONS DO NOT OPERATE ON MOLECULES ONLY. NOBLE GAS ATOMS, FOR INSTANCE, FEEL THE LONDON DISPERSION FORCE.

尽管它们被称为**分子**间力，但这些作用不只是发生在分子之间。例如，稀有气体原子也会受色散力影响。

FROM NOW ON, WE'LL BE A LITTLE LOOSE WITH LANGUAGE AND SOMETIMES REFER TO IMFs AS BONDS. BONDS OR IMFs: THEY'RE ALL ELECTRIC ATTRACTIONS BETWEEN PARTICLES!

从现在起，我们先假设将分子间作用力归为键力。键力或分子间作用力都是粒子之间的电荷作用！

THIS TABLE SUMMARIZES THE STRENGTHS OF DIFFERENT ATTRACTIVE FORCES. THE **STRENGTH** OF A BOND MEANS THE ENERGY REQUIRED TO BREAK IT.

此表总结了不同引力的强度。键的**强度**指破坏键所需的能量。

Strong attractions 强作用力

	STRENGTH 强
IONIC 离子键 ION-ION ATTRACTION 离子-离子吸引力	300 - 1000 kJ/mol
METALLIC 金属键 ELECTRON SHARING AMONG METAL IONS 金属离子间的电子共享	50 - 1000 kJ/mol
COVALENT 共价键 ELECTRON SHARING 电子共享	300-1000 kJ/mol

Moderate attractions 中等作用力

HYDROGEN BONDS AN EXPOSED PROTON IN ONE MOLECULE ATTRACTS A NEGATIVELY CHARGED ATOM IN A NEARBY MOLECULE 氢键 一个分子中裸露质子吸引了附近一个分子中的负电荷原子	20 - 40 kJ/mol
ION-DIPOLE 离子-偶极	10 - 20 kJ/mol

Weak attractions 弱作用力

DIPOLE-DIPOLE 偶极-偶极	1 - 5 kJ/mol
ION-INDUCED DIPOLE 离子-诱导偶极	1 - 3 kJ/mol
DIPOLE-INDUCED DIPOLE 偶极-诱导偶极	0.05 - 2 kJ/mol
INSTANTANEOUS DIPOLE-INDUCED DIPOLE (DISPERSION) 瞬时偶极-偶极诱导偶极（色散）	0.05 - 2 kJ/mol

NOTE: DISPERSION FORCES ARE GREATER BETWEEN LARGER ATOMS, WHICH HAVE MORE ELECTRONS TO PUSH AROUND AND WHERE ELECTRONS ARE FARTHER FROM THE NUCLEUS AND SO MORE EASILY PUSHED.

注：大原子之间的色散力更大，因为大原子有更多的电子，并且离原子核更远，所以更容易被推动。

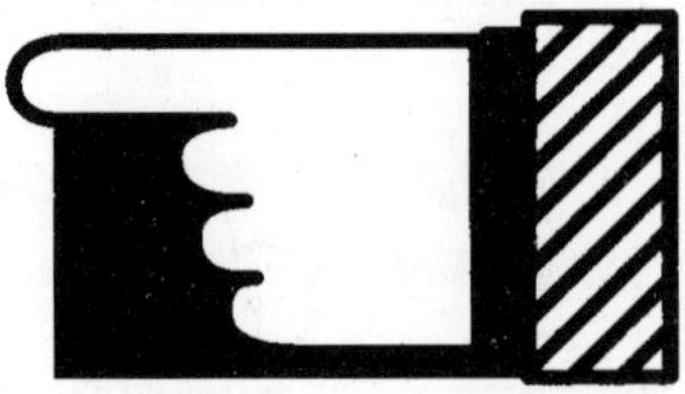

见过冰雪融化的人都知道**温度会影响物质状态**。只要温度提高到足够高，就可以使任何物质变成气体。需要多高的温度取决于物质的键力和分子间的作用力的强度。

SUBSTANCES WITH WEAK IMFs CAN BE SOLID OR LIQUID ONLY AT VERY LOW TEMPERATURES, WHEN PARTICLES MOVE SLUGGISHLY.

弱分子间作用力使物质只在低温下保持固体或液体，这时粒子移动缓慢。

AS TEMPERATURE RISES, MOLECULAR MOVEMENT STRAINS IMFs. IF THE FORCES ARE WEAK, THE SUBSTANCE MUST BECOME LIQUID OR GASEOUS.

随着温度的升高，分子运动并摆脱分子间作用力。如果力减弱，物质会变成液态或气态。

BY CONTRAST, STRONGLY BONDED SUBSTANCES CAN REMAIN SOLID EVEN AT THOUSANDS OF DEGREES CELSIUS.

相比之下，分子间作用力强的物质在几千摄氏度下还能保持固体状态。

IN OTHER WORDS, SUBSTANCES WITH WEAK IMFs MELT AND BOIL AT LOWER TEMPERATURES, WHILE THOSE WITH STRONG BONDS MELT AND BOIL AT HIGHER TEMPERATURES. WATER, WITH ITS HYDROGEN BONDS, IS SOMEWHERE IN BETWEEN.

换言之，分子间作用力弱的物质只需要较低的温度就能熔化和沸腾，而那些具有较强键作用的物质，需要在较高温度下才能熔化和沸腾。由于有氢键，水介于两者之间。

物质 SUBSTANCE		作用力 FORCE		BOND STRENGTH 键强度 (kJ/mol)	MELTING POINT 熔点 (°C)	BOILING POINT 沸点 (°C)
Ar	氩	DISPERSION	色散力	8	−189	−186
NH_3	氨气	HYDROGEN	氢键	35	−78	−33
H_2O	水	HYDROGEN	氢键	23	0	100
Hg	汞	METALLIC	金属	68	−38	356
Al	铝	METALLIC	金属	324	660	2467
Fe	铁	METALLIC	金属	406	1535	2750
NaCl	氯化钠	IONIC	离子	640	801	1413
MgO	氧化镁	IONIC	离子	1000	2800	3600
Si	硅	COVALENT	共价	450	1420	2355
C (DIAMOND)	碳（金刚石）	COVALENT	共价	713	3550	4098

THE SIMPLEST STATE OF MATTER HAS (ALMOST) NO IMFS AT ALL.

物质最简单的物态（几乎）没有分子间作用力。

Gases, Real and Ideal
气体，现实状态和理想状态

GAS PARTICLES ZOOM AROUND FREELY, OR NEARLY SO. WHEN THEY DO BUMP INTO EACH OTHER, THEY FEEL AN IMF, SO THEIR COLLISIONS ARE A BIT "STICKY" (I.E., SOME K.E. IS LOST IN OVERCOMING THE ATTRACTION).

我得走了！　想你……

气体粒子活动自由或几乎自由。当它们互相碰撞时，就能感受到分子间的作用力，因此，它们碰撞时会有些“黏滞”（例如，在克服吸引力时会失去一些动能）。

FOR THEORETICAL PURPOSES, CHEMISTS IGNORE THIS MINOR COMPLICATION AND THINK ABOUT AN **IDEAL GAS.** IN AN IDEAL GAS, ALL PARTICLES ARE IDENTICAL, THEY ZOOM AROUND FREELY, AND ALL COLLISIONS ARE PERFECTLY BOUNCY, OR **ELASTIC**—THAT IS, K.E. IS PRESERVED.

出于理论研究的目的，化学家忽略掉这个小问题，只考虑**理想气体**。在理想气体中，所有的粒子都是一样的，它们自由地运动，并且产生的所有碰撞都是完全弹性碰撞，也就是说，所有动能都得以保留。

ONE CAN DISCUSS CERTAIN PROPERTIES OF AN IDEAL GAS:

讨论一下理想气体的某些属性：

n THE NUMBER OF MOLES, A MOLE BEING 6.02×10^{23} PARTICLES

1摩尔的数目是6.02×10^{23}个粒子

V THE VOLUME

体积

T THE TEMPERATURE IN DEGREES KELVIN

开氏温度

P THE PRESSURE

压强

PRESSURE IS DEFINED AS **FORCE PER UNIT OF AREA.** A FORCE APPLIED TO A SMALL AREA CAN HAVE MORE EFFECT THAN A FORCE SPREAD OVER A LARGE AREA. THAT'S WHY YOU SIT ON A STOOL INSTEAD OF A NEEDLE! SAME FORCE (YOUR WEIGHT), DIFFERENT AREA.

GAS HAS PRESSURE BECAUSE ITS PARTICLES BUMP INTO THINGS.

气体有压强，因为气体的粒子会撞到物体。

$$\text{Pressure} = \frac{\text{Force}}{\text{Area}} \qquad 压强 = \frac{力}{面积}$$

SINCE DOUBLING AN AREA DOUBLES THE NUMBER OF COLLISIONS AND SO DOUBLES THE FORCE, FORCE AND AREA GO UP TOGETHER, SO THE PRESSURE IS CONSTANT THROUGHOUT THE GAS.

压强被定义为**物体单位面积上所受力的大小**。相同的力作用在小面积上产生的压强大于在大面积上产生的压强。这就是为什么你坐在凳子上而不是针上的原因！同样的力（你的体重）作用于不同的区域，效果不同。

由于面积增加一倍，碰撞的数量增加一倍，所以产生双倍的力。力和面积一起增加，所以贯穿整个气体的压强不变。

THE AIR AROUND US EXERTS ATMOSPHERIC PRESSURE. **ONE ATMOSPHERE** (1 atm) IS THIS PRESSURE (ON AVERAGE) AT SEA LEVEL. IN TERMS OF METRIC UNITS:

1 atm = **101,325** NEWTONS/m^2

= **10.1325** NEWTONS/cm^2

我们周围的空气产生大气压强。**1个大气压**（1atm）是在海平面（平均）的大气压强。在度量单位方面：

1atm=101325 牛顿/米^2

=10.1325 牛顿/厘米^2

ATMOSPHERIC PRESSURE IS **HUGE!** WE DON'T FEEL IT BECAUSE IT PUSHES FROM ALL DIRECTIONS, BUT RECALL GUERICKE'S EXPERIMENT WITH HORSES TO APPRECIATE ITS TRUE MAGNITUDE.

大气压力是**巨大的**！我们感觉不到它是因为它从各个方向袭来，回忆一下居里克的马拉铜球实验，以了解大气压的实际强度。

Gas Laws 气体定律

NOT SURPRISINGLY, n, T, V, AND P ARE ALL RELATED. FOR INSTANCE, YOU MIGHT EXPECT THAT MORE PARTICLES WOULD OCCUPY A GREATER VOLUME, ALL ELSE BEING EQUAL. AND SO THEY DO! IN FACT, IT'S A LAW, THE FIRST OF THREE GAS LAWS, WHICH WE LIST IN ALPHABETICAL ORDER.

毫不奇怪，摩尔数（*n*）、温度（*T*）、体积（*V*）和压强（*P*）是互相关联的。例如，在所有其他条件相同时，你可能认为更多的粒子将占据更大的体积。它们做到了！事实上，这是一个定律，三大气体定律之一，我们按字母顺序列出。

AVOGADRO'S LAW: IF T AND P ARE FIXED, THEN VOLUME IS PROPORTIONAL TO THE NUMBER OF MOLES.

阿伏加德罗定律(Avogadro's hypothesis)：温度（*T*）同压强（*P*）固定不变时，体积与摩尔数成正比。

$$\frac{n_1}{V_1} = \frac{n_2}{V_2}$$

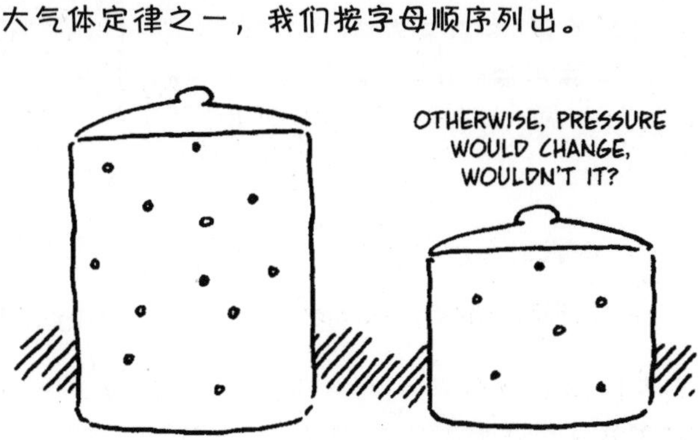

THIS IMPLIES THAT A SET VOLUME OF GAS (AT FIXED T AND P) ALWAYS HAS THE **SAME NUMBER OF MOLECULES**—NO MATTER WHAT WHAT GAS IT IS! THIS FACT ENABLED NINETEENTH-CENTURY CHEMISTS TO FIND ATOMIC WEIGHTS FOR THE FIRST TIME.

这意味着，同体积的气体（*T*和*P*恒定）具有**相同数目的分子**——不论什么气体！这一事实使19世纪的化学家们首次发现了原子量。

BOYLE'S LAW: IF n AND T ARE FIXED, THEN VOLUME IS INVERSELY PROPORTIONAL TO PRESSURE.

波义耳定律（*Boyle's law*，有时又称 *Mariotte's Law*）：摩尔数（n）和温度（T）固定不变时，气体的压力P和体积V成反比。

$$P_1V_1 = P_2V_2$$

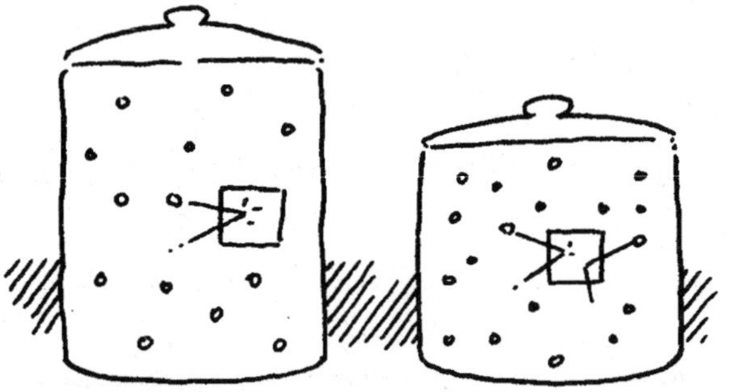

IN A LARGER VOLUME, FEWER PARTICLES HIT A UNIT OF AREA...

在一个较大的体积内，单位面积上碰撞的粒子数较少……

THAN IN A SMALLER VOLUME.

在一个较小的体积内，单位面积上碰撞的粒子数较多。

CHARLES'S LAW: WITH n AND P FIXED, VOLUME IS PROPORTIONAL TO TEMPERATURE.

查理定律（*Charles's law*）：摩尔数（n）和体积（V）固定不变时，气体的压强*P*与温度*T*成正比。

$$\frac{V_1}{T_1} = \frac{V_2}{T_2}$$

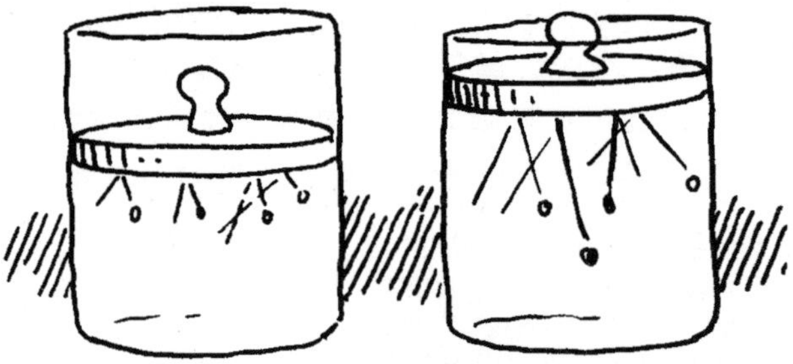

IF T RISES...

如果温度T升高……

MORE-ENERGETIC PARTICLES PUSH UP THE PISTON.

更多高能量的粒子推动活塞。

ALL THESE LAWS CAN BE ROLLED INTO A SINGLE EQUATION THAT COMBINES THE RELATIONSHIP AMONG ALL FOUR VARIABLES. IT'S CALLED THE **IDEAL GAS LAW**, AND IT GOES

所有这些定律可以由一个方程推导，这个方程结合了四个变量之间的关系。这就是所谓的**理想气体定律**，它就是

$$PV = nRT$$

HOLD ANY TWO VARIABLES FIXED, AND YOU SEE THE RELATIONSHIP BETWEEN THE OTHER TWO AS GIVEN IN THE A, B, C LAWS ON THE PREVIOUS PAGE.

如你在前面的A、B、C三大定律所见，保持任意两个变量固定，再考虑其他两个变量的变化关系。

R CAN BE FOUND AS FOLLOWS: FIRST, EXPERIMENTALLY DETERMINE THE VOLUME OF ONE MOLE OF GAS (ANY GAS, BY AVOGADRO!). AT 0°C (= 273°K) AND 1 ATM, IT TURNS OUT THAT **ONE MOLE OF GAS OCCUPIES 22.4 LITERS.** SO:

R可以由如下方法得到：首先，R是1摩尔气体（任何气体,阿伏加德罗说的！）在0°C（= 273°K）和1个大气压下的体积，事实证明，**1摩尔气体有22.4升**，所以：

n = 1 mol
T = 273°K
P = 1 atm
V = 22.4 L.

PLUG INTO THE GAS LAW EQUATION:

插入气体定律公式：

$$(1\ \text{atm})(22.4\ \text{L}) = (1\ \text{mol})R(273°\text{K})$$

SO 因此，

$$R = (22.4/273)\ \text{atm-L/mol°K}$$
$$= 0.082\ \text{atm-L/mol°K}$$

THE CONDITIONS 条件如下

T = 0°C AND 并且
P = 1 atm

ARE KNOWN AS **STANDARD TEMPERATURE AND PRESSURE (STP).**

被称为**标准温度和压强**（STP）。

Example: 例如：

WHAT VOLUME OF GAS IS RELEASED BY THE EXPLOSION OF ONE GRAM OF BLACK POWDER?

一克黑火药爆炸释放多少体积的气体？

$$4KNO_3(s) + 7C(s) + S(s) \longrightarrow 3CO_2\uparrow + 3CO\uparrow + 2N_2\uparrow + K_2CO_3(s) + K_2S(s)$$

3 + 3 + 2 = 8 mol GAS

摩尔气体

THE MOLAR WEIGHT OF THE LEFT SIDE IS 520 g, WHICH PRODUCES 8 mol GAS. SO ONE GRAM OF POWDER PRODUCES

(1/520)(8) = 0.015 mol GAS.

SO n = 0.015. P = 1 atm, AND EXPERIMENT SHOWS THAT THE TEMPERATURE T IS ABOUT 2250°K.

在左侧的摩尔质量是520克，产生8摩尔气体。因此，1克火药产生

(1/520) × (8) = 0.015摩尔气体

因此，n=0.015。P = 1atm时，实验表明，温度T大约是2250°K。

SOLVING FOR VOLUME,

$$V = \frac{nRT}{P}$$

$$= \frac{(0.015\ \text{mol})(0.082\ \text{atm-L/mol°K})(2250°)}{1\ \text{atm}}$$

= 2.8 LITERS

计算体积。

$$V = \frac{nRT}{P}$$

$$= \frac{(0.015\ \text{mol})(0.082\ \text{atm-L/mol°K})(2250°)}{1\ \text{atm}}$$

= 2.8升

A GRAM OF POWDER, WE MEASURE, OCCUPIES A TINY VOLUME, ABOUT 0.8 mL.

我们测量到1克火药的体积很小，大约为0.8毫升。

THE EVOLVED GAS EXPANDS TO $(2800)/(0.8) =$ **3,500** TIMES THAT VOLUME! IF WE WANTED TO CONFINE THE GAS IN A LITTLE PACKAGE 1 mL (= .001 L) IN VOLUME, IT WOULD BUILD UP A PRESSURE OF:

而产生的气体体积能够膨胀至2800/0.8=3500倍！如果我们限制气体体积只有1毫升(=0.001L)，这将产生一个压力：

$$P = \frac{nRT}{V}$$

$$= \frac{(0.015)(0.082)(2250)}{(0.001)}$$

OR ABOUT **2800** atm. 约2800大气压。

Liquids 液体

BECAUSE OF THEIR IMFs, LIQUIDS HAVE COMPLICATED BEHAVIOR. THERE ARE NO "IDEAL LIQUIDS."

由于液体的分子间作用力，所以，液体的情况很复杂。没有“理想液体”。

LIQUIDS BEHAVE AS IF THEY HAVE A SKIN. ATTRACTION AMONG SURFACE MOLECULES—**SURFACE TENSION**—KNITS THEM TOGETHER MORE TIGHTLY THAN INTERIOR MOLECULES. THAT EXPLAINS WHY BUGS CAN WALK ON WATER...

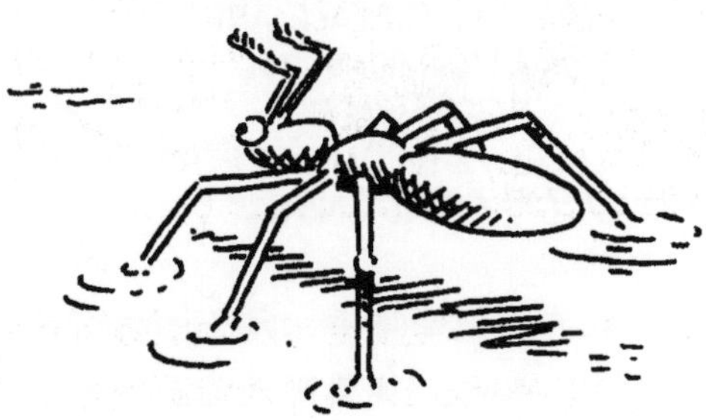

液体表面仿佛有一层皮肤，其分子之间的吸引力形成表面张力，比内部的分子结合得更紧密。这就解释了为什么虫子能在水上行走。

Evaporation and Condensation
蒸发和凝结

IN MOST LIQUIDS, MOLECULAR MOVEMENT CAN OVERCOME COHESIVE FORCES. IN THAT CASE, SOME MOLECULES BREAK THROUGH THE SURFACE AND **EVAPORATE.** CONVERSELY, LESS-ENERGETIC VAPOR MOLECULES MAY COLLECT INTO LIQUID, OR **CONDENSE.**

在大多数液体中，分子运动可以克服凝聚力。在这种情况下，一些分子冲破液体表面并**蒸发**。相反，能量降低的蒸气分子可以集成液体，即**凝结**。

WHEN A MOLECULE GOES GASEOUS, ENERGY MUST BE ABSORBED FROM THE SURROUNDINGS TO BREAK THE ATTRACTIVE FORCES (BONDS, IMFs) THAT EXIST WITHIN THE LIQUID. **EVAPORATION IS ENDOTHERMIC.**

当一个分子成为气态时，必须吸收周围环境中的能量以破坏液体内存在的引力（键力和分子间作用力）。**蒸发是吸热过程。**

$$\text{liquid} \longrightarrow \text{gas} \quad \Delta H > 0$$

液体 → 气体 $\Delta H > 0$

IN OTHER WORDS, GAS IS A **MORE ENERGETIC STATE OF MATTER** THAN LIQUID.

换句话说，气体是一种比液态物质**更有能量的状态。**

FOR EXAMPLE, WATER'S HEAT OF VAPORIZATION (AT 1 atm, 25°C) IS 44 kJ/mol. THAT IS THE ENTHALPY CHANGE OF THE "REACTION" $H_2O(l) \longrightarrow H_2O(g)$.

例如，水的汽化（1个大气压，25℃）吸热44kJ/mol摩尔。这是"反应" $H_2O(l) \rightarrow H_2O(g)$的焓变。

THIS IS WHY PERSPIRATION WORKS. EVAPORATING SWEAT DRAWS HEAT FROM YOUR BODY.

蒸发的汗水可从你体表吸热，这也是出汗的功能。

A BRILLIANTLY SIMPLE APPLICATION OF THIS 44 kJ/mol IS THE COOLING POT OF NIGERIAN POTTER **MOHAMMAD BAH ABBA.**

这个关于44kJ/mol的一个简单应用是尼日利亚的制陶工**波特·穆罕默德·巴哈·阿巴**的冷罐。

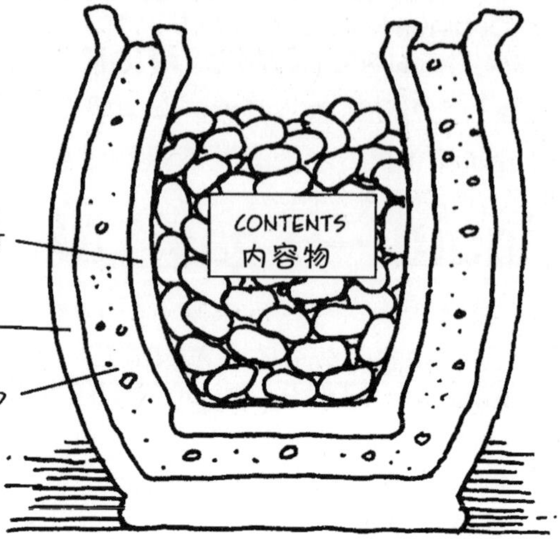

ONE CLAY POT SITS INSIDE ANOTHER, WITH A LAYER OF WET SAND IN BETWEEN. THE OUTER POT IS UNGLAZED AND POROUS.

一个陶罐在另一个陶罐内，之间用一层湿沙填充，外层的陶罐是无釉和多孔的。

IN A DRY ENVIRONMENT, THE WATER IN THE SAND LAYER EVAPORATES AND PASSES OUT THROUGH PORES IN THE OUTER POT. IN THE PROCESS, IT DRAWS HEAT FROM THE APPARATUS.

在干燥的环境中，沙层中的水分蒸发，并通过外层的孔排出。在这个过程中，它吸收了这个装置的热量。

THE TEMPERATURE INSIDE CAN FALL AS FAR AS 14°C (= 25°F) BELOW THAT OF THE OUTSIDE—A LIFESAVER IN DESERT COUNTRIES WHERE MOST PEOPLE CANNOT AFFORD A FRIDGE.

内部的温度可降至低于外部温度 14℃（= 25℉）。这个方法对于一个沙漠国家来说简直是救星，在这些国家里，大多数人都买不起冰箱。

NOW IMAGINE A LIQUID IN A CLOSED CONTAINER AT CONSTANT TEMPERATURE. AS LIQUID EVAPORATES, VAPOR BUILDS UP, AND SOON SOME OF THIS VAPOR BEGINS TO CONDENSE.

想象一下在恒温密闭容器中的液体，由于液体蒸发，蒸气逐渐增多，很快这种蒸气便开始凝结。

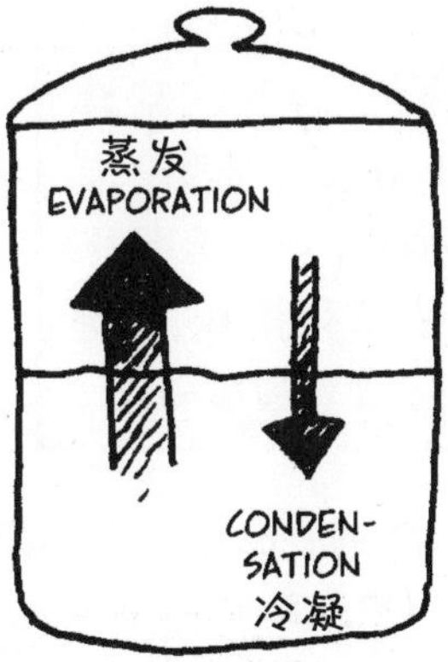

AT FIRST, EVAPORATION OUTPACES CONDENSATION, BUT EVENTUALLY, CONDENSATION MAY CATCH UP. WHEN THE TWO PROCESSES EXACTLY BALANCE, THERE IS NO NET CHANGE IN THE AMOUNT OF LIQUID OR GAS. THE TWO STATES ARE SAID TO BE IN **EQUILIBRIUM,** AND WE WRITE

起初，蒸发过程的速度快于凝结过程，但最终，凝结的速度会赶上。当两个进程正好平衡时，液体或气体量没有变化，就是说，这两种状态处于平衡，我们写成：

$$\textbf{liquid} \rightleftharpoons \textbf{vapor}$$

液体　　蒸气

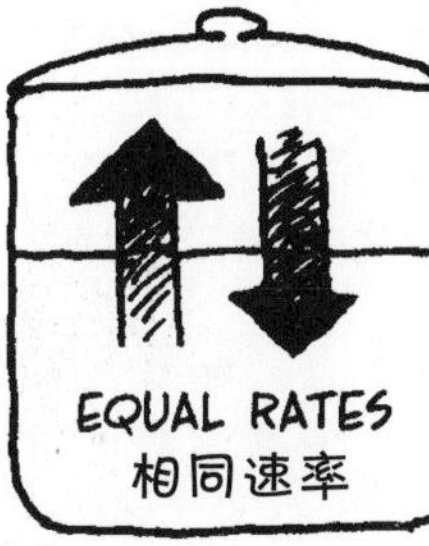

THE EXTRA PRESSURE DUE TO VAPOR ALONE IS CALLED ITS **PARTIAL PRESSURE.*** AS VAPOR BUILDS UP, ITS PARTIAL PRESSURE RISES STEADILY (BIGGER n, SAME V AND T!) UNTIL EQUILIBRIUM. AT EQUILIBRIUM, THIS PARTIAL PRESSURE IS CALLED THE

vapor pressure.

IT'S THE PRESSURE THE VAPOR "WANTS" TO ATTAIN.

额外的压强是由于单独蒸气的原因，就是所谓的**分压***。当蒸气积聚时，其分压便稳步上升（摩尔数n增大，体积V和温度T恒定！），直至平衡。在平衡状态，这被称为**蒸气压强**。这是蒸气“希望”得到的压强。

VAPOR PRESSURE (P_V) RISES WITH TEMPERATURE, SINCE MORE-AGITATED MOLECULES HAVE A GREATER "NEED" TO VAPORIZE.

蒸气压强（P_V）随温度升高，因为骚动的分子更加有“需要”蒸发。

VAPOR PRESSURE OF WATER
水的蒸气压

T (°C)	P_V ATM
0	0.006
20	0.023
40	0.073
60	0.197
80	0.467
90	0.692
100	1.00
200	15.34
300	84.8

*THE TOTAL PRESSURE OF A MIXTURE OF GASES IS THE SUM OF ALL THEIR PARTIAL PRESSURES.
*一个混合气体的总压强是其中所有气体分压的总和。

P_V IS THE PRESSURE AT WHICH VAPOR "WANTS" TO STABILIZE. BUT WHAT IF NO MATTER HOW MUCH VAPOR THE LIQUID SPEWS, ITS PRESSURE NEVER REACHES P_V? IN THAT CASE, VAPORIZATION GOES UNCHECKED, AND THE LIQUID **BOILS.**

P_V为蒸气“需要”稳定下来时的压力。但是，如果不管有多少液体喷出蒸气，压力永远达不到P_V将会怎样？在这种情况下，蒸发就会难以控制，液体会**沸腾**。

WHETHER A LIQUID BOILS DEPENDS ON THE TOTAL PRESSURE ABOVE THE LIQUID—THE **EXTERNAL PRESSURE.** CALL IT P.

液体是否沸腾取决于液体表面外部压力的总压力——外部压力记为P。

EQUILIBRIUM IS POSSIBLE WHEN VAPOR PRESSURE **P_V IS LESS THAN P,** BECAUSE THEN P_V CAN ACTUALLY BE REALIZED AS A PARTIAL PRESSURE OF VAPOR.

当蒸气压力P_V小于P时，平衡是可能的，因为实际上，P_V可以被认为是蒸气分压。

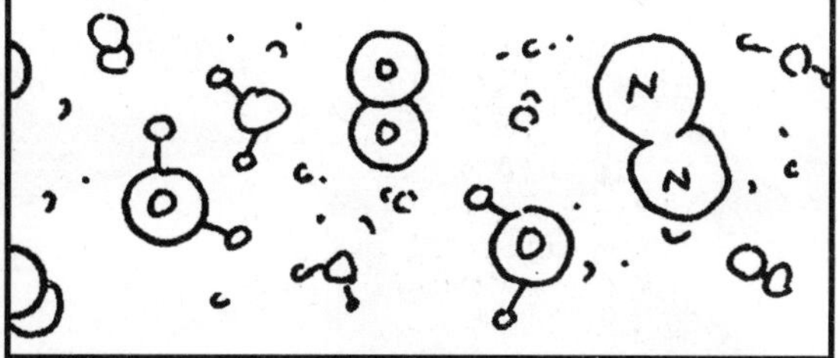

HERE H_2O MOLECULES ARE JUST PART OF THE AIR AND HAPPILY SO!

这里水分子只是快乐地作为空气的一部分！

IF P IS LESS THAN P_V, THE PARTIAL PRESSURE OF VAPOR MUST ALSO BE LESS THAN P_V, AND BOILING OCCURS.

如果P小于P_v，水蒸气分压也肯定比P_v少，这样就会发生沸腾。

THAT IS, BOILING BEGINS PRECISELY WHEN **VAPOR PRESSURE EQUALS EXTERNAL PRESSURE.**

也就是说，当**蒸气压等于外部压强**时，沸腾就会发生。

THE TEMPERATURE AT WHICH A LIQUID BOILS IS CALLED ITS
boiling point.
BOILING POINT DEPENDS ON EXTERNAL PRESSURE.
液体沸腾的温度称为它的**沸点**，沸点取决于外部压强。
AT SEA LEVEL (PRESSURE = 1 ATM), WATER BOILS AT 100° C, BUT AT HIGH ALTITUDE, WHERE AIR IS THIN, BOILING POINT CAN DROP BELOW 85°. IN THE VACUUM OF SPACE, WATER BOILS AT ANY TEMPERATURE.
在海平面（压强=1大气压），水的沸点是100℃；但在高海拔地区，那里空气稀薄，沸点可降到85℃；在太空的真空环境中，任何温度的水都会沸腾。
EXTERNAL PRESSURE, BY THE WAY, CAN INCLUDE LIQUID PRESSURE AS WELL AS GAS PRESSURE. IN THE DEEP OCEAN (PRESSURE VERY HIGH!) WATER NEAR VOLCANIC VENTS CAN REMAIN LIQUID ABOVE 350°C.
顺便说一下外部压强，包括液体压力以及气体压力。在深海（压力非常高），火山口附近的水可以使液态保持在350℃以上。
I'M COOKED!
我被煮熟了！

WE SUMMARIZE ALL THIS WITH A LIQUID-GAS MINI-DIAGRAM. THE HORIZONTAL AXIS IS TEMPERATURE; THE VERTICAL AXIS IS PRESSURE; AND AT EACH PAIR OF VALUES (T,P) WE SEE WHETHER A SUBSTANCE IS LIQUID OR GAS.

我们把这些总结在一张液-气小图上。横轴代表温度，纵轴代表压强，每对数值（温度、压强）代表我们看到的某种物质是否是液体或气体。

THE CURVE BETWEEN THEM INDICATES THE BOILING POINT FOR ANY PRESSURE.

它们之间的曲线代表任意压力下的沸点。

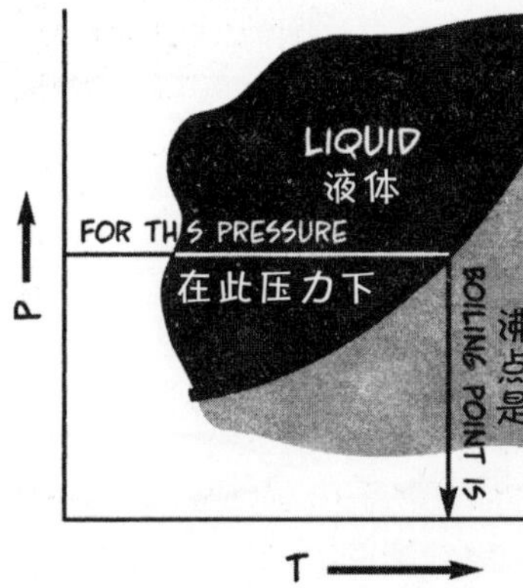

NOTE THAT PHASE TRANSITIONS CAN RESULT FROM CHANGING PRESSURE ALONE, OR TEMPERATURE ALONE, OR A COMBINATION.

注意：单独改变压强或温度或同时改变两者都会导致相变。

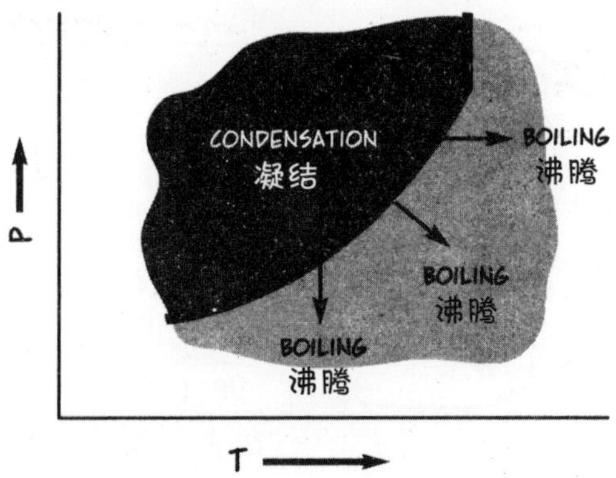

THE CURVE HAS ITS LIMITS. EVERY LIQUID HAS A CHARACTERISTIC **CRITICAL TEMPERATURE,** THE HIGHEST AT WHICH THE LIQUID STATE CAN EXIST. ABOVE THE CRITICAL TEMPERATURE, NO AMOUNT OF PRESSURE CAN STOP THE LIQUID FROM BOILING AWAY.

曲线也有其局限性。每种液体都有一个特征**临界温度**，即可以保持液体状态的最高温度。在临界温度以上，施加压力无法再阻止液体沸腾。

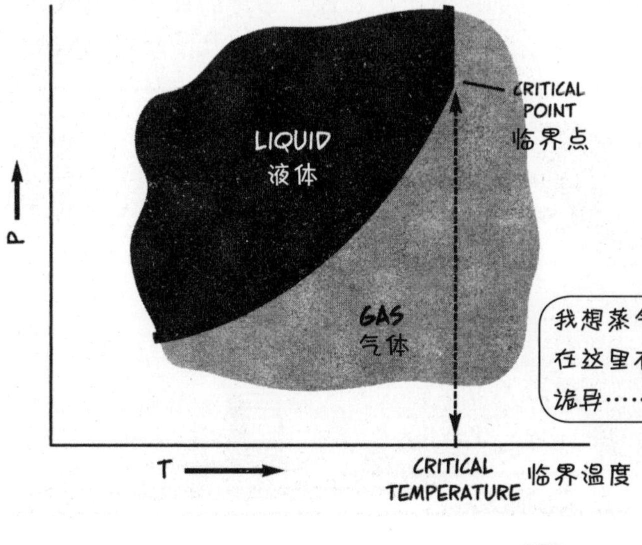

Melting Solids
固体熔化

IN THE OPEN AIR, MANY LIQUIDS SIMPLY EVAPORATE AWAY. SINCE THE VAPOR ESCAPES, IT BUILDS UP NO SIGNIFICANT PRESSURE ON THE SURFACE, AND EVAPORATION CONTINUES INDEFINITELY.

在露天环境中，很多液体容易蒸发。因为蒸汽逸出，所以表面没有很大的压力，因此蒸发不受限制地持续进行。

在液体表面的蒸气分压是P_V，但大气压$<P_V$，所以……

分子继续离开。

IN SOLIDS, BY CONTRAST, VERY FEW PARTICLES HAVE ENOUGH ENERGY TO ESCAPE. VAPOR PRESSURE IS LOW—THOUGH NOT SO LOW WE CAN'T SMELL MANY SOLIDS. IN SOME CASES, VAPOR PRESSURE IS VIRTUALLY NIL. DIAMONDS ARE FOREVER!

相比之下，固体中几乎没有粒子有足够的能量可以逃逸。固体的蒸气压力低但不至于低到我们闻不到所有的固体的味道。在某些情况下，蒸气压几乎是零。钻石的蒸气压永远是零，因此钻石恒久远！

AS WE ALL KNOW, SOLIDS **MELT***, AND THEY DO SO AT A SET TEMPERATURE, THE **MELTING POINT**, WHICH VARIES FROM SOLID TO SOLID.

众所周知，固体可以在一定的温度下**熔化***，不同的固体**熔点**不同。

AT THIS TEMPERATURE, ANY ADDED HEAT IS ENTIRELY CONSUMED IN BREAKING BONDS UNTIL THE SOLID IS COMPLETELY MELTED. **MELTING, LIKE EVAPORATION, IS ENDOTHERMIC.**

SOLID ⟶ LIQUID $\Delta H > 0$

THIS ENTHALPY CHANGE IS CALLED THE **HEAT OF FUSION**. FOR ICE AT STP, IT'S 6.01 kJ/mol.

在此温度下，任何吸收的热量都完全用于破坏键，直到固体完全融化。**熔化过程同蒸发过程一样，是吸热过程。**

固体 ⟶ 液体 $\Delta H > 0$

其中焓变称为**熔化热**。处于标准状态下的冰，其熔化热是6.01kJ/mol。

*USUALLY. SOME OF THEM **SUBLIME**, OR GO STRAIGHT TO THE GAS PHASE. MORE ON THAT SHORTLY.

*通常，固体中的一部分**升华**，即直接进入气相。后面很快会谈到这一现象。

EXTERNAL PRESSURE AFFECTS MELTING POINT: IN THIS SOLID-LIQUID MINI-DIAGRAM WITH P AND T AXES, THE CURVE SHOWS THE MELTING POINT FOR EACH VALUE OF P.

外部压强影响熔点：在这张带有P和T轴的固-液小图中，曲线代表不同压力值P下的熔点。

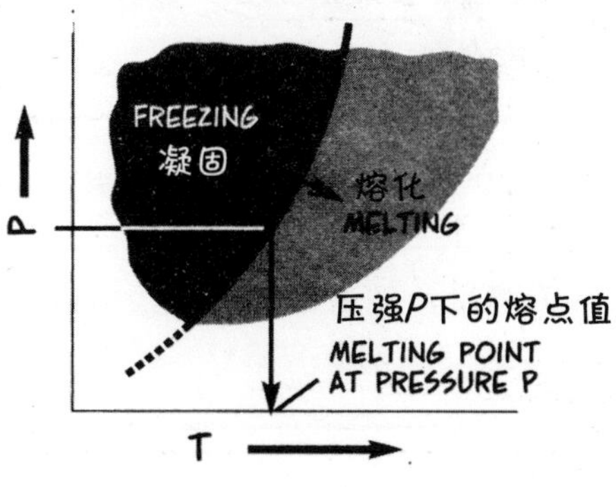

THE EFFECT IS LESS DRAMATIC THAN WITH BOILING POINT, HOWEVER, SO THE MELTING CURVE IS USUALLY PRETTY STEEP.

其效果远远低于沸点，即便如此，这样的熔化曲线通常也很陡峭。

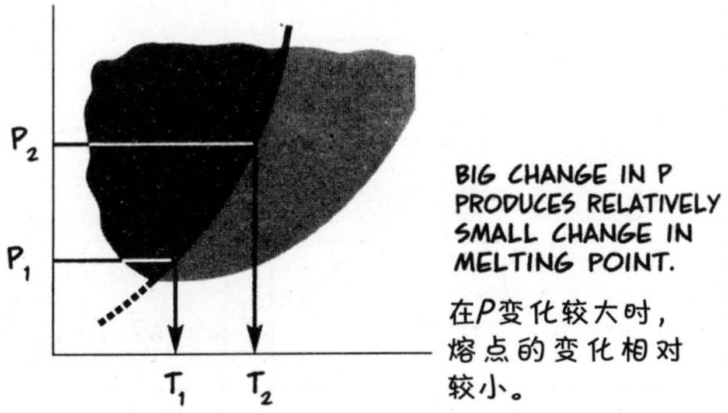

IN A FEW WEIRD MATERIALS, ADDED PRESSURE ACTUALLY DECREASES MELTING POINT. WATER IS ONE SUCH.

对于一些特殊的物质，增加压强实际上使熔点降低。水就是这样的。

THAT'S BECAUSE WATER **EXPANDS** WHEN IT FREEZES. THE CRYSTALLINE STRUCTURE OF ICE IS UNUSUALLY SPACIOUS.

这是因为水凝固时体积**膨胀**。冰的晶体结构是非常松散的。

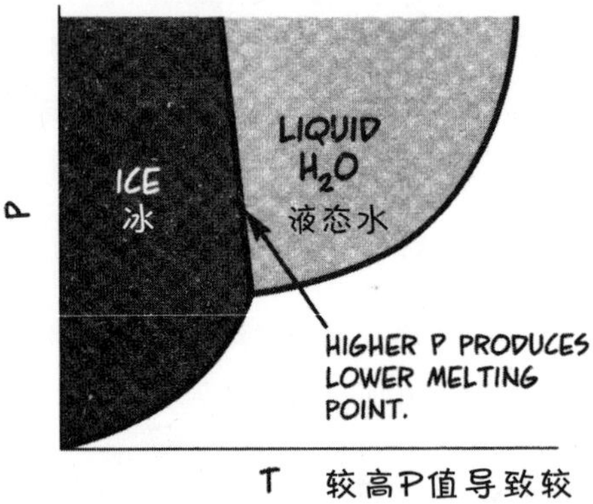

较高P值导致较低熔点。

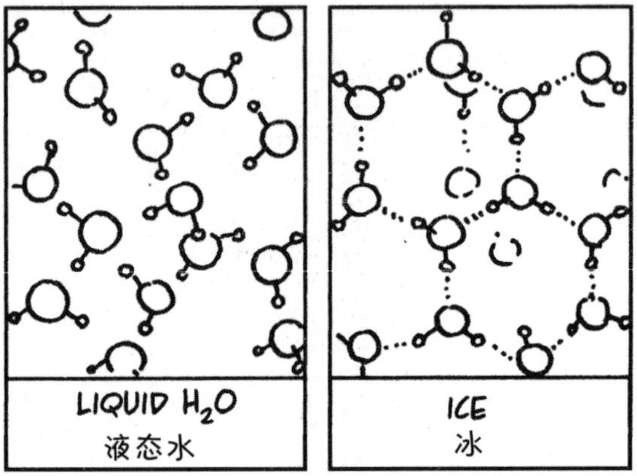

PRESSING ON AN ICE CUBE PUTS STRAIN ON THE BONDS AND DRIVES THE MOLECULES INTO A TIGHTER BUT MORE RANDOM CONFIGURATION, AND THE ICE MELTS AT THE POINT OF PRESSURE.

在冰块上施加压强，使键处于紧张状态并且使分子成为更加紧凑但更随意的结构，冰在此压强施加处融化。

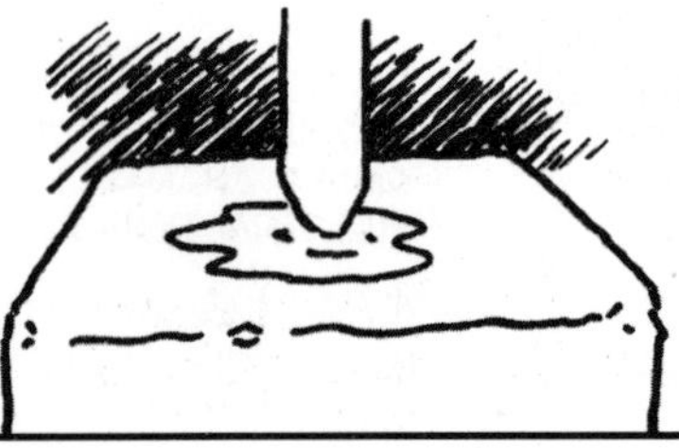

SO, UNLIKE MOST SOLIDS, ICE FLOATS ON ITS LIQUID FORM... THE EXPANSION OF FREEZING WATER CAN CRACK ROCKS... AND THIS ODD FEATURE HAS A PROFOUND IMPACT ON THE WORLD AROUND US.

因此，与大多数固体不同，冰浮在水面上……水结冰产生的膨胀甚至可以裂开石头……这些特征深刻地影响着我们周围的世界。

ICE-SKATING AS IT WOULD BE IF WATER FROZE LIKE A NORMAL SUBSTANCE.

如果水像正常的物质一样凝固，那么，滑冰就会像这个样子。

Phase Diagrams 相图

PUT OUR MINI-DIAGRAMS TOGETHER AND THEY SHOW A COMPLETE PICTURE OF THE THREE STATES OF MATTER IN TERMS OF T AND P. THE SOLID-LIQUID CURVE MEETS THE LIQUID-GAS CURVE AT A **TRIPLE POINT** WHERE ALL THREE PHASES ARE IN EQUILIBRIUM.

将小图放在一起，它们便表示在不同*T*和*P*下的三种物态的全貌，固-液曲线与液-气曲线相交于一个三相点，代表三相平衡。

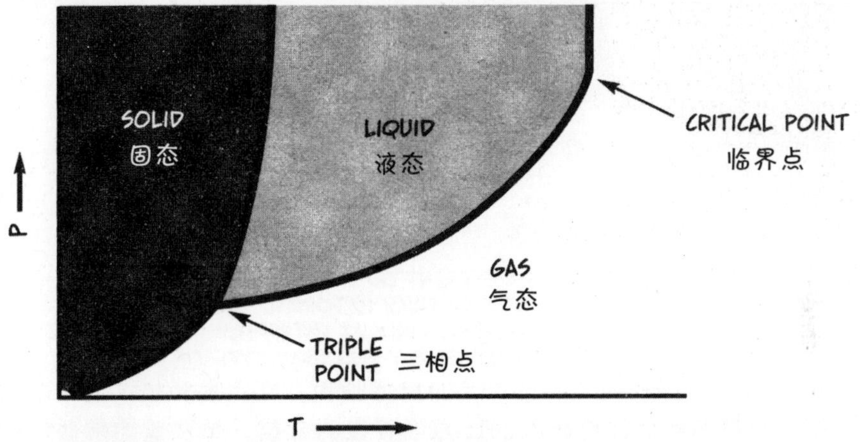

NOTE THAT THERE ARE ALSO CONDITIONS WHEN A SOLID CAN CHANGE DIRECTLY INTO A GAS, A PROCESS CALLED **SUBLIMATION**. THE REVERSE PROCESS, GAS ⟶ SOLID, IS **DEPOSITION**. THE BEST-KNOWN EXAMPLE AT NORMAL PRESSURE IS CO_2, "DRY ICE," THE STUFF USED IN THEATRICAL SMOKE MACHINES.

请注意：满足一定条件时，固体也可以直接变成气体，这一过程称为**升华**。相反的过程，即气体直接变为固体，称为**凝华**。最著名的例子是在常压下的二氧化碳舞台烟雾机所用的“干冰”。

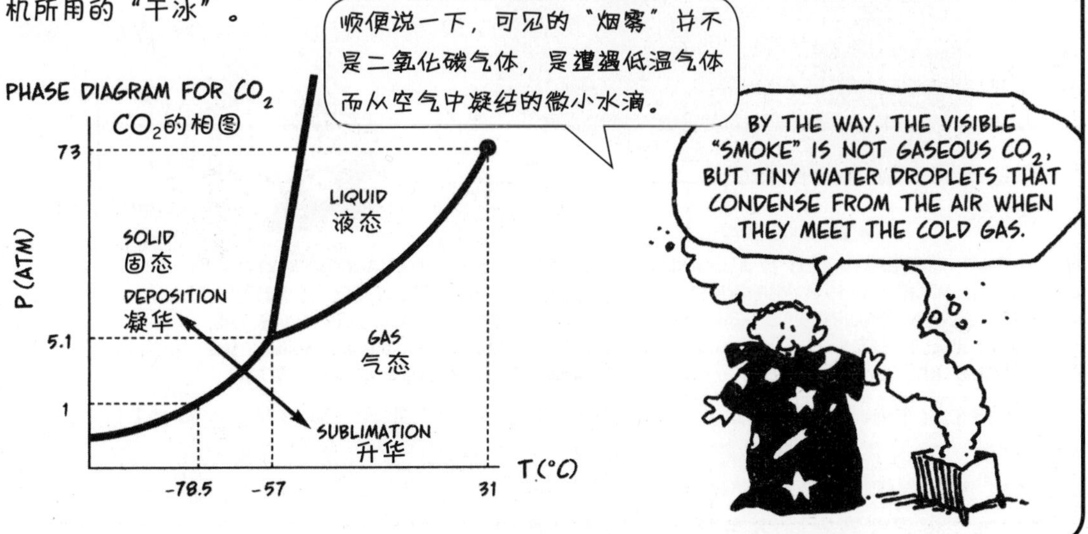

A COUPLE OF OTHER PHASE DIAGRAMS SHOW SOME MORE SUBTLE AND UNUSUAL FEATURES OF MATTER. HERE IS **CARBON.**

一些相图显示更多关于物质微妙和不寻常的特性。下图表示的是**碳**的特性。

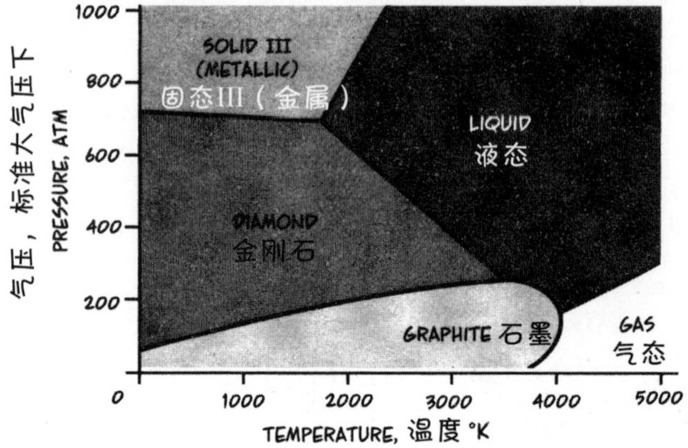

CARBON HAS **THREE SOLID FORMS,** WITH DIFFERENT CRYSTALLINE STRUCTURES: GRAPHITE, FOUND IN COAL AND PENCIL LEADS, DIAMOND, WHICH IS FORMED ONLY UNDER HIGH-PRESSURE CONDITIONS, AND METALLIC, WHICH EXISTS ONLY AT EXTREMELY HIGH PRESSURE. NOTE HOW THE MELTING CURVE SLOPES DIFFERENTLY FOR EACH TYPE OF CRYSTAL.

碳有**三种不同的固态形式**，对应不同的晶体结构：在煤炭和铅笔芯中的石墨、在高压条件下形成的钻石和只能在极高的压力下存在的金属。请注意熔解曲线对每个晶体类型的斜率也不同。

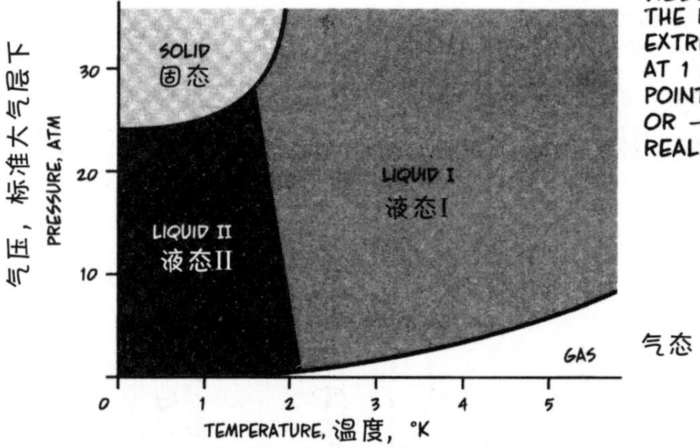

HELIUM, LIGHTEST OF THE NOBLE GASES, HAS EXTREMELY WEAK IMFs. AT 1 ATM, ITS BOILING POINT IS JUST OVER 4°K, OR -269°C. THAT'S REALLY **COLD!!!**

氦，最轻的稀有气体，在1个大气压时，只有极小的分子间作用力，它的沸点大约是4°K或-269°C。**够冷**！

BELOW THAT TEMPERATURE IT IS A LIQUID... AND BELOW 2.17°K—IT IS ANOTHER KIND OF LIQUID! THIS HELIUM II IS A "SUPERFLUID" WITH WEIRD PROPERTIES. IT FLOWS WITHOUT VISCOSITY (GOOPINESS)... IT WILL LEAK OUT THE TINIEST PORE... IT WILL EVEN CLIMB THE CONTAINER WALLS! SEE http://cryowwwebber.gsfc.nasa.gov/introduction/liquid_helium.html FOR DETAILS. HELIUM CAN ALSO BE SOLID, BUT ONLY AT PRESSURES ABOVE 25 ATM.

在那个温度以下它是液体……而在低于2.17°K时它是另一种液体！这种氦II是一种有奇怪特性的"超流体"。它流动起来没有黏度……它从微小的孔隙泄漏出来……它甚至会爬上容器壁！详情参阅http://cryowwwebber.gsfc.nasa.gov/introduction/liquid_helium.html。氦气也可以是固体，但只能在25个大气压的压强以上存在。

Heating Curves
升温曲线

FINALLY, LET'S RETURN TO THE HEATS OF FUSION AND EVAPORATION, AND SEE HOW THEY PLAY OUT WHEN WE HEAT A BLOCK OF ICE UNTIL IT MELTS AND THEN BOILS.

最后，让我们回到熔解热和蒸发热，看看当我们加热冰块直至融化然后沸腾，它们是如何起作用的。

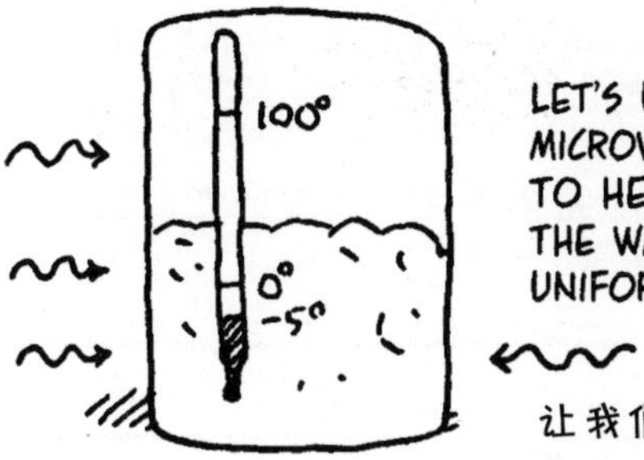

LET'S USE MICROWAVES TO HEAT THE WATER UNIFORMLY.

让我们用微波来均匀地加热水。

LET'S SUPPOSE THE ICE'S INITIAL TEMPERATURE IS -5°C. AS WE ADD HEAT, TEMPERATURE RISES TOWARD 0°C.

让我们假设冰的初始温度为-5°C。当我们加热时，温度向0°C升高。

AT THE MELTING POINT, TEMPERATURE STALLS AT 0°, EVEN THOUGH WE KEEP ADDING HEAT.

到达熔点时，尽管我们继续加热，但是温度仍停在0°C。

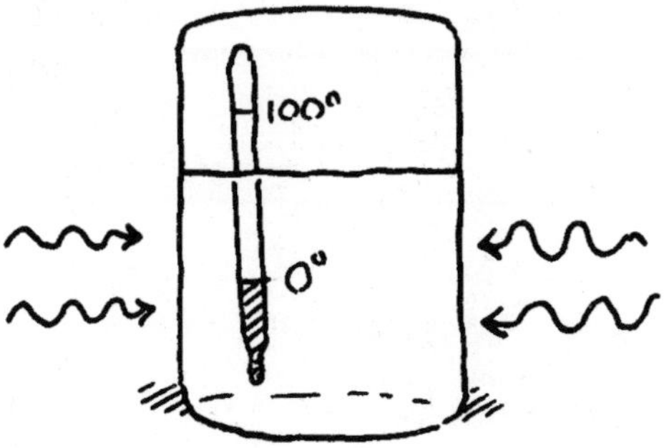

ALL THE ADDED HEAT GOES INTO BREAKING BONDS WITHIN THE ICE CRYSTAL.

所有增加的热量都用于破坏冰晶体内的键力。

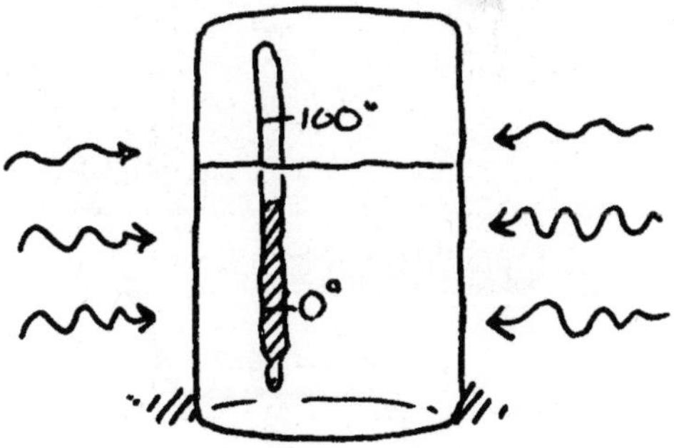

ONLY WHEN THE ICE IS FULLY MELTED DOES TEMPERATURE RISE AGAIN.

只有当冰块完全融化时，温度才会继续升高。

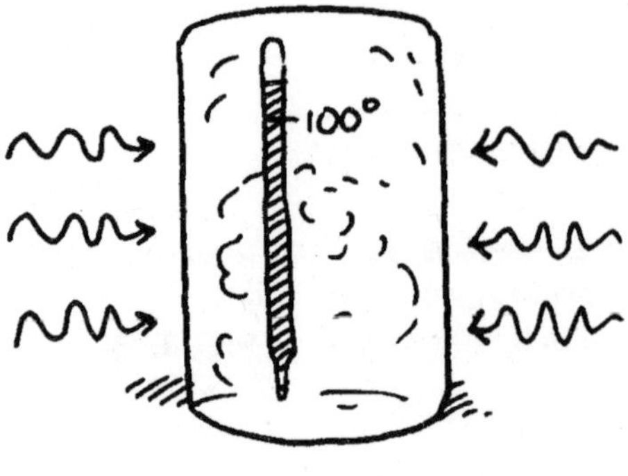

AT THE BOILING POINT, TEMPERATURE AGAIN STALLS, AS HEAT IS TAKEN UP BY PHASE CHANGE ALONE.

到达沸点时，温度再次停止升高，因为热量仅用于相变。

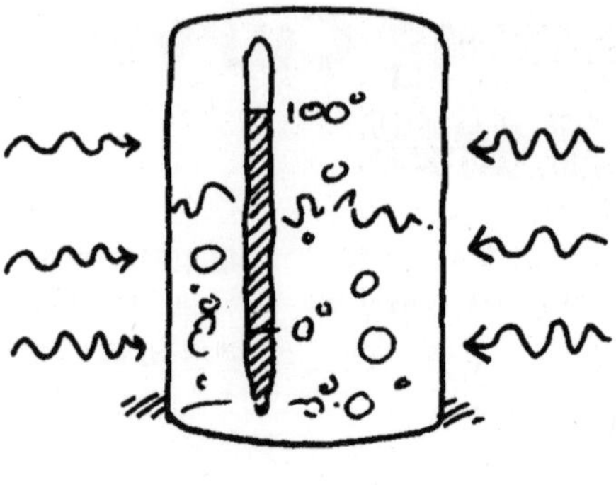

ONCE THE WATER IS FULLY VAPORIZED, THE STEAM'S TEMPERATURE RISES.

一旦水完全汽化，蒸气的温度就会上升。

THAT SIX-PANEL COMIC STRIP TRANSLATES INTO THIS **HEATING CURVE** THAT PLOTS TEMPERATURE AGAINST ADDED HEAT. T STOPS RISING DURING PHASE TRANSITIONS.

这6个有趣的转变过程可用这个**升温曲线**表示，曲线以温度和增加的热量为坐标来绘制。在相变过程中，即使继续加热，温度也不会上升。

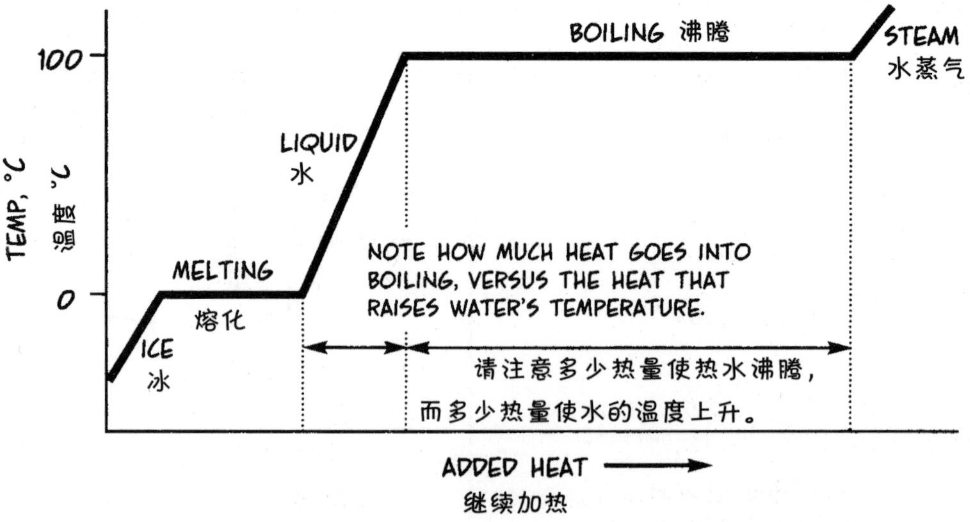

THE SPECIFIC HEAT OF WATER, RECALL, IS AROUND 4.18 J/g°C . SO TO RAISE THE TEMPERATURE OF ONE GRAM OF LIQUID WATER BY 100° REQUIRES AN ADDITION OF ABOUT

$$(4.18\ \text{J}/°\text{C})(100°\text{C})$$

$$= 418\ \text{Joules}$$

水的比热大约是4.18J/g°C，所以1克液体水提高100°C的温度需要约（4.18焦耳/°C）（100°C）=418焦耳。

BY CONTRAST, AT 100°C THE HEAT OF VAPORIZATION OF WATER IS ABOUT 41 **KILO**JOULES PER MOLE. SINCE A MOLE OF WATER WEIGHS 18 GRAMS, THIS IS

相比之下，100°C的水的汽化热约为41千焦耳/摩尔。由于1摩尔水分子重18克，因此：

$$\frac{41\ \text{kJ/mol}}{18\ \text{g/mol}} = 2.28\ \text{kJ/g}$$

$$= 2{,}280\ \text{J/g}$$

IN OTHER WORDS, IT TAKES ABOUT **FIVE TIMES AS MUCH HEAT** TO BOIL WATER COMPLETELY AWAY AS IT DOES TO HEAT IT ALL THE WAY FROM 0° TO 100°!!

换句话说，它需要约5**倍**烧开水的**热量**把水完全蒸发干！

IN THIS CHAPTER, WE REVIEWED THE THREE STATES OF MATTER, WHAT HOLDS THEM TOGETHER AND PULLS THEM APART. WE ALSO LEARNED THE GAS LAWS, WHICH EXPLAIN EVERYTHING FROM CALCULATING ATOMIC WEIGHTS TO RUNNING REFRIGERATORS.

在这一章中，我们回顾了物质的三种状态，以及什么使物质结合在一起或使它们分离。我们还了解到理想气体定律，解释了从如何计算原子量到冰箱如何运行的一切原理。

THERE EXISTS, BY THE WAY, A FOURTH STATE OF MATTER. AT VERY HIGH TEMPERATURE, ELECTRONS JUMP OFF THEIR NUCLEI; ALL BONDS BREAK; AND ALL SUBSTANCES TURN INTO A HOT PARTICLE SOUP CALLED **PLASMA.** LUCKILY, THIS IS NOT SOMETHING CHEMISTS HAVE TO THINK ABOUT VERY OFTEN…

顺便说一句，物质存在第四态。在非常高的温度下，电子脱离原子核，所有的键断裂，而所有的物质粒子成为“热汤”，称为等**离子体。**幸运的是，这不是化学家必须经常考虑的事情……

Chapter 7
Solutions

第七章
溶液

WE'VE JUST LOOKED AT STATES OF MATTER ONE AT A TIME... NOW LET'S COMBINE TWO OF THEM—OR RATHER, LET'S COMBINE SOMETHING, ANYTHING, WITH A LIQUID. FOR INSTANCE: ADD A PINCH OF TABLE SALT TO A FLASK OF WATER.

我们之前关注了物质状态的问题……现在让我们来同时考虑其中两个问题，让我们把一些东西甚至任何东西都加入液体中。例如，在一烧瓶水中添加少许食盐。

THE SALT, OF COURSE, COMPLETELY VANISHES.

盐当然完全消失。

THE SALT, AS WE SAY, **DISSOLVES** IN THE WATER.

正如我们所说，盐**溶解**在了水中。

WHEN A SUBSTANCE DISSOLVES IN A LIQUID, THE COMBINATION IS CALLED A **SOLUTION.** THE LIQUID IS THE **SOLVENT,** AND THE DISSOLVED MATERIAL IS THE **SOLUTE.***

当一种物质溶解在液体中，所得到的物质被称为**溶液**。该液体为**溶剂**，被溶解的物质是**溶质**.*

Solute + Solvent ⟶ Solution

溶质+溶剂⟶溶液

*ACTUALLY, A SOLUTION CAN BE SOLID OR GASEOUS TOO. ANY HOMOGENEOUS MIXTURE OF TWO OR MORE SUBSTANCES IS CONSIDERED A SOLUTION, WHATEVER ITS PHASE.

*其实，溶液也可以是固体或气体。任何两种或两种以上的物质均匀混合就可以是溶液，无所谓其状态。

A DISSOLVED SOLID FALLS APART INTO ITS INDIVIDUAL CONSTITUENT PARTICLES, EITHER IONS OR MOLECULES. GASES ALSO DISSOLVE MOLECULE BY MOLECULE. THIS EXPLAINS WHY SOLUTIONS ARE USUALLY TRANSPARENT.

固体溶解分散成独立而连续的粒子，可以是离子或分子。气体分子也可以被溶解。这可以解释为什么溶液通常是透明的。

FOR EXAMPLE, SODIUM CHLORIDE, NaCl, DISSOCIATES IN WATER INTO SINGLE Na^+ AND Cl^- IONS, WHICH BIND WITH THE WATER MOLECULES.

例如，氯化钠（NaCl）在水中解离成单独的Na^+和Cl^-离子，并与水分子结合。

SUGAR—SUCROSE, $C_{12}H_{22}O_{11}$—BREAKS INTO WHOLE MOLECULES. (WATER MOLECULES LIKE ITS OH GROUPS.)

糖-蔗糖，$C_{12}H_{22}O_{11}$，分解成无数个分子（水分子喜欢糖分子中的OH基团）。

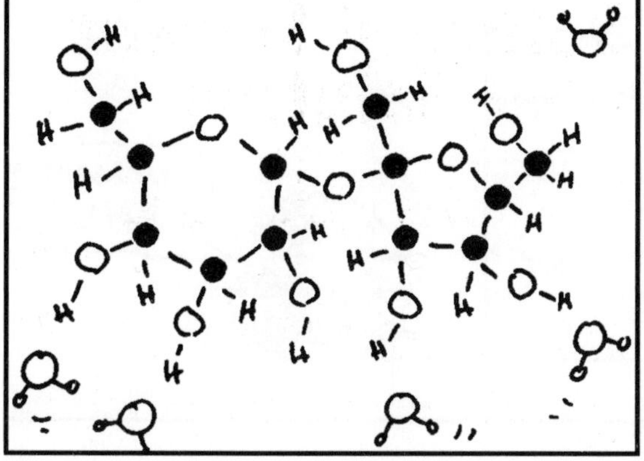

VINEGAR, A SOLUTION OF **ACETIC ACID,** CH_3CO_2H, CONTAINS HYDROGEN IONS, H^+, ACETATE IONS, $CH_3CO_2^-$, AND MUCH CH_3CO_2H STILL IN COMBINATION.

醋是**醋酸**的溶液，CH_3CO_2H包含氢离子H^+，醋酸根离子（$CH_3CO_2^-$）和许多以分子形式存在的CH_3CO_2H。

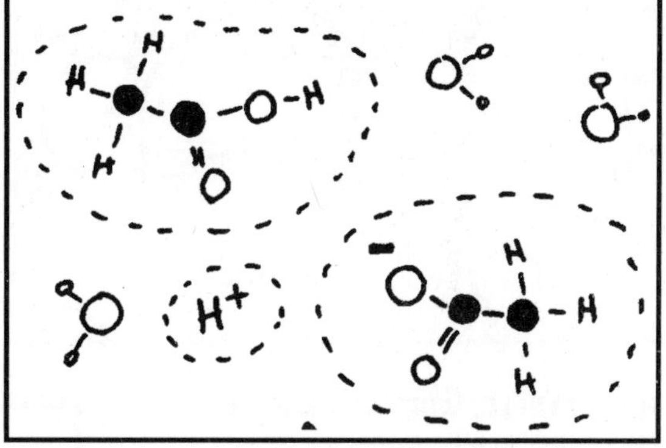

LET'S LOOK MORE CLOSELY AT THE DISSOLVING PROCESS. IMAGINE A CHUNK OF MATERIAL IMMERSED IN LIQUID. IN ORDER TO DISSOLVE, SOME OF ITS PARTICLES MUST BREAK THE BONDS THAT HOLD THEM TOGETHER AND FORM NEW BONDS WITH MOLECULES OF LIQUID. SIMILARLY, IMFs WITHIN THE LIQUID MUST ALSO BE OVERCOME.

让我们更深入地了解溶解过程。将大量的物质浸入液体之中。为了溶解，其中的一些粒子必须打破原有的化学键，并与液体分子一起产生新的化学键。同样，也必须克服分子间作用力。

EACH FREE SOLUTE PARTICLE ATTRACTS ONE OR MORE MOLECULES OF SOLVENT, WHICH CLUSTER AROUND IT IN A SOLVENT "CAGE." THIS PROCESS OF BREAKING AND FORMING BONDS IS CALLED **SOLVATION.**

每一个自由溶质粒子吸引一个或多个溶剂分子，围绕在它的周围形成一个溶剂"笼"。这种键的断裂和形成的过程称为**溶剂化**。

ALL THIS BOND REARRANGING MEANS THAT **DISSOLVING IS A CHEMICAL REACTION.** AMONG OTHER THINGS, THEN, IT HAS AN ASSOCIATED ENTHALPY CHANGE, WHICH MAY BE POSITIVE OR NEGATIVE.

键的重新组合意味着**溶解是一种化学反应**。除此之外，它还对应一个焓变化，值可能是正也可能是负。

FOR EXAMPLE, WHEN MAGNESIUM CHLORIDE, $MgCl_2$, DISSOLVES IN WATER, IT HAS AN ENTHALPY OF SOLVATION

ΔH = 119 kJ/mol

HIGHLY ENDOTHERMIC! A MERE 4g OF $MgCl_2$ (= .042mol) IN 50mL (= 50g) OF WATER DROPS THE WATER'S TEMPERATURE BY 23.9°C (BY THE BASIC CALORIMETRY EQUATION).

例如，当氯化镁（$MgCl_2$）溶解在水中时，它就有一个溶剂化焓

ΔH = 119 千焦耳/摩尔

这是个高度吸热的过程！仅仅4克（=0.042摩尔）氯化镁溶于50毫升（=50克）水，水的温度就降低了23.9°C（基本热力学方程）。

CHEMICAL COLD PACKS ARE IN FACT MADE FROM $MgCl_2$ AND OTHER SALTS THAT ABSORB HEAT WHEN DISSOLVED IN WATER.

化学冰袋就是利用氯化镁等盐类溶解在水中吸收热量的性质制作的。

SOME LIQUID MIXTURES ARE NOT SOLUTIONS:

一些液体是混合物而不是溶液：

WHEN I STIR POWDERED MILK INTO WATER, THE SOLID PARTICLES REMAIN IN VERY LARGE CLUMPS OF MOLECULES. A MIXTURE LIKE MILK IS CALLED A **SUSPENSION,** AND SUSPENSIONS ARE OPAQUE.

当我把水和奶粉一起搅拌时，固体粒子仍然保持非常大的分子团。像牛奶的混合物被称为**悬浊液**，悬浊液是不透明的。

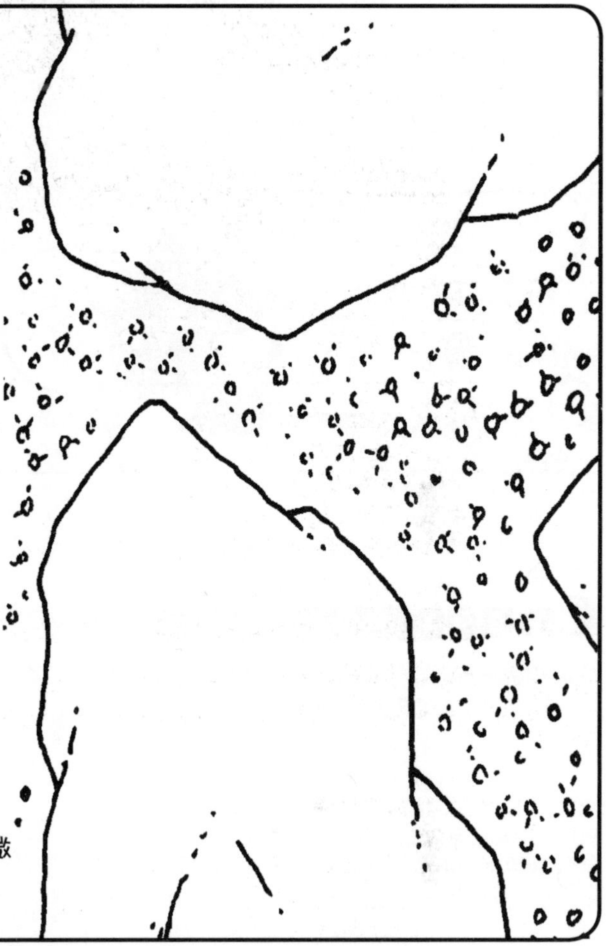

ANOTHER EXAMPLE WOULD BE PAINT, IN WHICH FLECKS OF PIGMENT ARE SUSPENDED IN OIL OR SOME GEL-LIKE MEDIUM.

另一个常见例子是油漆，这些颜料微粒是以油状或凝胶状分散在溶液中的。

AN **EMULSION** IS A SUSPENSION OF ONE LIQUID IN ANOTHER. MAYONNAISE, FOR EXAMPLE, MAINLY CONSISTS OF TINY DROPLETS OF OIL SUSPENDED IN VINEGAR. ORDINARILY, OIL AND VINEGAR WOULD SEPARATE, BUT THE ADDITION OF A SMALL AMOUNT OF MUSTARD AND EGG YOLK STABILIZES THE EMULSION.

乳化指一种液体悬浮在另一种液体中。例如蛋黄酱，主要是由悬浮在醋中的微小液滴所组成。在通常情况下，油和醋会分层，但加入少量的蛋黄和芥末就会乳化。

LONG MOLECULES FROM THE YOLK BURROW INTO OIL DROPLETS. A POLAR "TAIL" STICKS OUT AND ATTRACTS THE POLAR WATER MOLECULES IN VINEGAR, WHICH BLOCK THE DROPLETS FROM MERGING.

蛋黄中的长分子渗入油滴中。一端的"尾巴"伸出来吸引醋中的极性分子，阻止液滴合并。

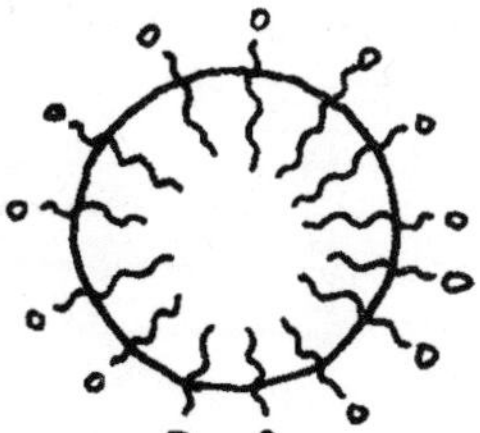

Concentration

IS A MEASURE OF HOW MUCH SOLUTE IS PRESENT IN A SOLUTION RELATIVE TO THE WHOLE.

浓度

是溶质在溶液中的相对含量。

FOR EXAMPLE, WEIGH OUT 35g OF NaCl. PUT IT IN A GRADUATED CONTAINER AND ADD WATER UNTIL THERE IS ONE LITER OF SOLUTION.

THE CONCENTRATION OF THIS SOLUTION IS 35 g/L AND MEASURES **MASS OF SOLUTE** PER **VOLUME OF SOLUTION.**

例如，称出35克的氯化钠放入一个容器内，并不断地加水，直到溶液达1升。

此溶液的浓度为35克/升，是通过测量**单位体积溶液中的溶质质量**得到的。

OTHER POSSIBLE MEASURES (ALL USED!):

MASS OF SOLUTE PER MASS OF SOLUTION

VOLUME OF SOLUTE PER VOLUME OF SOLUTION

MASS OF SOLUTE PER VOLUME OF SOLVENT
(NOT THE SAME THING AS VOLUME OF SOLUTION!)

MASS OF SOLUTE PER MASS OF SOLVENT

PARTS PER MILLION (PPM)
(A MASS-PER-MASS RATIO OF VERY DILUTE SOLUTIONS)

PARTS PER BILLION (PPB, EVEN MORE DILUTE)

其他可能的计量方式（所有可能使用到的）：

单位质量溶液中的溶质质量

单位体积溶液中的溶质体积

单位体积溶剂中的溶质质量（与溶液体积不同！）

单位质量溶剂中的溶质质量

百万分之一（ppm）（浓度非常稀的溶液的一种质量比）

十亿分之一（ppb，更稀的溶液）

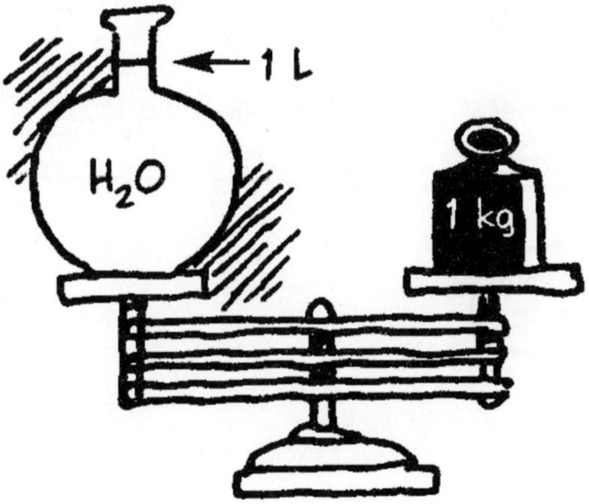

WHEN THE SOLVENT IS WATER, WE CAN EASILY CONVERT FROM A MASS-VOLUME RATIO TO A MASS-MASS RATIO, BECAUSE **ONE LITER OF WATER WEIGHS ONE KILOGRAM.** A LITER OF VERY DILUTE AQUEOUS SOLUTION, OF COURSE, WEIGHS THE SAME.

当溶剂是水时，我们可以很容易地把一个质量体积比转换成一个质量-质量比，**因为1升的水重1千克**，1升极稀的水溶液当然也是1千克。

OUR FAVORITE MEASURE OF CONCENTRATION ACTUALLY TELLS YOU HOW MANY MOLECULES ARE DISSOLVED RELATIVE TO VOLUME. **MOLARITY,** OR MOLAR CONCENTRATION, IS THE NUMBER OF MOLES OF SOLUTE PER LITER OF SOLUTION. WE WRITE

M = MOLES/LITER.

我们最喜欢的浓度单位可以告诉你有多少分子溶解于相对体积的溶液。**摩尔浓度**是每升溶液含有溶质的摩尔数，表示为：

M = 摩尔/升

WHAT'S THE MOLARITY OF OUR 35 g/L SALT SOLUTION? ONE MOLE OF NaCl WEIGHS 58.4 g, SO WE HAVE

$$\frac{35\ \text{g}}{58.4\ \text{g/mol}} = 0.6\ \text{mol NaCl}$$

IN A LITER OF SOLUTION. THE MOLARITY IS 0.6 M.

35克/升盐溶液的摩尔浓度是多少？1摩尔的氯化钠重58.4克，所以我们有。

$$\frac{35\ 克}{58.4克/摩尔} = 0.6\ \textbf{摩尔氯化钠}$$

在1升溶液中，该摩尔浓度是0.6 M。

WE USE SQUARE BRACKETS, [ ], TO DENOTE MOLAR CONCENTRATION OF ANY "SPECIES" (I.E., ANY PARTICULAR MOLECULE OR ION) IN SOLUTION. HERE, SINCE NaCl DISSOCIATES COMPLETELY IN SOLUTION,

$$[Na^+] = 0.6M$$
$$[Cl^-] = 0.6M$$

IN A 1 M SOLUTION OF Na_2SO_4, WHICH ALSO FULLY DISSOCIATES,

$$[Na^+] = 2\ M$$
$$[SO_4^{2-}] = 1\ M$$

THERE ARE TWO MOLES OF Na^+ FOR EACH MOLE OF Na_2SO_4.

* 英文中“摩尔”（mole）和“鼹鼠”词形相同。

我们使用方括号[]来表示任何“种类”溶液的摩尔浓度（即任何特定的分子或离子）。在这里，因为氯化钠在溶液中完全解离，

$$[Na^+] = 0.6\ M \qquad [Cl^-] = 0.6\ M$$

1M硫酸钠在溶液中也是完全解离，

$$[Na^+] = 2\ M \qquad [SO_4^{2-}] = 1\ M$$

每摩尔硫酸钠有两摩尔钠离子。

Solubility 溶解度

ANY SUBSTANCE WILL DISSOLVE IN ANY LIQUID—TO SOME DEGREE, THOUGH IT MAY BE VERY SMALL INDEED. FOR INSTANCE, NO MORE THAN .000006 g OF ELEMENTAL MERCURY (Hg) WILL DISSOLVE IN A LITER OF WATER AT ROOM TEMPERATURE. A MOLE OF Hg WEIGHS 200.6 g...

任何物质在任何液体中都会有一定程度的溶解，尽管有时它可能会非常小。例如，室温下低于0.000006克的单质汞（Hg）会溶解在1升水中。1摩尔的汞重200.6克……

BUT EVEN WHEN A SUBSTANCE IS HIGHLY SOLUBLE, THERE IS ALWAYS A LIMIT! YOU CAN THROW ONLY SO MUCH SALT INTO WATER BEFORE IT STARTS PILING UP ON THE BOTTOM, UNDISSOLVED.

但即使是有高度溶解性的物质，还是有一个限制！你抛进水里的盐，只能溶解这么多，然后剩下的开始沉到底部并不会溶解。

THIS LIMIT, A SUBSTANCE'S MAXIMUM POSSIBLE CONCENTRATION, IS CALLED ITS **SOLUBILITY.** A MAXIMALLY CONCENTRATED SOLUTION IS CALLED **SATURATED.**

这个限制，即一种物质可能浓度的最大值，叫作它的**溶解度**。最大浓度溶液叫做**饱和溶液**。

WE SAY A MATERIAL IS SOLUBLE IF IT DISSOLVES TO AN "APPRECIABLE" DEGREE, AND INSOLUBLE IF NOT—A FUZZY CONCEPT, CLEARLY.

如果物质溶解到“可观”的程度，我们就说它是可溶性物质，如果不溶解就是难溶物质，这是一个模糊的概念。

THE EQUIVALENT WORD FOR LIQUID-LIQUID INTERACTION IS **MISCIBILITY:** TWO LIQUIDS ARE MISCIBLE IF THEY DISSOLVE IN ONE ANOTHER AND IMMISCIBLE IF, LIKE OIL AND WATER, THEY SEPARATE.

液液相溶有一个同义词，即**互溶**：如果两种液体彼此溶解，它们就是互溶的，否则就是不互溶的，就像油和水会分层。

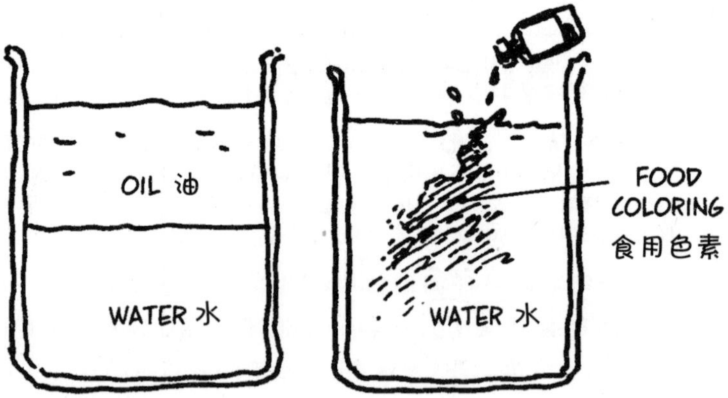

IMMISCIBLE 不互溶　　MISCIBLE 互溶

LIKE TENDS TO DISSOLVE LIKE. A POLAR SOLVENT (SUCH AS WATER) TENDS TO DISSOLVE (OR MIX WITH) OTHER POLAR COMPOUNDS. HERE DIPOLE-DIPOLE OR DIPOLE-ION ATTRACTIONS DRIVE SOLVATION. FOR INSTANCE:

相似相溶。极性溶剂（如水）往往易于溶解（或与之混合）其他极性化合物。这里偶极-偶极或偶极-离子作用会引起溶解。例如：

METHANOL, CH_3OH, IS POLAR AND FORMS A HYDROGEN BOND WITH WATER, WITH WHICH IT WILL MIX IN ANY AMOUNT.

甲醇：（CH_3OH）具有极性，并与水形成了氢键，它可以以任意比例与水互溶。

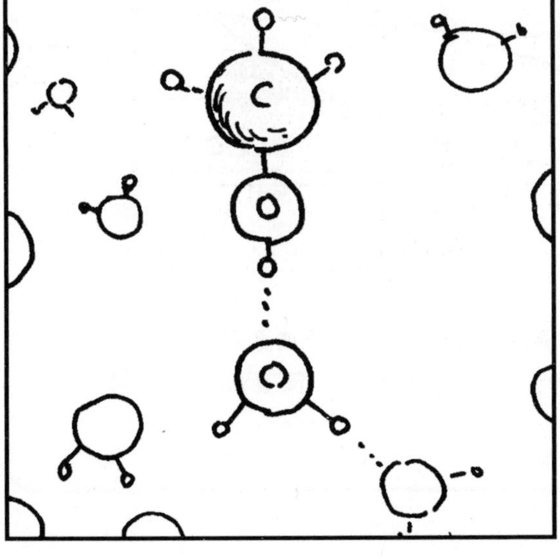

ITS COUSIN **METHANE,** CH_4, IS UTTERLY SYMMETRICAL AND NONPOLAR. WATER SHUNS IT, AND ITS SOLUBILITY IS VERY LOW (0.024 g/L OR 0.0015 M).

而它的堂兄**甲烷**（CH_4），是完全对称的并且没有极性。水分子避开它，它的溶解度很低（0.024克/升或0.0015M）。

MOLECULAR SIZE: BIG, HEAVY MOLECULES TEND TO BE LESS SOLUBLE THAN SMALL, LIGHT ONES. SOLVENT MOLECULES FIND IT HARD TO "CAGE" BIG PARTICLES.

分子大小：大而重的分子往往比小而轻的分子更难溶解，溶剂分子难以"笼罩"大个粒子。

TEMPERATURE ALSO AFFECTS SOLUBILITY. AS TEMPERATURE RISES, AGITATED MOLECULES OR IONS BREAK THEIR BONDS MORE EASILY, SO SOLUBILITY USUALLY GOES UP. EXCEPTIONS EXIST, HOWEVER, AND THE EFFECT IS SOMETIMES SLIGHT.

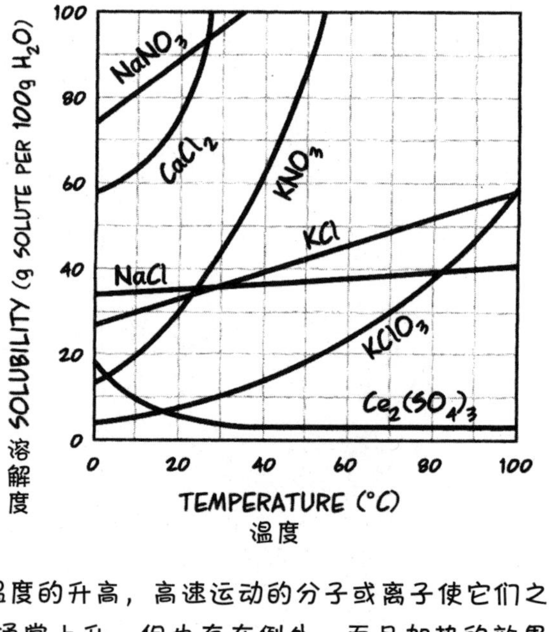

温度同样影响溶解度。随着温度的升高，高速运动的分子或离子使它们之间的键更容易被破坏，因此溶解度通常上升。但也存在例外，而且加热的效果有时并不显著。

FOR DISSOLVED GASES, PRESSURE AFFECTS SOLUBILITY. TO BE PRECISE, THE **PARTIAL PRESSURE** OF A GAS ABOVE THE SOLUTION AFFECTS THE AMOUNT OF GAS THAT WILL DISSOLVE. THE HIGHER THE PARTIAL PRESSURE, THE GREATER THE GAS'S SOLUBILITY.

对于气体溶解，压力则会影响溶解度。准确地说，溶液液面上气体的**蒸气分压**会影响气体溶解量。分压越高，气体的溶解度越大。

LOWER PRESSURE
LOWER CONCENTRATION
较低的压力
较低的浓度

HIGHER PRESSURE
HIGHER CONCENTRATION
较高的压力
较高的浓度

SOFT DRINKS, WHICH CONTAIN DISSOLVED CO_2, ARE BOTTLED AT HIGH PRESSURE TO INCREASE THE AMOUNT OF DISSOLVED GAS. WHEN THE CAP IS REMOVED, PRESSURE EASES, AND CO_2 FIZZES OUT OF SOLUTION.

软性饮料含有溶解的二氧化碳，在装瓶时用高压力使溶解气体的数量增加。当瓶盖被打开时，压力降低，二氧化碳从溶液中冒出。

Freezing 凝固

GENERALLY SPEAKING, DISSOLVED MATERIAL LOWERS THE FREEZING POINT. SOLUTE PARTICLES DISRUPT THE NORMAL COHESIVE FORCES WITHIN THE SOLVENT, MAKING IT HARDER FOR THE SOLUTION TO SOLIDIFY. THE HIGHER THE CONCENTRATION, THE LOWER THE FREEZING POINT.

一般来说，物质溶解降低了溶液的凝固点。溶质粒子在溶剂中扰乱正常的凝聚力，使其更难凝固。浓度越高，凝固点越低。

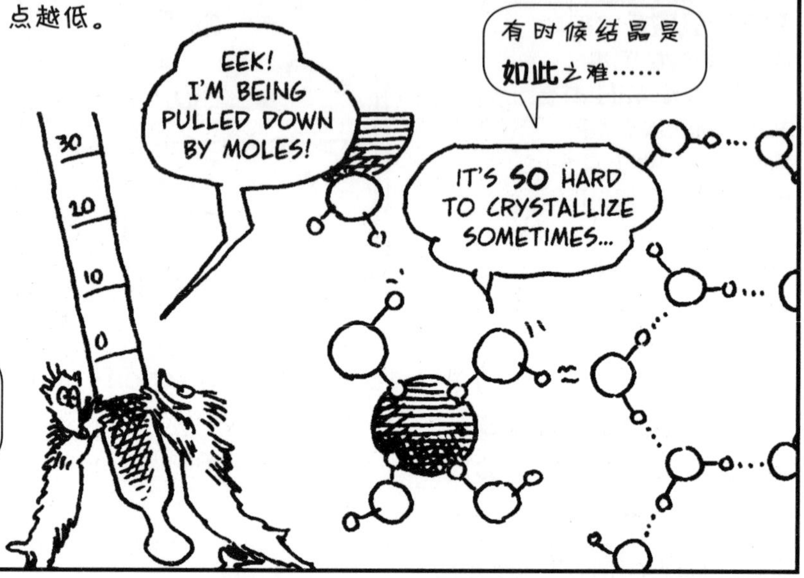

FOR EXAMPLE, IN AN ICE CREAM MAKER, A BUCKET OF CREAM, DISSOLVED SUGAR, AND FLAVOR IS SURROUNDED BY ICE, WHICH MAY BE AT -3° TO -5°C.

例如，在一个冰淇淋机里，一桶溶解了糖和色素的奶油，被-3°C到-5°C的冰所包围。

WHEN SALT IS ADDED, THE ICE MELTS. THE BELOW-ZERO SALT WATER NOW MAKES CONTACT WITH THE FULL SURFACE OF THE BUCKET.

当加入盐时，冰融化。在0°C下，盐水充分地接触到桶表面。

NOW THE CREAM CAN BE RAPIDLY COOLED BELOW 0°C. LIQUID WATER ALSO HAS A HIGHER HEAT CAPACITY THAN ICE, AND SO COOLS MORE EFFICIENTLY.

现在奶油也可以迅速冷却至0°C以下。液态水的热容量比冰高，所以冷却效率高。

ICE TOUCHES THE CREAM CONTAINER IN ONLY A FEW PLACES.
冰只接触奶油容器的一小部分。

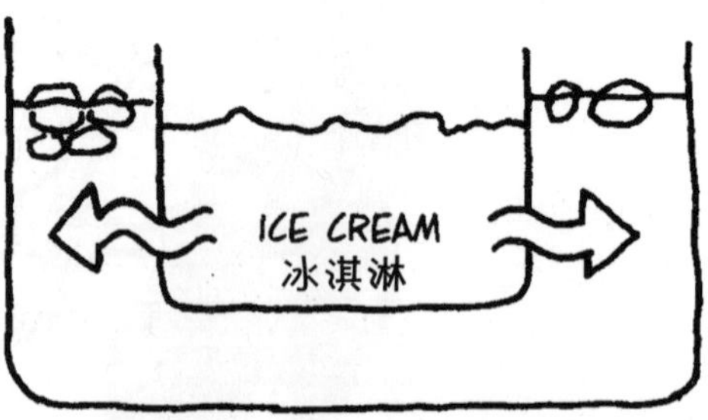

EFFICIENT HEAT TRANSFER
高效传热

ICE CREAM RARELY FREEZES TOTALLY. AS THE LIQUID FREEZES, SUGAR BECOMES MORE CONCENTRATED IN THE REMAINING SYRUP, SO ITS FREEZING POINT DROPS EVEN LOWER, AND SOME OF IT REMAINS UNFROZEN. THAT'S WHY ICE CREAM IS USUALLY SOFT.

冰淇淋很少完全凝固。由于液体凝固，糖在剩下的糖浆里浓度变得更高，所以它的凝固点下降得更低，另一部分仍然没有凝固。这就是为什么冰淇淋通常是软的。

Boiling 沸腾

DISSOLVED MATERIAL UPS THE BOILING POINT, AND THUS EXTENDS THE RANGE OF THE LIQUID STATE IN BOTH DIRECTIONS.

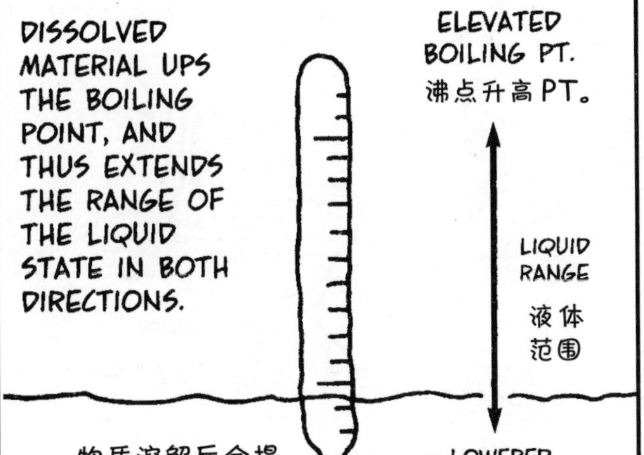

物质溶解后会提高溶液沸点，从而向两个方向扩展液相状态的温度范围。

THIS IS AGAIN A RESULT OF SOLUTE-SOLVENT INTERACTIONS. SOLVENT MOLECULES THAT ARE ATTACHED TO SOLUTE PARTICLES FIND IT HARDER TO ESCAPE INTO THE GAS PHASE.

这又是一个溶质与溶剂相互作用的结果。与溶质粒子相互作用的溶剂分子更难逃入气相。

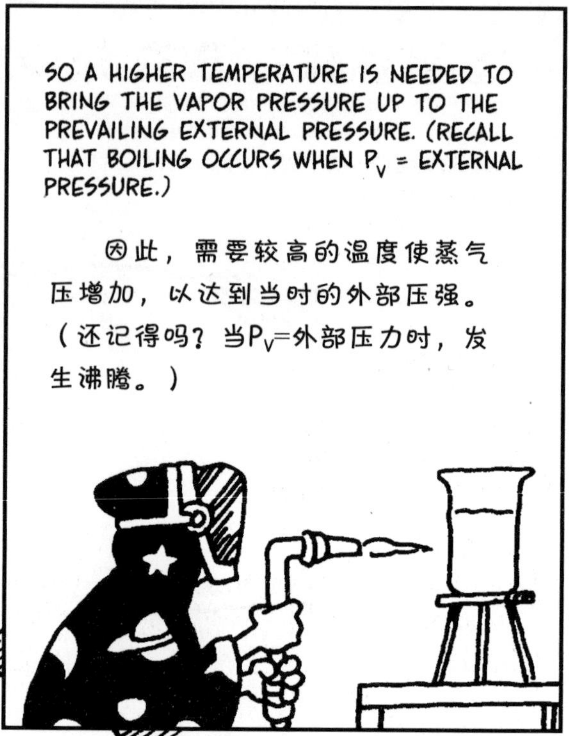

MAYBE THIS IS WHY CHEFS ADD SALT TO WATER FOR COOKING SPAGHETTI. THE SALT SOLUTION BOILS AT A TEMPERATURE ABOVE 100°C (AT ONE ATM, ANYWAY), AND THE SPAGHETTI IS DONE SOONER. ALSO, IT TASTES BETTER...

也许这就是为什么厨师在煮意大利面时会向水中加盐。在100°C以上的温度（一个大气压）时，盐水溶液沸腾，并更快地煮熟意大利面，同时，它的味道也更好……

*SEE CHAPTER 6, PAGE 156.
*见第6章，第156页。

So What? 那又如何？

AN ENORMOUS AMOUNT OF FAMILIAR AND IMPORTANT CHEMISTRY HAPPENS IN SOLUTION: COOKING, BREWING, FERMENTATION, DIGESTION, ELECTRIC BATTERY POWER, MEDICINE, ETCHING OF METAL AND GLASS, LAUNDRY AND OTHER WASHING, BLOOD CHEMISTRY, TOOTH DECAY, CALCIFICATION OF PIPES, ACID RAIN, OIL REFINING, WATER PURIFICATION, CELLULAR METABOLISM—JUST TO NAME A FEW!

溶液中发生了大量我们熟悉且很重要的化学现象：烹饪、酿酒、发酵、消化、电池发电、医药、电镀、玻璃蚀刻、洗涤、血液生化、蛀牙、管道钙化、酸雨腐蚀、炼油、水处理、细胞的新陈代谢，这仅仅是几个例子而已！

MUCH OF THE REST OF THIS BOOK WILL BE AN ATTEMPT TO UNDERSTAND SUCH PROCESSES IN MORE DETAIL. WE BEGIN BY LOOKING AT WHY SOME REACTIONS GO FAST, WHILE OTHERS GO SLOW...

这本书的其余部分将更详细地阐明这些过程。我们先看看为什么有些反应十分迅速，而其他反应却十分缓慢……

Chapter 8

Reaction Rate and Equilibrium

第八章

反应速率和平衡

IN CHEMISTRY, WE CARE ABOUT NOT ONLY **WHAT** REACTS, BUT ALSO **HOW FAST.** BLACK POWDER EXPLODES IN A FLASH, WHILE THE SUGAR IN YOUR COFFEE NEVER SEEMS TO DISSOLVE FAST ENOUGH. WE TRY TO SPEED UP ENVIRONMENTAL CLEANUP AND RETARD RUST AND AGING. IN OTHER WORDS, **RATES MATTER!**

在化学上，我们不仅关心是**何种**反应，而且关心反应的**快慢**。火药能在一瞬间爆炸，而在你咖啡里的糖的溶解速度似乎永远不够快。我们试图加快环境清理速度，并防止生锈和老化。换言之，这都是**速率的问题**！

"AT FIRST SIGHT, NOTHING SEEMS MORE OBVIOUS THAN THAT EVERYTHING HAS A BEGINNING AND AN END."

—SVANTE ARRHENIUS, 1903 NOBEL PRIZE WINNER IN CHEMISTRY

"乍一看，万事都有开头和结尾是再明显不过的了。"

——斯万特·阿伦尼乌斯，1903年诺贝尔化学奖获得者。

WHAT'S THE RATE OF A CHEMICAL REACTION? WE BEGIN WITH THE ULTRA-SIMPLE CASE OF ONLY ONE REACTANT:

A ⟶ PRODUCTS

HERE THE **REACTION RATE** r_A IS THE RATE AT WHICH REACTANT **A** IS USED UP OVER TIME. IT MAY BE EXPRESSED IN MOLES PER SECOND.

IF **A** IS IN SOLUTION, r_A USUALLY REFERS TO THE RATE AT WHICH CONCENTRATION [A] CHANGES, IN MOLES PER LITER PER SECOND, AND IF **A** IS A GAS, r_A MAY REFER EITHER TO CONCENTRATION OR PARTIAL PRESSURE P_A, WHICH AMOUNT TO THE SAME THING.

什么是化学反应速率？我们从只有一种反应物的最简单例子开始：

A → 产物

在这里，反应速率r_A是反应物A在单位时间消耗的速率，记为摩尔/秒。

如果A位于溶液中，r_A通常指浓度[A]的变化速率，记为摩尔/升/秒；如果A是一种气体，r_A则是浓度或分压P_A，它们的数值是一样的。

FOR EXAMPLE, IN THE LOWER ATMOSPHERE, SUNLIGHT FALLING ON NITROGEN DIOXIDE, NO_2, CAUSES IT TO BREAK INTO NITRIC OXIDE, NO, AND A LOOSE OXYGEN ATOM (CALLED A FREE RADICAL):

例如，在低层大气中，阳光照射二氧化氮（NO_2），使其分解成一氧化氮（NO）和一个未束缚的氧原子（称为自由基）：

$$NO_2 \xrightarrow[\text{日光照射}]{\text{LIGHT}} NO + O$$

(THE FREE OXYGEN GOES ON TO BIND WITH O_2 TO FORM OZONE, O_3. OZONE AND THE NITROGEN OXIDES ARE AMONG OUR NASTIER AIR POLLUTANTS.)

（氧自由基与氧气结合，变成臭氧（O_3）。臭氧和氮氧化物都是主要的大气污染物。）

AT MIDDAY, NO_2 MAKES UP ABOUT 20 PARTS PER BILLION OF THE AIR—20 MOL OF NO_2 PER BILLION MOL OF AIR—OR 20 MOL OF NO_2 IN 24.4×10^9 L OF AIR (AT 25°C). SO MOLAR CONCENTRATION IS $[NO_2] = 20/(24.4 \times 10^9) = 8.2 \times 10^{-10}$ MOL/L. LET'S TAKE AN AIR SAMPLE, AND MEASURE $[NO_2]$ EVERY 40 SECONDS AS IT DECOMPOSES. WE WRITE $[A]_t$ FOR THE CONCENTRATION OF NO_2 AT TIME t.

中午，二氧化氮在空气中大约占十亿分之二十——意思是每十亿摩尔空气中有20摩尔二氧化氮——或者说在25°C时24.4×10^9 升的空气中有20摩尔二氧化氮。因此，摩尔浓度$[NO_2] = 20/(24.4\times10^9) = 8.2 \times 10^{-10}$摩尔/升，让我们拿一些空气样品，测量$[NO_2]$每40秒分解多少。我们用$[A]_t$表示在某一时间$t$的二氧化氮浓度。

t (SEC.)	$[A]_t$ ($\times 10^{-10}$ MOL/L)	
0	8.20	$[A]_0$
40	5.80	
80	4.10	$([A]_0)/2$
120	2.90	
160	2.05	$([A]_0)/4$
200	1.45	
240	1.02	$([A]_0)/8$
280	.72	
320	.51	$([A]_0)/16$
360	.36	

THE REACTION CERTAINLY SLOWS OVER TIME. IN 10^{10} LITERS OF AIR, **2.4** MOL ($[A]_0 - [A]_{40}$) WERE USED UP IN THE FIRST **40** SEC., BUT ONLY **0.21** MOL IN THE 40 SECONDS BETWEEN t = 280 AND t = 320 ($[A]_{280} - [A]_{320}$).

反应速度随着时间的推移肯定会降低。在10^{10}升空气中，前40秒消耗了**2.4摩尔**（$[A]_0-[A]_{40}$），但在t = 280到t = 320之间的40秒内只消耗了**0.21摩尔**（$[A]_{280}-[A]_{320}$）。

THE DECLINE HAS A PATTERN: **HALF THE REMAINING REACTANT IS CONSUMED EVERY 80 SECONDS.** AT t = 80 SEC., HALF THE NO_2 IS LEFT... AT 160 SEC., A FOURTH REMAINS... AT 240, AN EIGHTH, ETC. WE SAY THE REACTION HAS A **HALF-LIFE,** h, OF 80 SECONDS. DURING **ANY** INTERVAL OF LENGTH h, HALF THE REACTANT IS CONSUMED. IN n HALF LIVES, THEN:

这一下降过程有一个模式：**每80秒会消耗剩下反应物的一半**。在t = 80秒时，只剩一半的二氧化氮了……在160秒时，剩下四分之一……在240秒时，剩下八分之一……我们认为这个反应有一个**半衰期**，*n*=80秒。在**任意**时间间隔*h*内，反应物消耗一半。在一个半衰期*h*内，则：

$$[A]_{nh} = (1/2)^n [A]_0$$

n HALF LIVES　*n*个半衰期

A SIMPLE MODEL ACCOUNTS FOR THIS BEHAVIOR. START WITH A BIG BUNCH OF MOLECULES OF REACTANT **A**, AND IMAGINE THAT EVERY MOLECULE HAS THE SAME PROBABILITY OF DECOMPOSING. THEN A FIXED FRACTION OF THE WHOLE WILL REACT IN EACH UNIT OF TIME.

一个简单模型可以解释这个现象。由一个含有大量分子的反应物**A**开始，假使每个分子具有相同的分解概率。在单位时间内反应物的消耗量的比例是固定的。

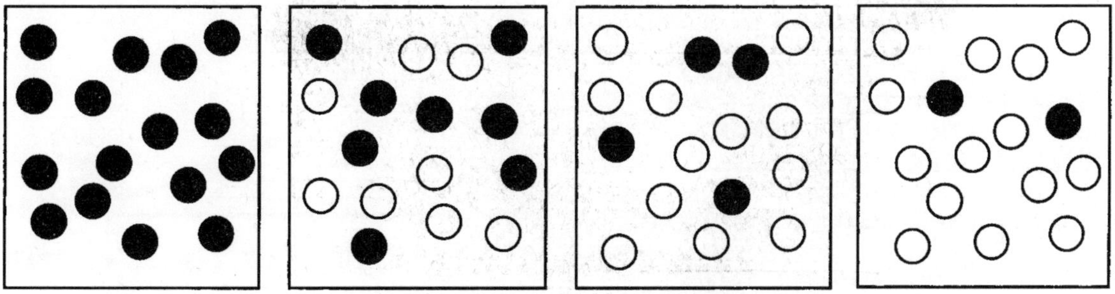

t=1　　t=2　　t=3　　t=4

IN OTHER WORDS, THE REACTION **RATE** (NUMBER OF MOLES OR MOL/L DECOMPOSING PER UNIT TIME) IS PROPORTIONAL TO THE **QUANTITY OF REACTANT** PRESENT (NUMBER OF MOLES OR MOL/L). SO WE CAN WRITE A SECOND FORMULA FOR THE REACTION RATE: AT ANY GIVEN TIME,

换句话说，反应速率（每单位时间内分解的摩尔数或摩尔/升）与**反应物的量**（摩尔数或摩尔/升）成正比。我们因此可以写出反应速率的第二个公式：即在任何特定时间，

$$r_A = -k[A]$$

k IS A CONSTANT CALLED THE **RATE CONSTANT.** BY CONVENTION, k IS ALWAYS A POSITIVE NUMBER, SO THE MINUS SIGN IS NECESSARY TO MAKE r NEGATIVE, MEANING [A] IS DECREASING.

k是一个常数，称为**速率常数**。按照惯例，k总是正数，所以负号是必要的，说明[A]在减少。

NOTE: MATH-AVERSE READERS MAY SKIP THIS PAGE. OTHERWISE, KEEP READING.

注：厌恶数学的读者可以跳过这几页。爱好者请继续阅读。

WE CAN EVALUATE k FROM THE DATA. START WITH THE FIRST EQUATION

我们可以用数据计算k。从第一个方程开始

$$[A]_{nh} = 2^{-n}[A]_0$$

[A] DECREASES EXPONENTIALLY (AS THE EXPONENT OF 2 IN THIS EQUATION). IN PARTICULAR, [A] **NEVER REACHES ZERO.** THEORETICALLY, THE REACTION NEVER ENDS!

[A]以**指数方式降低**（在这个等式中指数为2）。特别指出，[A]**永远不会为零**。从理论上讲，反应永远不会结束！

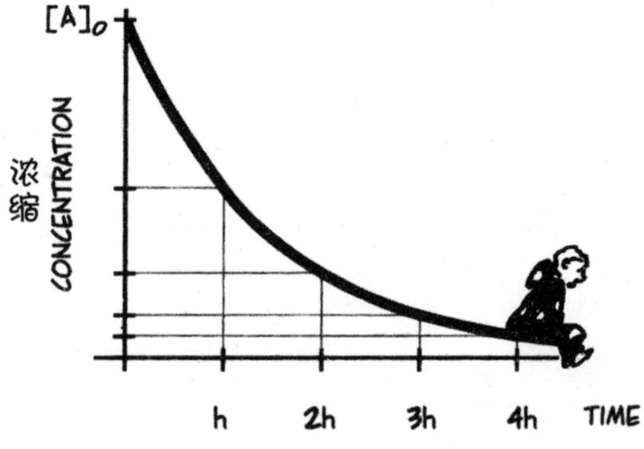

h IS AN AWKWARD TIME UNIT—IT VARIES FROM ONE REACTION TO ANOTHER. WE WANT A FIXED UNIT OF TIME, t (DAYS, SECONDS, WHATEVER'S APPROPRIATE). THEN

*h*是一个尴尬的时间单位，它在每一个反应中都不相同，我们希望有一个固定的时间单位t（天，秒，只要是适当的）。接着

$t = nh$, or $n = t/h$

$t = nh$，或$n = t/h$

AND WE CAN WRITE

我们可以得到

$$[A]_t = 2^{-t/h}[A]_0$$

TAKING THE NATURAL LOG OF BOTH SIDES,

取等式两边的自然对数，

$$\ln [A]_t = \frac{-1}{h}(\ln 2)t + \ln[A]_0$$

SETTING $k = (1/h) \ln 2$, WE FIND:

设定$k = (1/h) \ln 2$，我们发现：

$$\ln [A]_t = -kt + \ln[A]_0$$

THAT IS, THE PLOT OF $\ln[A]_t$ AGAINST t IS A **STRAIGHT LINE** WITH SLOPE $-k$. ONE CAN SHOW (USING CALCULUS) THAT THIS IS THE SAME k AS IN $r_A = -k[A]$. IN OUR NO_2 EXAMPLE, THEN,

也就是说，$\ln[A]_t$对t的曲线是一个斜率为k的**直线**。我们可以用计算器计算证明，这与$r_A = -k[A]$中的k相同。在二氧化氮的例子中，

$k = (1/80 \text{ SEC})(\ln 2) = (1/80 \text{ SEC})(0.693) = 0.0087 \text{ SEC}^{-1}$. THAT IS,

0.87% OF THE NO_2 GAS IS CONSUMED EVERY SECOND.

$k = (1/80$ 秒$)(\ln 2) = (1/80$ 秒$)\times(0.693) = 0.0087$ 秒$^{-1}$。也就是说，每秒消耗0.87%的二氧化氮气体。

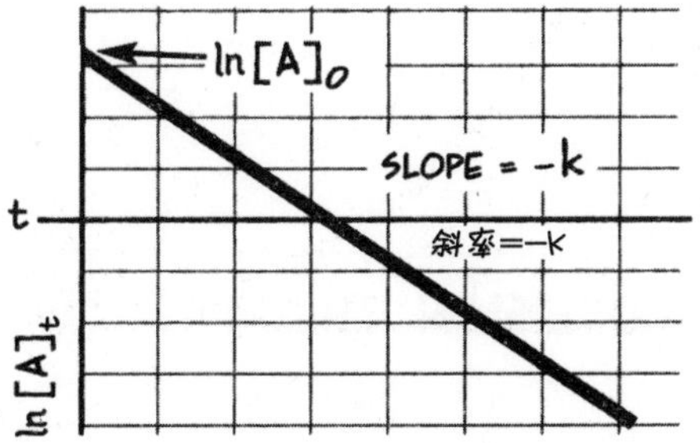

A REACTION WITH $r = -k[A]$ IS CALLED A **FIRST-ORDER REACTION:** IT GOES AS THE FIRST POWER OF A SINGLE CONCENTRATION. YOU CAN CHECK EXPERIMENTALLY IF A REACTION IS FIRST-ORDER BY GRAPHING $\ln[A]_t$ AGAINST t AND SEEING IF IT'S A STRAIGHT LINE. IF SO, THE RATE CONSTANT IS THE NEGATIVE OF THE SLOPE.

一个用$r = -k[A]$表示的反应称为**一级反应**：它由一个单一浓度驱动。通过$\ln[A]_t$对t的曲线是不是一条直线，你可以判断这个反应是否是一级反应。如果是这样，速率常数是斜率的负值。

Collision Course 碰撞过程

HOW ABOUT A SECOND-ORDER REACTION? THAT MIGHT LOOK LIKE

A + B ⟶ PRODUCTS

HERE $r_A = r_B$ BECAUSE THE REACTION REMOVES MOLECULES OF A AND B TOGETHER IN PAIRS. THE REACTION RATE r IS THEN TAKEN TO BE

$$r = r_A = r_B$$

二级反应是什么样的？这可能看起来像

A + B ⟶ 产物

这里$r_A = r_B$，因生成物为反应同时成对消耗A分子和B分子。反应速率是

$$r = r_A = r_B$$

TO ANALYZE r, THE FIRST THING WE NOTICE IS THAT TWO MOLECULES CAN COMBINE ONLY IF THEY FIRST **COLLIDE.**

分析r时，我们首先注意到的是，两个分子只有在第一次**碰撞**时才能结合。

THIS BRILLIANT OBSERVATION IS THE START OF **COLLISION THEORY.**

这伟大的观测是**碰撞理论**的开端。

HOW OFTEN DO PARTICLES COLLIDE? IT DEPENDS ON THEIR CONCENTRATION (OR PARTIAL PRESSURE).

粒子碰撞的概率是多大？这取决于粒子的浓度（或者分压）。

IMAGINE THAT A VOLUME OF GAS OR SOLUTION IS DIVIDED INTO COUNTLESS TINY COMPARTMENTS. IF TWO PARTICLES SHARE A COMPARTMENT, WE'LL CALL THAT A COLLISION.

试想想，如果气体或溶液的体积被分成无数的小单位，其中两个粒子在同一个空间单位中，我们称其为碰撞。

IF [B] IS CONSTANT, THEN CHANGING [A] CHANGES THE NUMBER OF A-B COLLISIONS PROPORTIONALLY. (HERE **A** ARE BLACK AND **B** ARE WHITE.)

如果[*B*]是常数，那么[*A*]的变化会成比例地改变A与B的碰撞数（这里**A**是黑色，**B**是白色）。

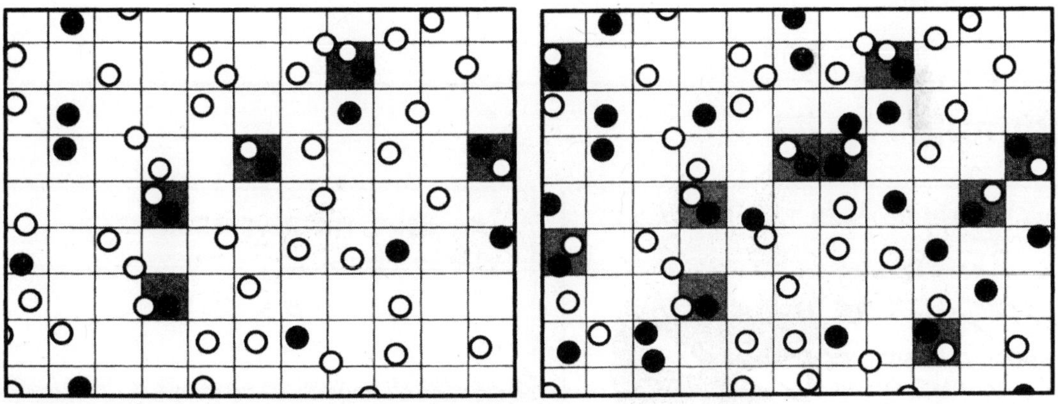

THE SAME IS TRUE WHEN [B] IS CHANGED, SO THE FREQUENCY OF COLLISIONS MUST BE PROPORTIONAL TO [A][B], OR P_AP_B, IF A AND B ARE GASES.

当[*B*]改变，同样如此。因此，碰撞频率必须与[*A*][*B*]成比例，如果*A*和B是气体，则是P_AP_B。

NOT ALL COLLISIONS RESULT IN REACTION. THE ONES THAT DO ARE CALLED **EFFECTIVE.** WE ASSUME THAT THE RATIO OF EFFECTIVE COLLISIONS TO TOTAL COLLISIONS IS CONSTANT (AT A FIXED TEMPERATURE).

并非所有的碰撞都导致反应发生，只有**有效**碰撞才会如此。我们假设有效碰撞在碰撞总数中的比率是不变的（在一个固定的温度下）。

SO: REACTION RATE EQUALS RATE OF EFFECTIVE COLLISIONS, WHICH IS PROPORTIONAL TO RATE OF TOTAL COLLISIONS, WHICH IS PROPORTIONAL TO [A][B] OR $P_A P_B$. CONCLUSION:

因此：反应速率等于有效碰撞率，与[A] [B]或$P_A P_B$成正比。结论是：

$$r = -k[A][B]$$

k A POSITIVE CONSTANT

k是一个正的常数。

WE SAY THE REACTION IS FIRST ORDER IN **A,** FIRST ORDER IN **B,** AND SECOND ORDER OVERALL.

我们说这个反应对**A**是一级反应，对**B**也是一级反应，整体是二级反应。

Example 范例

WE'VE ALREADY SEEN THAT IN DAYLIGHT

我们已经知道，在白天

$$NO_2 \rightarrow NO + O$$

AND THE MONATOMIC OXYGEN GOES ON TO MAKE OZONE

而单原子氧继续产生臭氧

$$O + O_2 \rightarrow O_3$$

SO OVERALL

因此，整个反应是

$$NO_2 + O_2 \rightarrow NO + O_3$$

AT NIGHT, THE REVERSE REACTION TAKES PLACE:

到了晚上，发生逆反应：

THIS REACTION HAS RATE r = RATE OF CONSUMPTION OF NO = RATE OF CONSUMPTION OF O_3 AND IS GIVEN BY

反应速率r = 一氧化氮的消耗速率=臭氧的消耗速率，并给出

$$r = -k[NO][O_3] \qquad k = 1.11 \times 10^7 \ M^{-1}SEC^{-1}$$

A TYPICAL NO CONCENTRATION IS AROUND 24 PPB*, WHICH AS BEFORE GIVES MOLAR CONCENTRATION [NO] AS (24 MOL NO/24.4 X 10^9 L OF AIR) = 10^{-9} M. [O_3] IS AROUND TWICE THAT, OR 2 X 10^{-9} M.

一氧化氮的浓度大约是24ppB*，用之前的摩尔浓度[NO]表示为（24摩尔的一氧化氮/24.4×10^9升的空气）= 10^{-9} M，[O_3]大约是[NO]的两倍，约为2×10^{-9} M。

*PARTS PER BILLION

*十亿分之一

A BIT OF CALCULUS PRODUCES THIS PLOT OF THE CONCENTRATIONS. THE REACTION GOES QUICKLY: IT'S ESSENTIALLY OVER IN FIVE OR SIX MINUTES.

简单计算得出浓度曲线。这个反应很快：五六分钟基本就结束。

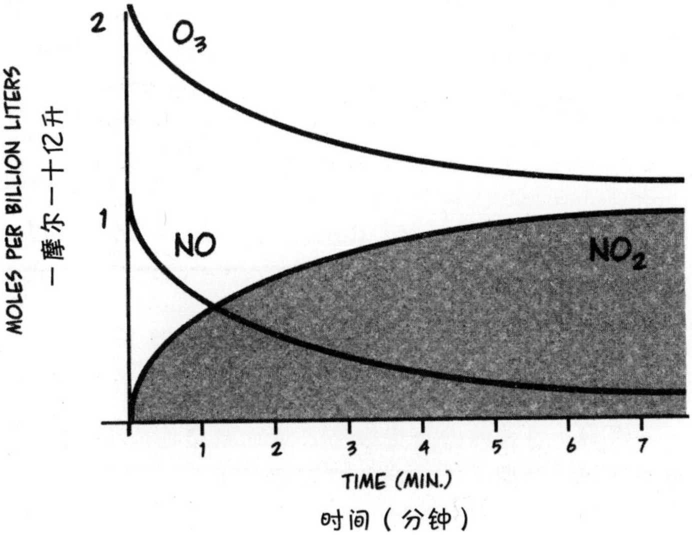

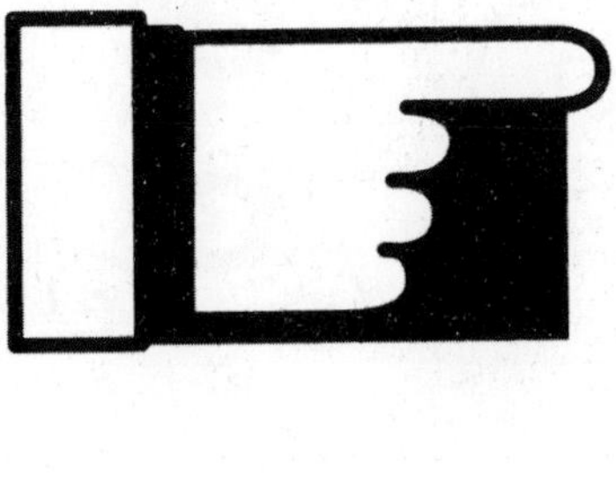

NOTE: THIS GRAPH IS GOOD ONLY FOR AN ISOLATED SAMPLE. TO PREDICT CONCENTRATIONS IN THE ENVIRONMENT, WE NEED TO KNOW THE RATES OF ALL REACTIONS THAT CONSUME AND PRODUCE NO AND O_3, AS WELL AS HOW MUCH ENTERS THE AIR FROM OUTSIDE SOURCES.

注：本图只是一个孤立的例子。为了预测在环境中的浓度，我们需要知道所有消耗和产生一氧化氮和臭氧的反应速率以及有多少从外部源进入了空气。

Reactions Up Close
近距离反应

WHY ARE SOME COLLISIONS EFFECTIVE, AND SOME ARE NOT?

为什么有些碰撞是有效的，有些则不是？

ONE REASON IS PARTICLES' RELATIVE **ORIENTATION.** TWO MOLECULES MAY NEED TO PRESENT A CERTAIN "FACE" TO EACH OTHER BEFORE THEY CAN COMBINE. FOR EXAMPLE, WHEN A HIGHLY POLAR MOLECULE OF HCl MEETS ETHENE, CH_2CH_2, A LOT OF ANGLES DON'T WORK.

原因之一是粒子的相对**位置**，两个分子可能需要摆出特定的“面”，才可以互相结合。例如，当一个高极性分子盐酸（HCl）碰到乙烯（CH_2CH_2）时，在很多角度都不会发生反应。

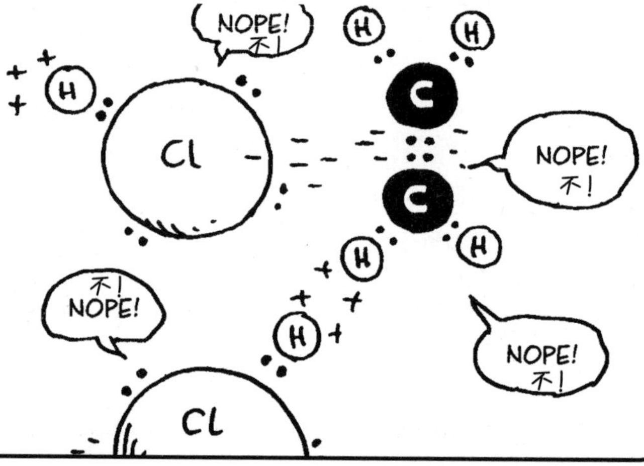

BUT WHEN THE POSITIVE POLE OF HCl MEETS CH_2CH_2'S VERY NEGATIVE DOUBLE BOND, ELECTRONS SHIFT—FIRST, ONE GOES TO HYDROGEN (IT'S CLOSER).

但是，当盐酸的正电极与乙烯的电负性很强的双键相遇时，有一个电子会首先转移到氢原子上（因为它更接近）。

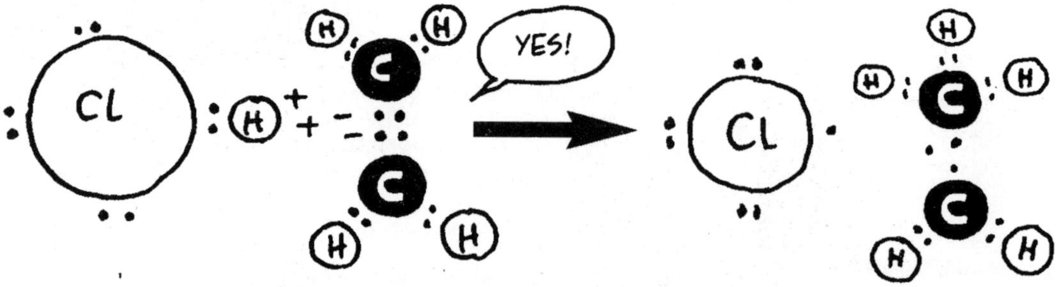

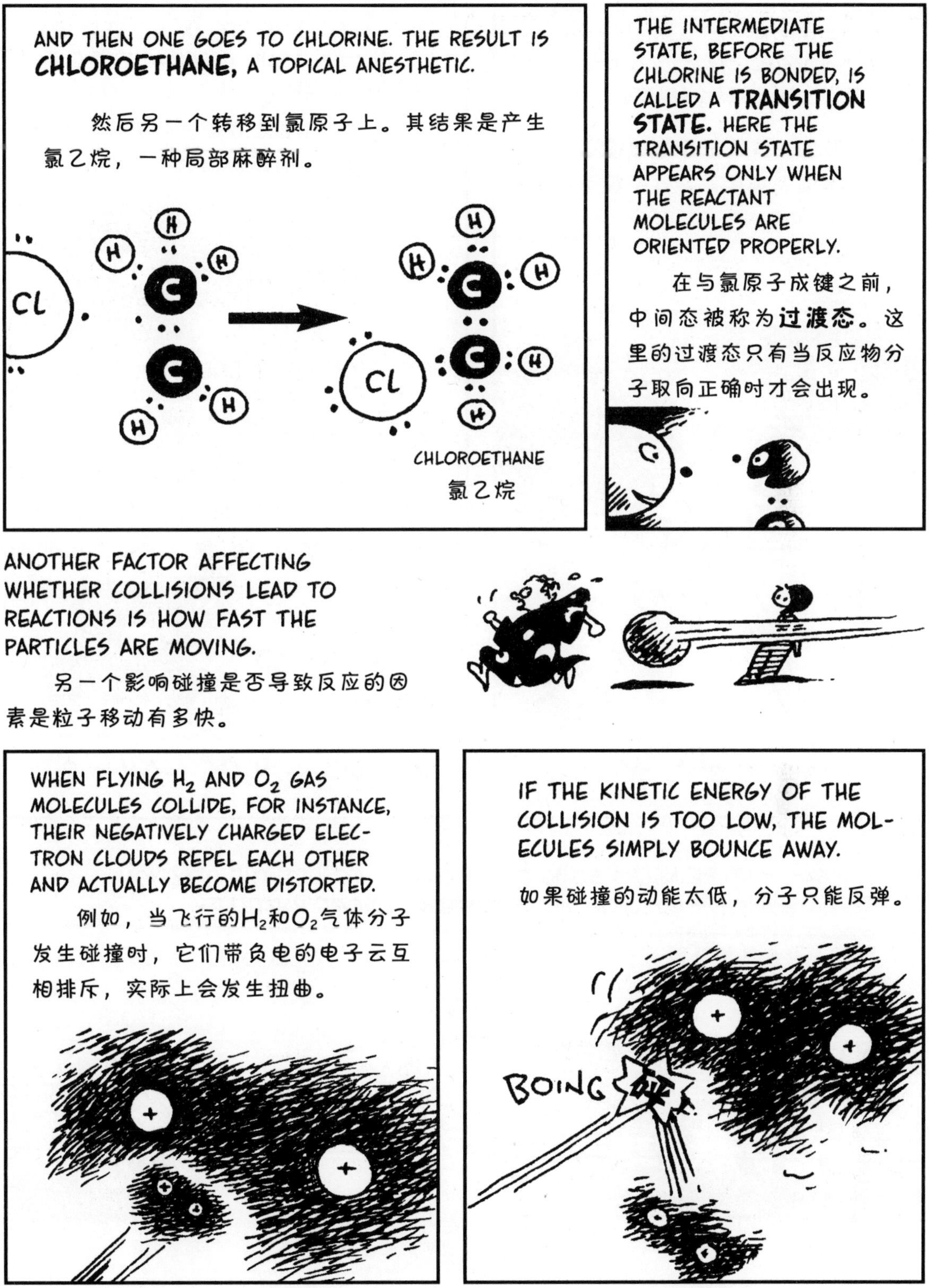
AND THEN ONE GOES TO CHLORINE. THE RESULT IS **CHLOROETHANE,** A TOPICAL ANESTHETIC.
然后另一个转移到氯原子上。其结果是产生氯乙烷，一种局部麻醉剂。
CL
H
C
CHLOROETHANE
氯乙烷
THE INTERMEDIATE STATE, BEFORE THE CHLORINE IS BONDED, IS CALLED A **TRANSITION STATE.** HERE THE TRANSITION STATE APPEARS ONLY WHEN THE REACTANT MOLECULES ARE ORIENTED PROPERLY.
在与氯原子成键之前，中间态被称为**过渡态**。这里的过渡态只有当反应物分子取向正确时才会出现。
ANOTHER FACTOR AFFECTING WHETHER COLLISIONS LEAD TO REACTIONS IS HOW FAST THE PARTICLES ARE MOVING.
另一个影响碰撞是否导致反应的因素是粒子移动有多快。
WHEN FLYING H_2 AND O_2 GAS MOLECULES COLLIDE, FOR INSTANCE, THEIR NEGATIVELY CHARGED ELECTRON CLOUDS REPEL EACH OTHER AND ACTUALLY BECOME DISTORTED.
例如，当飞行的H_2和O_2气体分子发生碰撞时，它们带负电的电子云互相排斥，实际上会发生扭曲。
IF THE KINETIC ENERGY OF THE COLLISION IS TOO LOW, THE MOLECULES SIMPLY BOUNCE AWAY.
如果碰撞的动能太低，分子只能反弹。
BOING

BUT IF INITIAL K.E. IS HIGH ENOUGH TO OVERCOME ELECTRIC REPULSION, THINGS CAN BREAK APART.

但是，如果一开始动能高到足以克服电子排斥力，分子就可以被分开。

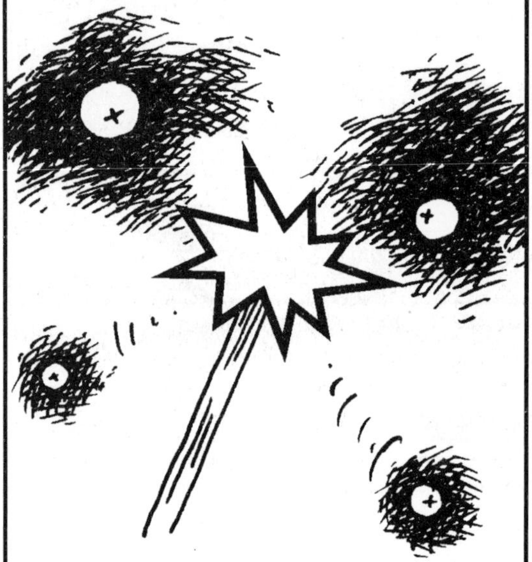

$H_2 + O_2 \longrightarrow 2H + 2O$

IF A FREE O MEETS AN H_2, ELECTRIC REPULSION AGAIN DEFORMS THE ELECTRON CLOUDS.

如果一个自由氧原子碰到一个H_2分子，电子排斥力再次使电子云变形。

IF THE **COLLISION ENERGY** IS SUFFICIENT, ELECTRONS ARE REARRANGED, A WATER MOLECULE FORMS, AND ENERGY ESCAPES (THE REACTION IS EXOTHERMIC).

如果有足够的**碰撞能量**，电子会发生重排。产生一个水分子，并放出能量（反应是放热）。

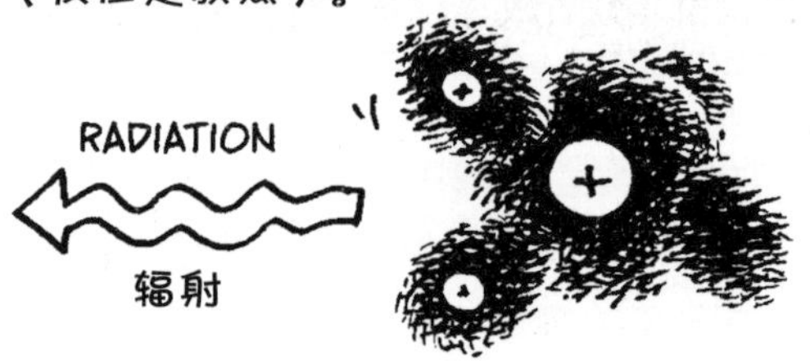

$H_2 + O \longrightarrow H_2O \qquad \Delta H < 0$

SO—THE GAS MIXTURE NEEDS SOME EXTRA ENERGY TO GET THE REACTION STARTED: A SPARK OR A FLAME, SAY, TO ENERGIZE SOME PARTICLES.

因此，混合气体需要一些额外的能量才能发生反应：火花或火焰，或者说，使一些粒子活跃起来。

BUT ONCE IT STARTS, $H_2 + O \rightarrow H_2O$ IS SO **EXOTHERMIC** THAT IT EXCITES THE PARTICLES AROUND IT, AND THE WHOLE REACTION RUSHES FORWARD WITH A SUDDEN, LOUD—

但是，一旦反应开始，$H_2 + O \rightarrow H_2O$就是**强烈放热**的反应，以至于它可以激发周围的粒子，并且使整个反应快速进行……

THIS IS ONE REASON WHY CHEMISTS ARE ALWAYS HEATING THINGS... WE HAVE TO SUPPLY THAT INITIAL ENERGY KICK TO GET REACTIONS "OVER THE HUMP."

这是化学家总喜欢加热东西的一个原因……我们必须为反应提供“渡过难关”的最初能量。

NEARLY EVERY COMBINATION REACTION WORKS THE SAME WAY: IT NEEDS AN ADDED ENERGY PUSH TO BRING THE REACTANTS TOGETHER. THIS BOOST IS CALLED THE **ACTIVATION ENERGY** OF THE REACTION, E_A. IN OTHER WORDS, CHEMICAL REACTIONS ARE NOT JUST LIKE FALLING DOWNHILL!

几乎所有化合反应的反应方式都相同：它需要一个额外的能量，推动反应物聚在一起。这就是**反应活化能**，记为E_A。换言之，化学反应可不像下坡那么简单！

FALLING DOWNHILL
下坡

CHEMICAL REACTION
化学反应

THE OBVIOUS WAY TO GET A REACTION MOVING FASTER, THEN, IS TO MAKE MORE OF THE PARTICLES EXCEED THE ACTIVATION ENERGY—IN OTHER WORDS, BY **RAISING TEMPERATURE.** THEN A HIGHER FRACTION OF COLLISIONS WILL BE EFFECTIVE.

使一个反应加快最显而易见的方式，就是让更多的粒子超过活化能。换句话说，通过**提高温度**，然后得到更高的有效碰撞比。

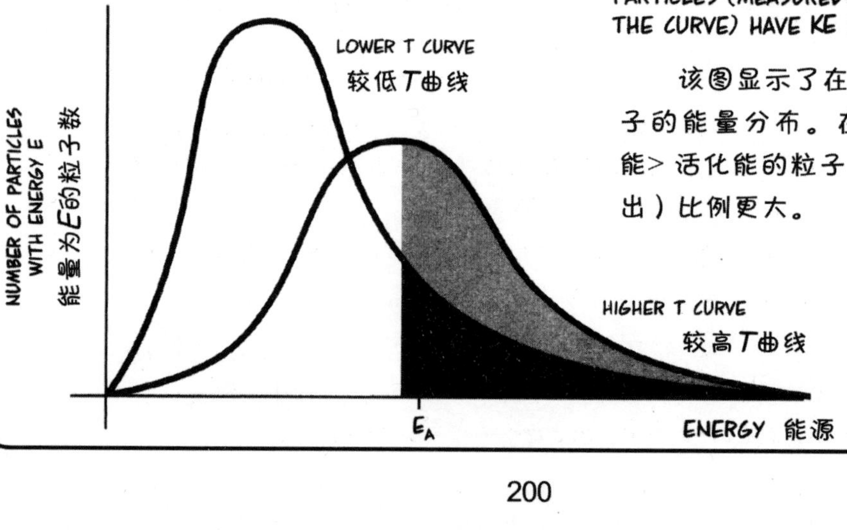

THIS GRAPH SHOWS THE ENERGY DISTRIBUTION OF A GROUP OF PARTICLES AT TWO DIFFERENT TEMPERATURES. AT HIGHER TEMPERATURE, A GREATER PROPORTION OF PARTICLES (MEASURED BY THE AREA UNDER THE CURVE) HAVE $KE > E_A$.

该图显示了在两个不同温度下粒子的能量分布。在较高的温度，动能>活化能的粒子（由曲线下面积得出）比例更大。

Catalysts, or Raising k

催化剂或提高反应速度k

YOU'RE PROBABLY NOT SURPRISED TO HEAR THAT RAISING TEMPERATURE ACCELERATES REACTIONS.* AFTER ALL, WE'VE ALL SEEN IMAGES OF CHEMISTS COOKING THINGS UP. MAYBE WE'VE EVEN TURNED UP THE FLAME A FEW TIMES OURSELVES.

你听到通过升温加速反应*，很可能不会感到惊讶。毕竟，我们对化学家煮东西都有一定印象，也许我们甚至在下厨时也曾自发地调大火焰。

*英语“催化剂”（catalyst）和“牛名单”（cattlelist）发音近似。

NOW, HOWEVER, WE CAN BE MORE PRECISE. SINCE $r = -k[A][B]$ FOR OUR SECOND-ORDER REACTION, WE CAN SAY THAT BOOSTING TEMPERATURE RAISES k, THE REACTION CONSTANT.

尽管如此，现在我们可以更精确一些。因为$r=-k[A][B]$作为我们的二级反应，我们就可以说升高温度可以增大K，使反应持续发生。

*WITHIN LIMITS. WHEN T RISES TOO HIGH, EVERYTHING SHAKES APART, AND THE REACTION IS DISRUPTED.

*在一定限度内。当温度T升得太高时，所有的物质都会分开，反应中断。

ARE THERE OTHER WAYS TO RAISE k? BASED ON THE PRECEDING DISCUSSION, WE MIGHT WONDER IF IT'S POSSIBLE TO REDUCE A REACTANT'S UNFAVORABLE ORIENTATIONS, OR LOWER THE ACTIVATION ENERGY. THIS IS WHERE **CATALYSTS** COME IN.

有没有别的方法提高k？基于前面的讨论，我们可能会想知道，有什么方法可以减少反应物的不利取向，或降低活化能。这是**催化剂**的研究领域。

A **CATALYST** IS A SUBSTANCE THAT SPEEDS UP A REACTION BUT ITSELF EMERGES FROM THE REACTION UNCHANGED.

催化剂是一种加快反应速度但本身在反应前后不发生变化的物质。

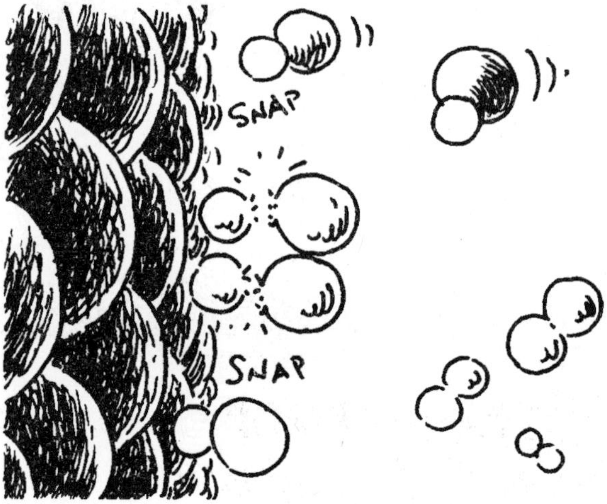

FOR EXAMPLE, THE **CATALYTIC CONVERTOR** IN A CAR ENGINE SPEEDS THE DETOXIFICATION OF EXHAUST GASES. ONE SUCH REACTION BREAKS CAUSTIC NITRIC OXIDE TO N_2 AND O_2:

例如，汽车发动机中的**催化转化器**，会加速降解产生的废气。这个反应把腐蚀性气体一氧化氮分解为N_2和O_2：

$$2NO \longrightarrow N_2 + O_2$$

IN THE CONVERTOR CHAMBER, PLATINUM, RHODIUM, AND PALLADIUM SCREENS BIND TO THE GAS MOLECULES VIA VARIOUS IMFs.

在转化器中，铂、铑、钯通过各种分子间作用力来筛选被结合的气体分子。

THE CATALYST BOTH ALIGNS THE **NO** MOLECULES FAVORABLY AND CUTS ACTIVATION ENERGY BY PULLING AGAINST THE N-O BOND—**PROBABLY.** THE EXACT MECHANISM IS UNKNOWN.

该催化剂帮助NO分子正确排列，并通过拉动N-O键，降低其活化能——**可能是这样**。但其确切的机理尚不清楚。

CATALYSTS ALSO PROBABLY ENABLED THE **ORIGIN OF LIFE.** THE CHEMICALS OF LIFE (OR PRE-LIFE) WERE TOO BIG AND UNGAINLY TO MAKE PROGRESS BY RANDOM COMBINATION... BUT IF (AS SEVERAL THEORIES SUGGEST) THEY WERE ANCHORED AT ONE END TO A CHARGED SURFACE, SUCH AS CLAY ON THE OCEAN FLOOR, THEY WOULD BE MUCH MORE LIKELY TO ENGAGE IN "GOOD" REACTIONS!

催化剂也可能与**生命的起源**有重大关系。生命体（或生命体的前体）的分子很巨大，几乎不可能随机组合产生……但如果它们固定在一个带电表面的一端，如海底的黏土，则它们将更有可能发生“好”的反应（几个理论认为）！

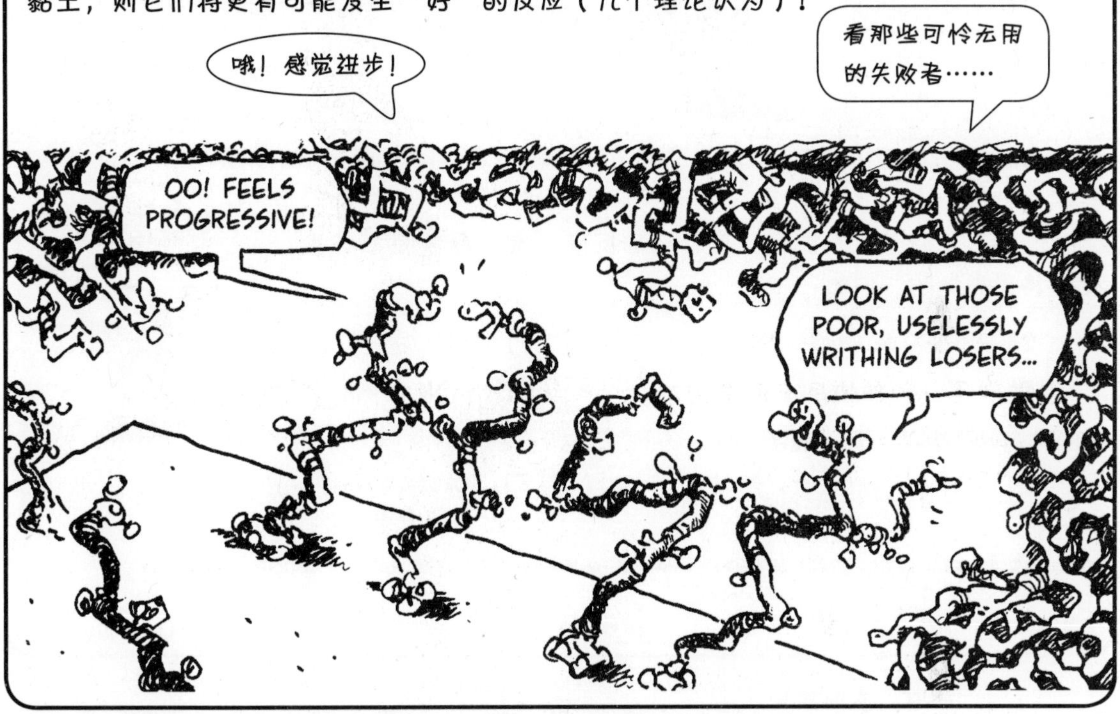

Higher-order Reactions, Maybe
多级反应，也许

WE SAW THAT

A + B ⟶ PRODUCTS

IS A SECOND-ORDER REACTION WITH RATE $r = -k[A][B]$. THIS, BY THE WAY, INCLUDES THE SPECIAL CASE WHEN **A AND B ARE THE SAME.** THE REACTION

A + A ⟶ PRODUCTS

HAS A RATE $-k[A]^2$.

我们看到，

A + B ⟶ 产物

是一个速度为$r = -k[A][B]$的二级反应。顺便说一下，这包括另一种特殊情况，即**当A和B相同时**，反应

A + A ⟶ 产物

其速度为$-k[A]^2$。

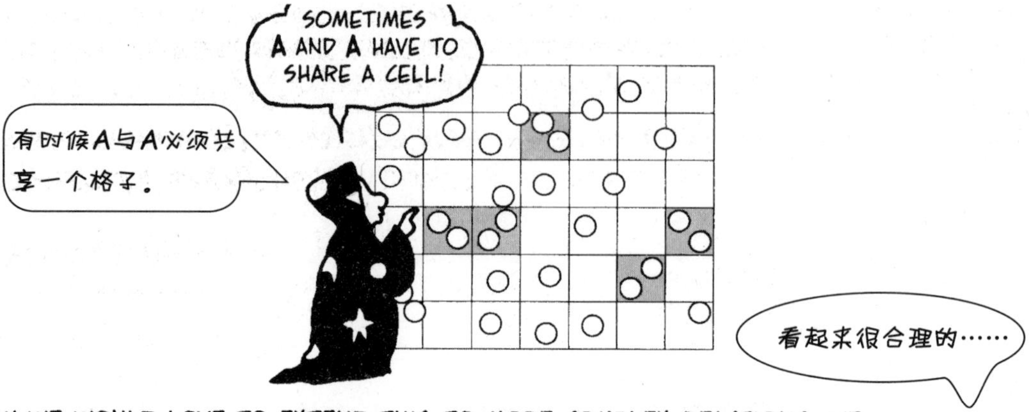

NOW WE WOULD LOVE TO EXTEND THIS TO MORE COMPLEX REACTIONS. WE MIGHT **HOPE,** FOR EXAMPLE, THAT RATE LAWS WOULD BE ANALOGOUS:

现在，我们很乐于扩展到更复杂的反应。我们**希望**反应速度定律将是类似的，例如：

$2A + B \longrightarrow$ PRODUCTS 产物 $r = -k[A]^2[B]$ (THIRD ORDER) (三级)

$2A + 3B \longrightarrow$ PRODUCTS 产物 $r = -k[A]^2[B]^3$ (FIFTH ORDER) (五级)

AND GENERALLY 而一般的

$aA + bB \longrightarrow$ PRODUCTS 产物 $r = -k[A]^a[B]^b$ (ORDER $a + b$)*
($a + b$级)

*WE HAVE TO BE A LITTLE CAREFUL ABOUT WHAT WE MEAN BY r. IT'S THE RATE AT WHICH $aA + bB$ IS CONSUMED. THAT IS, $r = (1/a)r_A = (1/b)r_B$.

*我们必须小心对待我们给予r的含义，r是$aA+bB$消耗的速率。也就是说，$r = (1/a)r_A = (1/b)r_B$。

WE WOULD LOVE TO SAY IT, READER, BUT UNFORTUNATELY WE CAN'T, BECAUSE IT'S **false.** RATES OF REAL-LIFE REACTIONS CAN'T BE PREDICTED FROM THEORY, BUT **MUST BE MEASURED EXPERIMENTALLY.**

我们很愿意这么说，读者们，但不幸的是我们不能，因为它尚不正确。现实生活中的反应速率从理论上是无法预测的，必须**通过实验测量**。

IN FACT, EVEN THE REACTION (A + B ⟶ PRODUCTS) SOMETIMES DOESN'T BEHAVE AS WE CLAIMED. YES, READER, MUCH OF THE FIRST HALF OF THIS CHAPTER IS SIMPLY **UNTRUE!**

事实上，反应（A + B → 产物）有时也不会像我们所说的那样进行。是的，读者们，本章的前半部分不一定是**真实**的！

FOR INSTANCE, WHEN WE WRITE 2A + B, ARE WE REALLY TO IMAGINE THREE PARTICLES COLLIDING AT ONCE? NOT LIKELY... MORE PROBABLY, **A** MEETS **B** TO FORM **AB**, THEN ANOTHER **A** COMES ALONG...

例如，当我们写2A+B时，难道我们真地认为三个粒子会同时碰撞？这不太可能……更可能的是，A碰到B形成AB，然后又一个A继续发生反应……

ONE-STEP REACTIONS ARE CALLED **ELEMENTARY**... AND IT **IS** TRUE THAT AN ELEMENTARY REACTION $aA + bB \longrightarrow$ PRODUCTS HAS A REACTION RATE OF

单步反应被称为**基元反应**……这是真的，一个基元反应 $aA + bB \rightarrow$ 产物的反应速率是

$$r = -k[A]^a[B]^b.$$

IN A MULTI-STEP REACTION, INTERMEDIATE STEPS ARE OFTEN UNCLEAR... THINGS GO BY TOO FAST TO OBSERVE. BUT THIS MUCH IS TRUE: THE **SLOWEST INTERMEDIATE REACTION RATE** DETERMINES THE OVERALL RATE.

在多步反应中，中间步骤往往不清楚……反应发生太快难以观察。但是有一点是真的：最慢的中间反应的速率决定着整体反应速率。

TO SEE THIS, IMAGINE A WASHER-DRYER COMBO THAT PROCESSES A LOAD OF DIRTY CLOTHES IN EXACTLY **24 HOURS**. LET'S LIFT THE LID AND SEE HOW IT WORKS...

为了明白这一点，想象一台在**24小时**内清洗一大堆脏衣服的集洗涤与烘干功能于一体的洗衣机。让我们揭开盖子，看看它是如何工作的……

WASHING, IT SEEMS, IS DONE MANUALLY BY ILL-TRAINED, UNCOOPERATIVE WEASELS WHO TAKE **23.999 HOURS** TO DO A LOAD. THE DRYER IS A NUCLEAR BLAST FURNACE THAT CRISPS YOUR CLOTHES IN A MILLISECOND.

洗涤就像用一群缺乏训练、无协作精神的黄鼠狼，它们可以用23.999小时人工洗完一桶衣服。烘干机则是一个核高炉能在一毫秒内烘干你的衣服。

PROCESS 1: RATE = ONE LOAD/DAY

过程1：速度=1桶/天

PROCESS 2: RATE = 86.4 MILLION LOADS/DAY

过程2：速度= 8640万桶/天

OVERALL PROCESS: RATE = ONE LOAD/DAY

整个过程：速度=1桶/天

NOW IS IT CLEAR THAT THE OVERALL RATE IS THE RATE OF THE SLOWEST STEP? WHEN THE WEASELS ARE DONE, THE "REACTION" IS ALL BUT OVER!

现在你明白整体速度取决于最慢的一步的速率了吧？当黄鼠狼完成工作时，“反应”几乎是结束了！

CHEMICAL EXAMPLE: IODIDE ION REDUCES PEROXYDISULFATE

化学的例子：碘离子还原硫酸盐

$$S_2O_8^{2-} + 2I^- \longrightarrow 2SO_4^{2-} + I_2$$

LOOKS THIRD-ORDER, BUT EXPERIMENT SAYS SECOND-ORDER, WITH

看起来是三级反应，但实验说明是二级反应，

$$r = -k[S_2O_8^{2-}][I^-]$$

CHEMISTS PROPOSE TWO ELEMENTARY STEPS:

化学家提出两个基本步骤：

$$S_2O_8^{2-} + I^- \longrightarrow 2SO_4^{2-} + I^+$$

$$I^+ + I^- \longrightarrow I_2$$

THE FIRST REACTION'S THEORETICAL RATE

第一个反应的理论速率

$$r = -k[S_2O_8^{2-}][I^-]$$

MATCHES THE OBSERVED RATE OF THE OVERALL REACTION. THE SECOND REACTION PRESUMABLY HAPPENS VERY FAST.

与观察到的整体反应速率相匹配。第二个反应大概发生得非常快。

Equilibrium...

化学平衡……

IS A STATE OF **DYNAMIC BALANCE.** IN NATURE, WE OFTEN FIND TWO PROCESSES THAT **UNDO** EACH OTHER—EVAPORATION AND CONDENSATION, FOR INSTANCE. WHEN THE PROCESSES UNDO EACH OTHER AT THE **SAME RATE,** NOTHING APPEARS TO BE CHANGING. THAT'S **EQUILIBRIUM.**

化学平衡是一种**动态的平衡**。在大自然中。我们常常发现两个过程互相**抵消**，比如蒸发和冷凝。当两个过程以**同样的速率**彼此抵消时，似乎没有什么发生改变。这就是化学平衡。

IF I SOIL MY CLOTHES AT THE SAME RATE THEY'RE WASHED AND DRIED, I ALWAYS HAVE THE SAME AMOUNT OF CLEAN CLOTHES.

如果我弄脏我的衣服和衣服被清洗和甩干的速度相同，那么，我总是有同样数目的干净衣服。

MANY CHEMICAL REACTIONS ARE **REVERSIBLE.**

许多化学反应是**可逆的**。

$$aA + bB \rightleftharpoons cC + dD$$

REACTANTS **A** AND **B** COMBINE TO MAKE **C** AND **D**... BUT IF EVERYTHING REMAINS MIXED TOGETHER, **C** AND **D** CAN FIND EACH OTHER TO MAKE **A** AND **B.**

反应物A和B结合起来，产生C和D……但如果一切仍然混合在一起，C和D可以遇到对方，再次生成A和B。

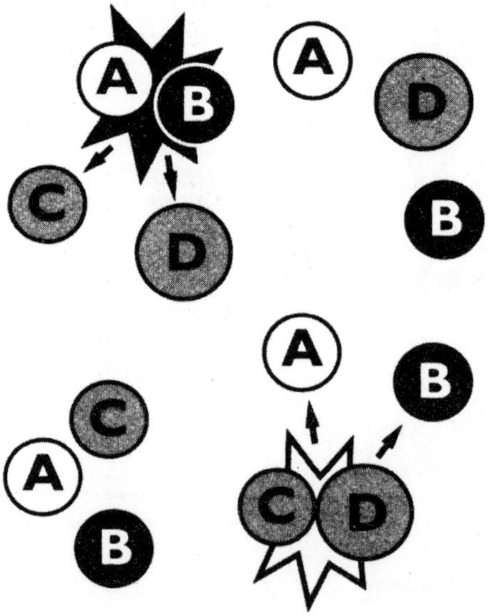

WE SAW AN EXAMPLE IN CHAPTER 4:

我们看看第四章中的例子：

$$CaCO_3(s) \rightleftharpoons CaO(s) + CO_2\uparrow$$

LIMESTONE WAS COOKED TO FORM QUICKLIME AND CARBON DIOXIDE GAS. LATER, THE WHITEWASH MADE FROM CaO REACTED WITH CO_2 FROM THE ATMOSPHERE TO MAKE $CaCO_3$ AGAIN.

石灰石被加工形成生石灰和二氧化碳气体，后来，从氧化钙制得的白浆与大气中的二氧化碳反应又重新生成碳酸钙。

IF THE CO_2 HAD NOT BEEN ALLOWED TO ESCAPE IN THE ORIGINAL REACTION (I.E., IF THE REACTION HAD OCCURRED IN A CLOSED VESSEL), SOME OF THE GAS WOULD HAVE RECOMBINED THEN AND THERE.

如果二氧化碳不允许从原反应中脱离（也就是说，如果反应在密闭容器中发生），那么有些二氧化碳会发生重组产生原料。

NOW IMAGINE A REACTION VESSEL CONTAINING THE REACTANTS **A** AND **B**.

现在想象一下，一个反应容器中含有反应物A和B。

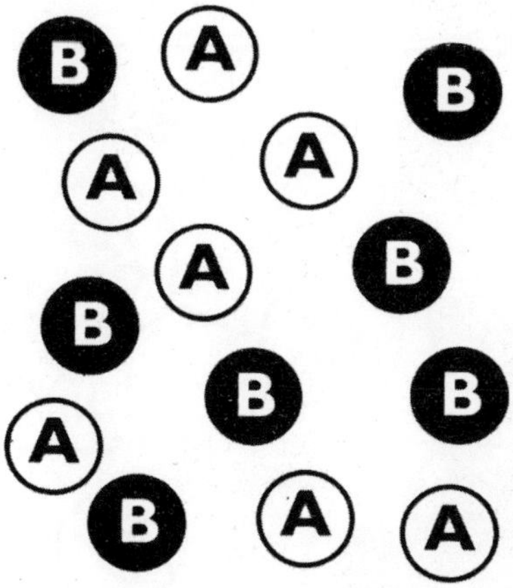

THE **FORWARD REACTION** BEGINS AND MAKES **C** AND **D** AT A RATE r_F. AS **C** AND **D** BUILD UP, A FEW OF THEM FIND EACH OTHER, AND THE REVERSE REACTION BEGINS AT A LOW RATE r_{REV}.

正反应开始，生成C和D的速率为r_F，C和D产生之后，其中有些分子找到对方，并以低的速率r_{REV}发生逆反应。

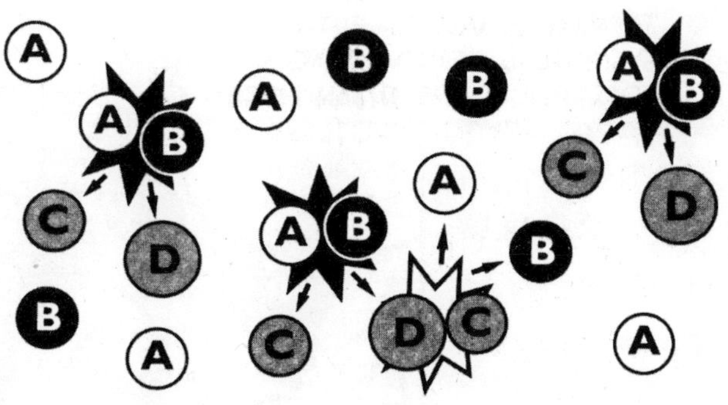

AT FIRST, $r_F > r_{REV}$, AND THE REACTION "GOES TO THE RIGHT." **A** AND **B** ARE CONSUMED FASTER THAN THEY ARE REPLENISHED, AND **C** AND **D** BUILD UP FASTER THAN THEY ARE CONSUMED.

起初，$r_F > r_{REV}$，反应“向右进行”。A和B消耗的速率大于产生的速率，C和D产生的速率大于消耗的速率。

IN OTHER WORDS, AS LONG AS $r_F > r_{REV}$, [A] AND [B] FALL AND [C] AND [D] RISE.

换句话说，只要$r_F > r_{REV}$，[A]和[B]就会降低，而[C]和[D]则会升高。

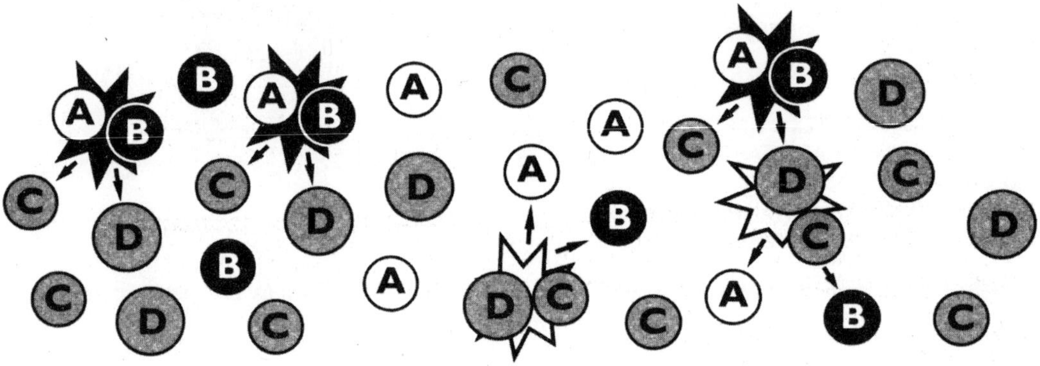

BUT RATES ARE PROPORTIONAL TO (POWERS OF) CONCENTRATIONS. SO AS LONG AS $r_F > r_{REV}$, r_F MUST FALL AND r_{REV} MUST RISE. **THE REACTION CONTINUES UNTIL**

但是速度与（驱动能力的）浓度成正比。所以，只要$r_F > r_{REV}$，r_F就一定会下降，r_{REV}则会上升。**反应继续进行，直到：**

$$r_F = r_{REV}$$

AT THIS POINT, EACH SUBSTANCE IS BEING CONSUMED AT THE SAME RATE IT IS BEING REPLENISHED. THE CONCENTRATIONS [A], [B], [C], AND [D] NO LONGER CHANGE. THE REACTION HAS REACHED **EQUILIBRIUM.**

在这一点，每一种物质被消耗的速率都等于产生它的速率。浓度[A]、[B]、[C]和[D]不再改变。该反应达到**平衡**状态。

NOW WE MAKE AN UNWARRANTED ASSUMPTION: SUPPOSE THE REACTION ORDERS ARE GIVEN BY THE STOICHIOMETRIC COEFFICIENTS a, b, c, AND d. THAT IS:

现在我们作一个大胆的假设：假设反应级数设为化学计量学系数a、b、c和d，那么：

$$r_F = -k_F[A]^a[B]^b$$
$$r_{REV} = -k_{REV}[C]^c[D]^d$$

(HERE k_F AND k_{REV} ARE THE FORWARD AND REVERSE RATE CONSTANTS.)

（这里，k_F和k_{REV}是正向和逆向的速度常数。）

AT EQUILIBRIUM, THEN, THE RATES ARE EQUAL:

在平衡状态，速率是相等的：

$$k_F[A]^a[B]^b = k_{REV}[C]^c[D]^d$$

REARRANGING,

重新排列得到，

$$\frac{[C]^c[D]^d}{[A]^a[B]^b} = \frac{k_F}{k_{REV}} = K,$$

WHERE K IS A **CONSTANT.**

其中k是一个**常数**。

BUT WHAT IF OUR ASSUMPTION IS WRONG, AND THOSE ARE NOT THE REAL RATES? NO PROBLEM! BY SOME MIRACLE, ALL INTERMEDIATE STEPS CAN BE SHOWN TO COMBINE PERFECTLY TO **VALIDATE THE USE OF THE STOICHIOMETRIC COEFFICIENTS.** THAT IS, THERE REALLY IS A CONSTANT K, SUCH THAT AT EQUILIBRIUM:

但是，如果我们的假设是错误的，这些都不是真正的速率会怎样？没问题！由于某种奇迹，所有中间体反应步骤显示出与**化学计量学系数**完美的结合，也就是说，在化学平衡状态，真实存在一个常数K：

$$\frac{[C]^c[D]^d}{[A]^a[B]^b} = K$$

TO PUT IT ANOTHER WAY, NO MATTER WHERE THE REACTION STARTS OR HOW MUCH OF ANY INGREDIENT IS PRESENT AT ANY TIME, THE CONCENTRATIONS **AT EQUILIBRIUM** ALWAYS SATISFY THE EQUATION:

换一种方式，无论反应从什么浓度开始，有多少成分是在任何时候出现的，浓度在平衡状态始终满足方程：

$$\frac{[C]^c[D]^d}{[A]^a[B]^b} = K$$

THIS FACT IS CALLED THE **law of mass action,** AND K IS THE REACTION'S **equilibrium constant.**

这其实就是**质量作用定律**，k为反应**平衡常数**。

Example: Ionization of water

范例：水的电离

CONSIDER $H_2O \rightleftharpoons H^+ + OH^-$. WATER MOLECULES OCCASIONALLY BREAK APART, AND H^+ AND OH^- REACH AN EQUILIBRIUM CONCENTRATION.

考虑$H_2O \rightleftharpoons H^+ + OH^-$。水分子偶尔断裂开，与$H^+$和$OH^-$的浓度达到平衡。

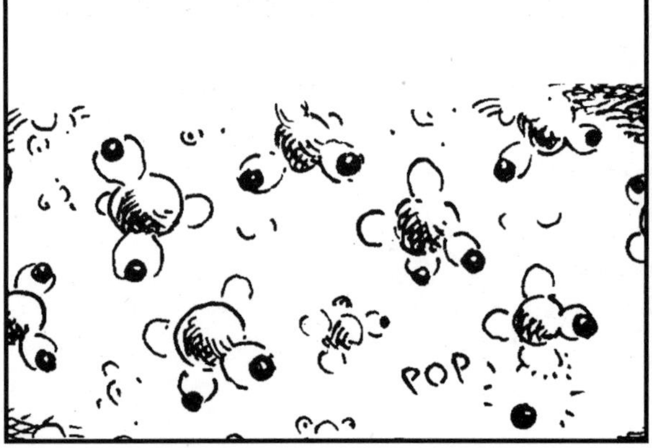

PRECISE MEASUREMENT OF PURE WATER AT 25° C SHOWS $[H^+]$ AND $[OH^-]$ TO BE ALMOST EXACTLY 10^{-7} M — NOT MUCH!

精确测量在25°C时纯净水的$[H^+]$和$[OH^-]$大约是10^{-7}M，没有多少！

H^+ IONS ALWAYS ATTACH THEMSELVES TO A WATER MOLECULE TO MAKE H_3O^+.

H^+离子总是依附一个水分子，使其成为H_3O^+。

WE PLUG IN THOSE VALUES AND CALCULATE THE EQUILIBRIUM CONSTANT.

我们插入这些值，并计算出平衡常数。

$$K = \frac{[H^+][OH^-]}{[H_2O]} = \frac{(10^{-7})(10^{-7})}{[H_2O]} = \frac{10^{-14}}{[H_2O]}$$

WHAT'S $[H_2O]$? BEFORE DISSOCIATION, IT'S 55.6 M. (1 L OF WATER WEIGHS 1000g; 1 MOL WATER WEIGHS 18 g; 1000/18 = 55.6.) AFTER DISSOCIATION, IT'S

55.6 - 0.0000001

BARELY DIFFERENT. SO WE CAN SAY

$[H_2O]$是多少？解离前，它是55.6M（1升水重1000克，1摩尔水重量18克。1000/18 = 55.6）。解离后为55.6 -0.0000001，几乎没有差别。因此，我们可以说

$$K = \frac{10^{-14}}{55.6}$$

NOW SUPPOSE 0.1 MOL OF **HYDROCHLORIC ACID,** HCl, DISSOLVES IN A LITER OF WATER. HCl, A POLAR MOLECULE, ALMOST COMPLETELY DISSOCIATES INTO H^+ AND Cl^- IONS. SUDDENLY, $[H^+]$ RISES TO 0.1 M. THEN WHAT?

现在假设0.1摩尔的盐酸（HCl）在1升水中溶解。盐酸是极性分子，几乎完全解离成H^+和Cl^-离子。突然，$[H]^+$升至0.1M，然后呢？

THEY DON'T CALL IT A CONSTANT FOR NOTHING! WE IMMEDIATELY WRITE

$$10^{-14} = 55.6K = [H^+][OH^-]$$
$$= (0.1)[OH^-]$$

SOLVING FOR $[OH^-]$,

$$[OH^-] = 10^{-13}$$

THAT IS, THE ADDED H^+ IONS GOBBLED UP EXACTLY ENOUGH OH^- IONS TO MAINTAIN THE PRODUCT $[H^+][OH^-]$ AT A CONSTANT 10^{-14}.

常数这个名字当然不是白叫的！我们立即写

$$10^{-14} = 55.6K = [H^+][OH^-]$$
$$= (0.1)\ [OH^-]$$

对$[OH^-]$求解

$$[OH^-] = 10^{-13}$$

也就是说，将H^+离子增加，使OH^-离子不断减少，以维持$[H^+][OH^-]$保持在10^{-14}的常数上。

Le Chatelier's Principle
勒夏特列原理

YOU CAN THINK OF EQUILIBRIUM AS A BALANCED SEESAW WITH REACTANTS ON ONE SIDE AND PRODUCTS ON THE OTHER. IN THE LAST EXAMPLE, H_2O WAS ON THE LEFT, OH^- AND H^+ ON THE RIGHT.

你可以把平衡状态想象为一个反应物对另一个反应物平衡的跷跷板。在上一个例子中，水分子是在左边，OH^-和H^+在右边。

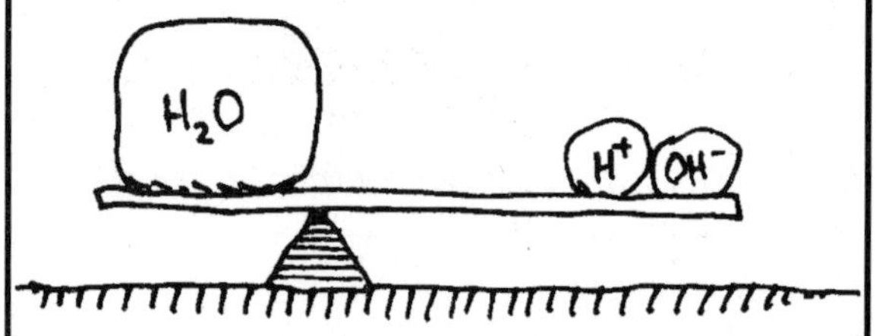

IN THAT EXAMPLE, THE EQUILIBRIUM WAS DISTURBED BY ADDING H^+ TO THE RIGHT SIDE. WHAT HAPPENS THEN?

在那个例子中，在右侧加入H^+，平衡状态便被打破。会发生什么呢？

THE FRENCH CHEMIST **HENRY LE CHATELIER** HAS LEFT US A GENERAL PRINCIPLE FOR ANALYZING WHAT HAPPENS WHEN CHEMICAL EQUILIBRIUM IS DISTURBED.

法国化学家**亨利·勒夏特列**提出一个原理，帮助我们分析化学平衡被打破时会发生什么。

When an external stress is applied to a system at equilibrium, the process evolves in such a way as to reduce the stress.

当一个外部压力被施加到一个平衡系统时，某种有利于减少这个压力影响的过程就会演变开来。

FOR EXAMPLE, IF $aA + bB \rightleftharpoons cC + dD$ IS IN EQUILIBRIUM, THEN ADDING REACTANT **A** DRIVES THE REACTION TO THE RIGHT—CONSUMING MORE **A.**

例如，如果$aA + bB \rightleftharpoons cC + dD$处于平衡状态，那么，添加反应物A使其向右移动，则此间消耗了更多的A。

IN OUR EXAMPLE, ADDING LOADS OF H^+ TO THE RIGHT-HAND SIDE OF $H_2O \rightleftharpoons H^+ + OH^-$ DROVE THE REACTION TO THE LEFT.

在我们的例子中，向$H_2O \rightleftharpoons H^+ + OH^-$这个反应的右边添加$H^+$，能够驱使反应向左侧移动。

$[OH^-]$ FELL SHARPLY, AND EVERY OH^- ION THAT DISAPPEARED TOOK AN H^+ WITH IT, THEREBY LOWERING $[H^+]$.

$[OH^-]$大幅降低，每一个OH^-离子携带一个H^+一起消失，从而降低了$[H^+]$。

LE CHATELIER VERY CLEVERLY APPLIED HIS OWN PRINCIPLE TO THE SYNTHESIS OF **AMMONIA,** NH_3, A KEY INGREDIENT OF COUNTLESS PRODUCTS, FROM FERTILIZER TO EXPLOSIVES.

勒夏特列很巧妙地运用他的原理合成无数产品（从化肥到炸药）都会用到的一个关键成分——氨（NH_3）。

$$N_2(g) + 3H_2(g) \rightleftharpoons 2NH_3(g)$$

INCREASING **PRESSURE,** SAID HIS PRINCIPLE, WILL DRIVE THE REACTION IN THE DIRECTION THAT **REDUCES PRESSURE.**

他的原理提出，加**压**将促进反应向**压强减小**的方向进行。

THERE ARE FOUR MOLES OF GAS ON THE LEFT, BUT ONLY TWO ON THE RIGHT. BY THE GAS LAW, PRESSURE IS DIRECTLY PROPORTIONAL TO THE NUMBER OF MOLES. SO PRESSURE IS RELIEVED WHEN THE REACTION GOES IN THE DIRECTION OF **FEWER MOLES,** THAT IS, TO THE RIGHT.

左侧有4摩尔气体，但右侧只有2摩尔。根据气体定律，压强与摩尔数直接成正比。所以，当反应向**少摩尔数**方向进行时，也就是向右，压强便会减小。

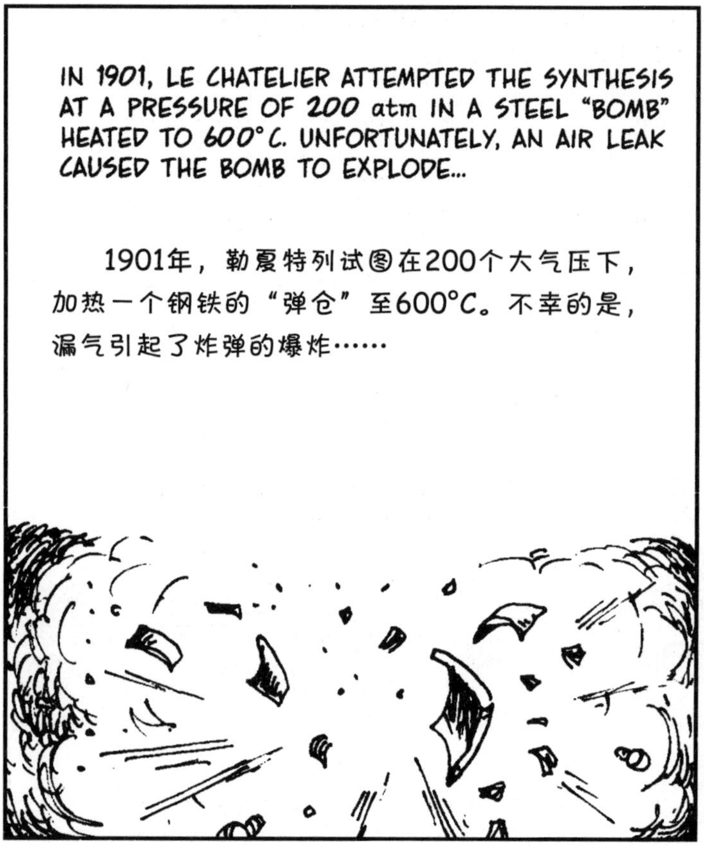

FIVE YEARS LATER, THE GERMAN **FRITZ HABER** SUCCEEDED WHERE LE CHATELIER HAD FAILED, AND EVER SINCE, AMMONIA SYNTHESIS HAS BEEN KNOWN AS THE

Haber process.

5年后，德国的**弗里茨·哈伯**成功地实现了勒夏特列已经失败的实验。从那时起，合成氨便以“**哈伯法**”闻名。

"I LET THE DISCOVERY OF THE AMMONIA SYNTHESIS SLIP THROUGH MY HANDS. IT WAS THE GREATEST BLUNDER OF MY SCIENTIFIC CAREER."
—LE CHATELIER

“合成氨从我手中溜过。这是我的科学生涯中最大的败笔。”

——勒夏特列

IN THIS CHAPTER, WE SAW HOW A NUMBER OF FACTORS AFFECTED REACTION RATES:

在这一章中，我们看到了影响反应速率的众多因素：

CONCENTRATION: RAISING CONCENTRATION UPS THE RATE.

浓度：提高浓度可加快反应速率。

TEMPERATURE: RAISING TEMPERATURE UPS THE RATE.

温度：升高温度可加快反应速率。

ACTIVATION ENERGY: LOWERING IT, BY MEANS OF A CATALYST, UPS THE RATE.

活化能：使用催化剂可降低活化能并加快反应速率。

WE ALSO SAW HOW A BUILDUP OF REACTION PRODUCTS COULD START A REVERSE REACTION THAT OVERTAKES THE FORWARD REACTION AT **EQUILIBRIUM.**

我们还看到了反应产物的生成如何使反应逆向进行，并在**平衡**状态与正向反应势均力敌。

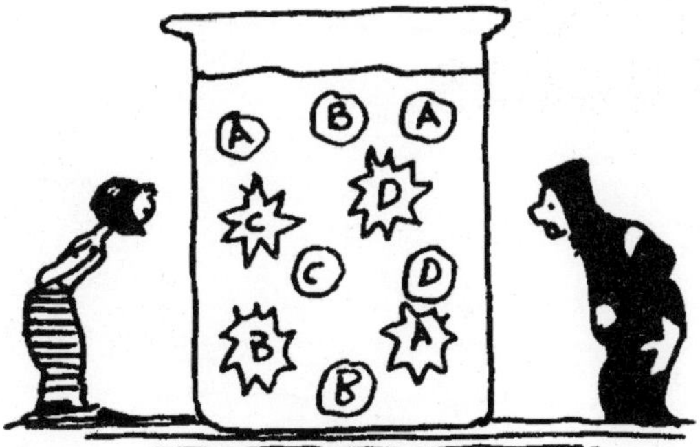

IN THE NEXT CHAPTER, WE'LL EXPLORE SOME GREAT USES OF THE CONCEPT—AND THE CONSTANT—OF EQUILIBRIUM, AND IN THE CHAPTER AFTER THAT, WE'LL DIG DEEP AND DISCOVER WHAT EQUILIBRIUM **REALLY MEANS.**

在下一章中，我们将探讨一些关于平衡状态的概念和常数极其广泛的应用。并且在之后的章节中，我们将深掘化学平衡的**真正意义**。

Chapter 9 第九章
Acid Basics 酸碱

ACIDS, SOUR AND AGGRESSIVE, ARE EVERYWHERE: IN SALAD DRESSING, RAINWATER, CAR BATTERIES, SOFT DRINKS, AND YOUR STOMACH. THEY CAN BURN, CORRODE, DIGEST, OR ADD A PLEASANT TANG TO FOOD AND DRINK...

BASES, BITTER AND SLIPPERY, MAY BE LESS FAMILIAR, BUT ARE EXACTLY AS COMMON AS ACIDS. YOU'LL FIND THEM IN BEER, BUFFERIN, SOAP, BAKING SODA, AND DRAIN CLEANERS...

ACIDS AND BASES ARE SOMETIMES USEFUL, OFTEN HARMFUL, AND ALWAYS A GREAT OPPORTUNITY TO PLAY WITH EQUILIBRIUM CONSTANTS!

有酸味和有腐蚀性的酸，是无处不在的：比如在沙拉酱、雨水、车载电池、软饮料，甚至你的胃里。它们可以燃烧、腐蚀、消化或在食品和饮料中添加一种令人愉快的味道……

我们可能不太熟悉苦涩和滑的碱，但正如酸一样，你会发现它们存在于啤酒、缓冲液、肥皂、小苏打、下水道清洁剂中……

酸和碱有时是有用的，但通常都是有害的，它们也是改变平衡常数的一个很好的方法！

稍微有点酸

ACIDS AND BASES ARE INTIMATELY CONNECTED VIA PROTONS, I.E., HYDROGEN IONS, H^+.

酸和碱通过质子紧密连接，质子也就是氢离子（H^+）。

AN **ACID** IS ANY SUBSTANCE THAT THROWS OFF PROTONS. THE STRONGER THE ACID, THE MORE EASILY IT SHEDS H^+.

酸是指任何可以失去质子的物质。酸性越强，就越容易失去H^+。

SINCE NAKED PROTONS ARE WILD, AGGRESSIVE CREATURES, STRONG ACIDS ARE VERY REACTIVE.

由于裸露的质子就像很凶猛的野生动物，所以强酸非常活泼。

A **BASE** IS ANY SUBSTANCE THAT TAKES UP PROTONS. BASES GENERALLY HAVE AN EXPOSED ELECTRON PAIR WHERE A PROTON CAN NESTLE.

碱是指任何可以接受质子的物质。碱一般都有一个裸露的电子对，可以把质子吸引过来。

THE STRONGER THE BASE, THE MORE STRONGLY IT WANTS TO BOND TO A PROTON.

碱性越强，越需要与一个质子成键。

AS YOU CAN SEE, AN ACID IS JUST A PROTON ATTACHED TO A BASE! AN ACID AND BASE PAIRED IN THIS WAY ARE CALLED **CONJUGATE** TO EACH OTHER.

正如你所看到的，酸仅仅是连接了一个质子的碱！以这种方式形成的一对酸和碱被称为互为**共轭**酸碱。

BY DEFINITION, THE STRONGER AN ACID, THE WEAKER ITS CONJUGATE BASE, AND VICE VERSA.

根据定义，酸性越强，其共轭碱越弱，反之亦然。

STRONG ACID, WEAK CONJUGATE BASE, LOOSE PROTON
强酸、弱共轭碱，易失去质子

WEAK ACID, STRONG CONJUGATE BASE, TIGHTLY BOUND PROTON
弱酸、强共轭碱，紧紧结合质子

SOME CONJUGATE ACID-BASE PAIRS:
一些共轭酸-碱对：

ACIDS, STRONGEST TO WEAKEST 酸，最强到最弱		BASES, WEAKEST TO STRONGEST 碱，最弱到最强	
SULFURIC, H_2SO_4	硫酸	BISULFATE, HSO_4^-	硫酸氢根
HYDROIODIC, HI	氢碘酸	IODIDE, I^-	碘离子
HYDROBROMIC, HBr	氢溴酸	BROMIDE, Br^-	溴离子
HYDROCHLORIC, HCl	盐酸	CHLORIDE, Cl^-	氯离子
NITRIC HNO_3	硝酸	NITRATE, NO_3^-	硝酸根
HYDRONIUM, H_3O^+	水合氢离子	WATER H_2O	水
BISULFATE, HSO_4^-	硫酸氢根	SULFATE, SO_4^{2-}	硫酸根
SULFUROUS, H_2SO_3	亚硫酸	BISULFITE, HSO_3^-	亚硫酸根
PHOSPHORIC, H_3PO_4	磷酸	$H_2PO_4^-$	磷酸二氢根
HYDROFLUORIC, HF	氢氟酸	FLUORIDE, F^-	氟离子
NITROUS HNO_2	亚硝酸	NITRITE NO_2^-	亚硝酸根
ACETIC (VINEGAR), CH_3CO_2H	乙酸(醋酸)	ACETATE, $CH_3CO_2^-$	乙酸根
CARBONIC H_2CO_3	碳酸	BICARBONATE, HCO_3^-	碳酸氢根
AMMONIUM NH_4^+	铵离子	AMMONIA NH_3	氨
HYDROCYANIC, HCN	氢氰酸	CYANIDE, CN^-	氰根
BICARBONATE, HCO_3^-	碳酸氢根	CARBONATE, CO_3^{2-}	碳酸根
WATER, H_2O	水	HYDROXIDE, OH^-	氢氧根

NOTE: BOTH ACIDS AND BASES CAN BE EITHER CHARGED OR NEUTRAL.

注：酸和碱都可以带电或呈中性……

Acids and Bases in Water
水中的酸碱

NOW WE WOULD LIKE A **NUMERICAL MEASURE** OF AN ACID'S STRENGTH. THIS IS EASIEST FOR ACIDS DISSOLVED IN WATER. (MOST ACIDS WE ENCOUNTER IN THE WORLD AND IN THE LAB ARE WATER SOLUBLE.)

现在我们得到酸强度的**测量数据**。酸很容易在水中溶解（我们在世界和实验室中遇到的大多数酸是水溶性的）。

重要安全事项：一定要**把酸加入水中**，而不是把水加到酸中。手拿强酸时要戴上手套。

WHEN A **STRONG** ACID DISSOLVES IN WATER, THE ACID COMPLETELY **IONIZES,** OR DISSOCIATES. HYDROCHLORIC ACID, FOR EXAMPLE, DOES THIS:

当**强**酸溶解在水中时，酸会完全**电离**或游离。如盐酸，就会这样：

$$HCl \longrightarrow H^+ + Cl^-$$

BUT THAT PROTON CAN'T FLOAT AROUND FREELY: ITS CHARGE SOON DRAWS A CLUSTER OF WATER MOLECULES.

但是，质子不能自由漂浮：它的电荷很快就绘制了一个水分子簇。

FOR CONVENIENCE, WE ASSIGN IT TO ONE OF THESE H_2O MOLECULES, AND WE CALL THE CLUSTER A **HYDRONIUM** ION, H_3O^+. IN SHORT,

为方便起见，我们将它分配给这些水分子之一，称其为**水合氢**离子（H_3O^+）。

$$HCl + H_2O \longrightarrow H_3O^+ + Cl^-$$

THAT IS, ADDING STRONG ACID TO WATER RAISES THE CONCENTRATION OF H_3O^+. H_3O^+ IS A STRONG ACID, AND THE HIGHER ITS CONCENTRATION, THE MORE ACIDIC THE SOLUTION.

也就是说，向水中加入强酸提高了H_3O^+的浓度。H_3O^+是一种强酸，它的浓度越高，溶液的酸性就越强。

WE CAN DESCRIBE THIS IN TERMS OF BASES, TOO.

我们也可以这样描述碱。

H_2O TAKES A PROTON FROM HCl TO FORM H_3O^+. THAT IS TO SAY, H_2O IS A **STRONGER BASE** THAN Cl^-.

水从盐酸得到质子形成H_3O^+，这就是说，水是一种比Cl^-**更强的碱**。

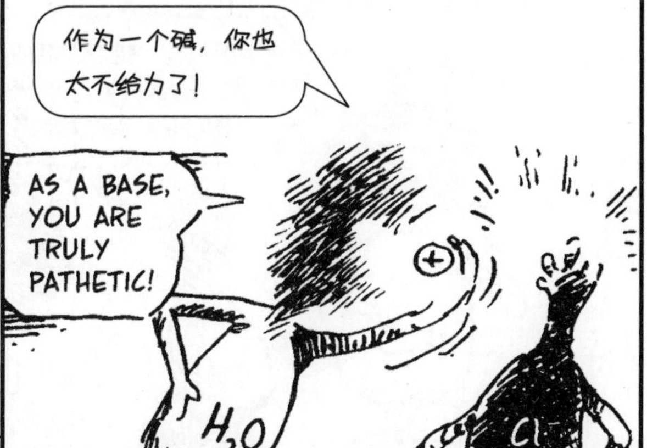

WHAT'S TRUE OF HCl IS TRUE OF ALL STRONG ACIDS. THEIR CONJUGATES (NO_3^-, ETC.) ARE ULTRA-WEAK BASES—WEAKER THAN WATER, WHICH IS VERY WEAK!

HCl是所有强酸的典范。它们的共轭碱（NO_3^-等）是超弱碱，比水的还要弱！

WHAT IF WE ADD A STRONGER BASE B^- TO WATER? THEN H_3O^+ WILL SURRENDER A PROTON TO IT:

如果我们将一个强碱B^-加入水中会怎样？然后，H_3O^+将移交给这个碱一个质子：

$$H_3O^+ + B^- \longrightarrow HB + H_2O$$

THAT IS, DISSOLVED BASES **REDUCE THE CONCENTRATION** OF H_3O^+.

也就是说，溶解碱**降低**H_3O^+的**浓度**，

TO SUM UP: IN AQUEOUS SOLUTION, ACIDS **INCREASE** $[H_3O^+]$, AND BASES **DECREASE** IT. $[H_3O^+]$ IS A **MEASURE OF A SOLUTION'S ACIDITY.**

总结：在水溶液中，酸**增加**$[H_3O^+]$，碱**减少**$[H_3O^+]$。$[H_3O^+]$**可以衡量溶液的酸性。**

pH

HOW HIGH IS $[H_3O^+]$? LET'S REVIEW THE DISCUSSION ON PAGE 212 IN CHAPTER 8. WATER ALWAYS IONIZES ITSELF A LITTLE:

$[H_3O^+]$有多少？让我们回顾一下在第8章第212页上的讨论。水自身的电离很小：

$$H_2O + H_2O \rightleftharpoons H_3O^+ + OH^-$$

AT EQUILIBRIUM, IN PURE WATER AT 25°C, THE MOLAR CONCENTRATIONS OF H_3O^+ AND OH^- ARE BOTH 1.0×10^{-7} M.

在平衡状态，纯净水在25℃时，H_3O^+和OH^-的摩尔浓度都是1.0×10^{-7}M。

THE EQUILIBRIUM CONSTANT FOR THIS REACTION IS

此反应的平衡常数是

$$K_{eq} = \frac{[H_3O^+][OH^-]}{[H_2O]^2}$$

BUT THE DENOMINATOR IS CONSTANT, OR NEARLY SO. ONLY ABOUT ONE WATER MOLECULE IN 556,000,000 IONIZES! THEREFORE THE NUMERATOR IS A CONSTANT TOO. WE CALL IT THE **WATER CONSTANT.**

分母是不变的，或者基本不变。556000000个离子中仅有1个水分子！因此，分子也是一个常数。我们把它叫作**水常数**。

$$K_w = [H_3O^+][OH^-]$$
$$= (10^{-7})(10^{-7})$$
$$= 10^{-14}.$$

A STRONG ACID GIVES ALL ITS PROTONS TO WATER TO MAKE H_3O^+. FOR INSTANCE, A 1 M SOLUTION OF HNO_3 HAS

强酸把它所有的质子给了水并形成H_3O^+。例如，硝酸溶液有：

$$[H_3O^+] = 1\ M = 10^0\ M$$

SO 因此，

$[OH^-]$ DROPS TO $K_w/[H_3O^+]$

$= 10^{-14}$

$[OH^-]$降至$K_w/[H_3O^+] = 10^{14}$

ON THE OTHER HAND, A BASIC COMPOUND LIKE NaOH DISSOCIATES FULLY IN WATER AND RAISES $[OH^-]$. $[H_3O^+]$ FALLS ACCORDINGLY. A 1 M SOLUTION OF NaOH HAS

另一方面，像氢氧化钠的碱性化合物在水中完全解离，提高$[OH^-]$，则$[H_3O^+]$相应下降。在1M氢氧化钠溶液中有：

$$[OH^-] = 1$$

$$[H_3O^+] = 10^{-14}.$$

FOR MOST PRACTICAL PURPOSES, THEN, $[H_3O^+]$ FLUCTUATES BETWEEN 1 AND 10^{-14}.

为了大多数实际目的，$[H_3O^+]$在1和10^{-14}之间波动。

NOW WHEN CHEMISTS SEE 10^x, THEY OFTEN FIND IT SIMPLER TO TALK ABOUT x, THE LOGARITHM. THEY DEFINE

现在，当化学家看到10^x，他们发现考虑对数x更简单。化学家们定义

$$pH = -\log[H_3O^+]$$

pH STANDS FOR POWER OF HYDROGEN. pH RANGES APPROXIMATELY FROM 0 TO 14. THE LOWER THE pH, THE MORE ACIDIC THE SOLUTION. FOR INSTANCE, A 0.01 M SOLUTION OF HCl HAS $[H_3O^+] = .01 = 10^{-2}$, SO pH = 2.

pH值代表氢的强度。pH值的范围约从0到14。pH值越低，溶液的酸性越强。例如，在0.01M的HCl溶液中，$[H_3O^+] = 0.01 = 10^{-2}$，因此pH= 2。

WHEN DEALING WITH BASES, IT CAN BE MORE CONVENIENT TO USE pOH. THIS IS DEFINED AS

当处理碱时，就可以更方便地使用pOH，定义为

$$pOH = -\log[OH^-]$$

AND WE HAVE

我们得到

$$pH + pOH = 14$$

pH	SUBSTANCE	pH	物质
0	5% SULFURIC ACID	0	5%的硫酸
1	STOMACH ACID	1	胃酸
2	LEMONS VINEGAR	2	柠檬 醋
3	APPLES, GRAPEFRUIT COCA-COLA, ORANGES	3	苹果、葡萄柚 可乐、橙子
4	TOMATOES, ACIDIFIED LAKES	4	西红柿、酸性湖泊
5	COFFEE BREAD POTATOES	5	咖啡 面包 土豆
6	NATURAL RIVERS	6	天然河流 牛奶
	MILK		
7	PURE WATER, SALIVA TEARS, BLOOD	7	纯净水、唾液 眼泪、血液
8	SEA WATER	8	海水
	BAKING SODA		小苏打
9		9	
10	WATER IN MONO LAKE	10	莫诺湖湖水
	MILK OF MAGNESIA		含镁牛奶
11		11	
12		12	
	LIME WATER		石灰水
13		13	
14	LYE, 4% SODIUM HYDROXIDE	14	碱液、4%氢氧化钠

* 莫诺湖是美国加利福尼亚州的一个盐湖。

Weak Ionization 弱电离

IN WATER, STRONG ACIDS IONIZE... WELL... STRONGLY. WHEN HCl DISSOLVES, IT RELEASES VIRTUALLY ALL ITS HYDROGEN AS H^+, AND pH IS GIVEN DIRECTLY BY HOW MUCH HCl IS IN SOLUTION.

在水中，强酸完全电离……当盐酸溶解时，会将几乎所有的氢以H^+的形式释放出来，pH值直接由溶液中的盐酸含量得出。

BUT A COMPLICATION ARISES WITH H_2SO_4, A STRONG ACID WITH TWO PROTONS TO GIVE. ONLY THE FIRST ONE IONIZES COMPLETELY:

但H_2SO_4就有些复杂，强酸有两个质子。只有第一个完全电离：

$$H_2SO_4 + H_2O \longrightarrow H_3O^+ + HSO_4^-$$

HSO_4^- IS A WEAKER ACID, WHICH PARTS WITH ITS PROTON LESS WILLINGLY.

硫酸氢根是一个较弱的酸，不情愿与它的质子分开。

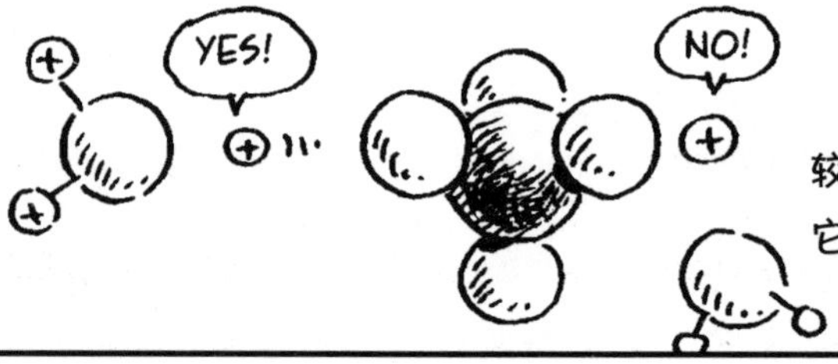

HOW DO WE SPECIFY THE "ACIDITY" OF WEAK ACIDS? THESE ACIDS IONIZE ONLY PARTWAY IN WATER. THAT IS, IF HB IS ANY WEAK ACID IN AQUEOUS SOLUTION, IT SOMETIMES HANDS OFF ITS H^+ TO H_2O, AND SOMETIMES THE PROTON COMES BACK:

我们如何鉴定弱酸的“酸度”？这些酸在水中只部分电离。也就是说，如果HB是在水溶液中溶解的任何弱酸，它有时会将自身的质子给水分子，有的时候质子又会回来。

$$HB + H_2O \rightleftharpoons H_3O^+ + B^-$$

THE REACTION'S EQUILIBRIUM CONSTANT EXPRESSES THE EXTENT OF IONIZATION:

这个反应的平衡常数表示电离程度：

$$\frac{[H_3O^+][B^-]}{[HB][H_2O]}$$

AS USUAL, $[H_2O]$ IS CONSTANT, SO WE CAN REMOVE IT FROM THE EXPRESSION. THEN THE **ACID IONIZATION CONSTANT** K_a IS DEFINED BY

像往常一样，$[H_2O]$是恒定的，所以，我们可以将其从表达式中删除。然后，**酸电离常数**Ka表示为

THE MORE IONIZED THE ACID, THE BIGGER I AM!

$$K_a = \frac{[H^+][B^-]}{[HB]}$$

酸电离越多，我越大！

WITH H_3O^+ ABBREVIATED AS H^+...

这里H^{3+}被简写为H^+…

HERE ARE K_a VALUES FOR A FEW WEAK ACIDS. A HIGH VALUE FOR K_a MEANS A LARGE NUMERATOR, THAT IS, A LOT OF IONS RELATIVE TO THE NON-IONIZED SPECIES IN THE DENOMINATOR. THAT IS, HIGHER K_a MEANS STRONGER ACID.

以下是一些弱酸的K_a值。K_a值越高表示分子越大，即相对于分母中非离子化的组分来说，存在着大量离子。也就是说，高K_a代表酸性更强。

		K_{a1}	K_{a2}
ACETIC 醋酸	CH_3CO_2H	1.75×10^{-5}	
CARBONIC 碳酸	H_2CO_3	4.45×10^{-7}	4.7×10^{-11}
FORMIC 甲酸	HCO_2H	1.77×10^{-4}	
HYDROFLUORIC 氢氟酸	HF	7.0×10^{-4}	
HYPOCHLOROUS 次氯酸	HOCl	3.0×10^{-8}	
NITROUS 亚硝酸	HNO_2	4.6×10^{-4}	
SULFURIC 硫酸	H_2SO_4	STRONG	1.20×10^{-2}
SULFUROUS 亚硫酸	H_2SO_3	1.72×10^{-2}	6.43×10^{-8}

ACIDS THAT SHED MORE THAN ONE PROTON WILL HAVE MORE THAN ONE IONIZATION CONSTANT. FOR EXAMPLE, H_2CO_3, WHICH CAN SHED TWO PROTONS, HAS K_{a1} FOR

可释放不止一个质子的酸，有一个以上的电离常数。例如，H_2CO_3，它可以释放两个质子，K_{a1}为

$$H_2CO_3 \rightleftharpoons H^+ + HCO_3^-$$

AND K_{a2} FOR　　K_{a2}为

$$HCO_3^- \rightleftharpoons H^+ + CO_3^{2-}$$

NOTE ALSO: IN WATER SOLUTION, SOME METAL IONS CAN ACT AS ACIDS. BY GRABBING OH^- FROM WATER, THEY GENERATE H_3O^+. Fe^{3+} IS AN EXAMPLE:

另注：在水溶液中，某些金属离子可以作为酸。通过捕获水中的OH^-，它们生成H_3O^+。Fe^{3+}就是一个例子：

$$Fe^{3+} + 2H_2O \rightleftharpoons FeOH^{2+} + H_3O^+$$
$$FeOH^{2+} + 2H_2O \rightleftharpoons Fe(OH)_2^+ + H_3O^+$$
$$Fe(OH)_2^+ + 2H_2O \rightleftharpoons Fe(OH)_3 + H_3O^+$$

ACID MINE DRAINAGE CONTAINS Fe^{3+}. WHEN IT ENTERS A RIVER WITH HIGHER pH, IT PRECIPITATES OUT AS AN UGLY SLIME CALLED "YELLOW BOY."

矿山的酸性废水含有Fe^{3+}。当它流入一个pH值较高的河流时，会沉淀出肮脏的黄泥，称为“黄色小子”*。

*译注：“黄色小子”是一种化学胶体沉淀，可以做黄色颜料。

Example 范例

K_a CAN BE USED TO FIND THE pH OF A WEAK ACID SOLUTION.

K_a可以用于得出某种弱酸溶液的pH值。

VINEGAR IS A 5% SOLUTION OF ACETIC ACID. THIS WORKS OUT TO ABOUT 0.8 MOL/L. WHAT IS THE pH OF AN 0.8 M SOLUTION OF CH_3CO_2H IN WATER?

醋是5%的醋酸溶液。这大约是0.8摩尔/升，0.8MCH_3CO_2H水溶液的pH值是多少？

$CH_3CO_2H \rightleftharpoons CH_3CO_2^- + H^+$ (ABBREVIATING H_3O^+ AS H^+) (缩写H_3O^+为H^+)

THE CONCENTRATION OF ACID BEFORE IONIZATION IS 0.8 M. SUPPOSE IONIZATION REDUCES THIS VALUE BY AN AMOUNT x. THEN WE CAN MAKE A TABLE:

电离前的酸浓度为0.8M，假设电离降低浓度值为x，那么我们可以列一个表：

	CH_3CO_2H	$CH_3CO_2^-$	H^+
CONC. BEFORE IONIZATION 电离前的浓度	0.8	0.0	0.0
CHANGE IN CONC. 浓度的变化	−x	x	x
EQUILIBRIUM CONC. 平衡点的浓度	0.8 − x	x	x

ASSUMPTION 1: H^+ IONS FROM WATER ARE NEGLIGIBLE COMPARED TO H^+ IONS FROM ACID.

假设1：水中的H^+离子与酸形成的H^+离子相比，可以忽略不计。

PLUG IN THE EQUILIBRIUM VALUES IN THE EQUATION FOR K_a

在方程中代入平衡常数K_a

$$K_a = \frac{[CH_3CO_2^-][H^+]}{[CH_3CO_2H]} = \frac{(x)(x)}{(0.8-x)} = 1.75 \times 10^{-5}$$ (FROM THE TABLE) (从表得知)

$$\frac{x^2}{0.8} = 1.75 \times 10^{-5}$$

ASSUMPTION 2: x IS NEGLIGIBLY SMALL COMPARED TO 0.8, SO WE CAN IGNORE IT IN THE DENOMINATOR.

假设2：与0.8相比，x是微乎其微的，所以我们可以忽略分母中的x。

$$x^2 = = (0.8)(1.75)10^{-5} = 14 \times 10^{-6}$$

$$x = (14)^{1/2} \times 10^{-3} = 3.74 \times 10^{-3}$$

ASSUMPTION 2 WAS JUSTIFIED. x REALLY IS MUCH SMALLER THAN 0.8.

假设2是合理的。x真的比0.8小多了。

BUT $x = [H^+]$, SO

但$x = [H^+]$，所以

$$pH = -\log(3.74 \times 10^{-3}) = 3 - \log(3.74) = 3 - 0.57$$

$$= \mathbf{2.43}$$

THIS ALSO TELLS US THE FRACTION OF MOLECULES THAT IONIZE.

这也告诉我们，一部分分子产生离子化。

$$\frac{[CH_3CO_2^-]}{[CH_3CO_2H]} = \frac{3.74 \times 10^{-3}}{0.8} = 4.7 \times 10^{-3}$$

A LITTLE LESS THAN 5 MOLECULES IN A THOUSAND.

少于千分之五的分子。

TRY DOING THE SAME CALCULATION WITH A 0.08 M SOLUTION. MAKE THE SAME TWO SIMPLIFYING ASSUMPTIONS. YOU SHOULD FIND pH = 2.93, AND ALSO THAT THE FRACTION OF IONIZED MOLECULES GOES UP AS CONCENTRATION GOES DOWN.

试着用相同方法计算0.08M的溶液。作出同样两个简单假设。您应该发现pH = 2.93，当浓度下降时，离子化的分子增加。

REACTIONS SUCH AS

$$Fe^{3+} + 2H_2O \rightleftharpoons FeOH^{2+} + H_3O^+$$

ARE CALLED **HYDROLYSIS,** OR WATER-SPLITTING. HERE IT INVOLVES AN ACID, BUT IT'S ALSO VERY COMMON WITH BASES.

如$Fe^{3+}+2H_2O \rightleftharpoons FeOH^{2+}+H_3O^+$的反应被称为**水解**，或水分解。在这里，它涉及一种酸，但它也非常普遍地涉及碱。

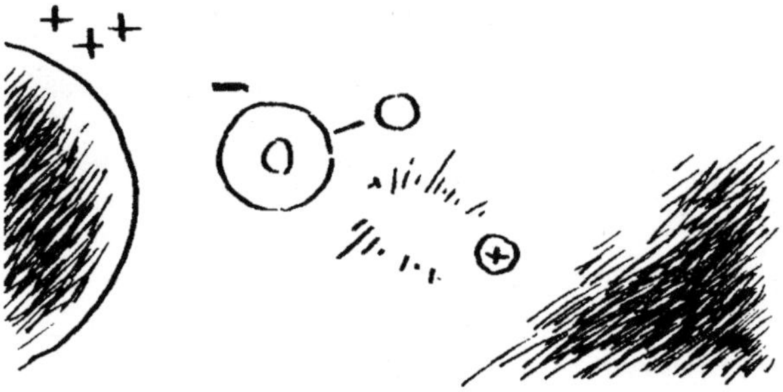

WHEN A BASE B^- (OTHER THAN OH^-) IS DISSOLVED IN WATER, B^- TAKES H^+ FROM H_3O^+.

当一个碱B^-（非OH^-）溶解于水时，B^-从H_3O^+得到一个H^+。

$[H_3O^+]$ DROPS... $[OH^-]$ MUST RISE TO MAINTAIN K_w...

$[H_3O^+]$降低……$[OH^-]$必须升高以维护K_W……

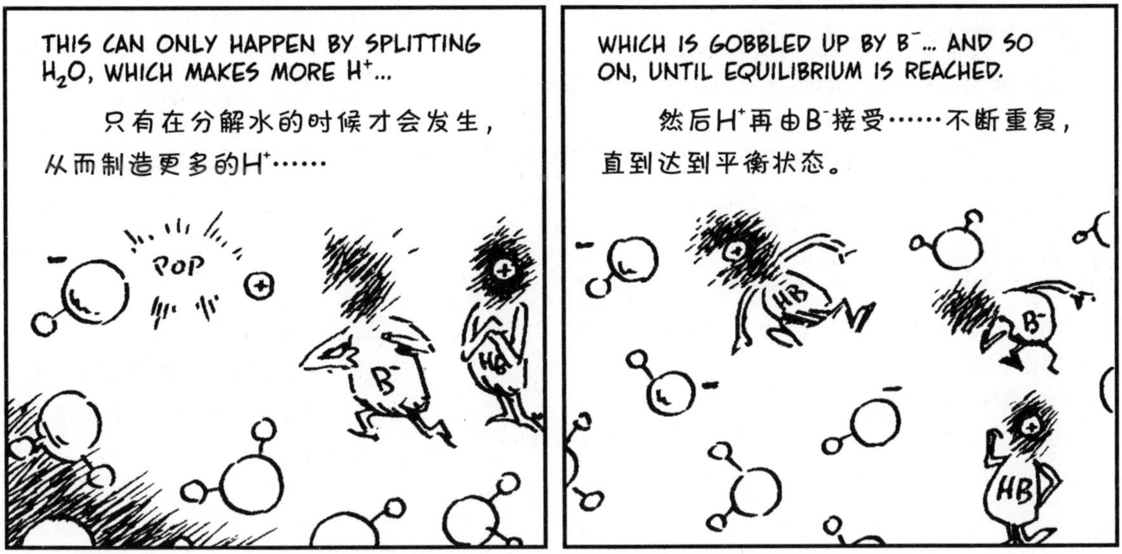

IN OTHER WORDS, B^- **HYDROLYZES** WATER AND CAUSES A RISE IN OH^-.

换句话说，B^-**分解**水造成OH^-增加，

$$H_2O + B^- \rightleftharpoons HB + OH^-$$

AND WE GET A NEW EQUILIBRIUM CONSTANT, THE **BASE IONIZATION CONSTANT** K_b.

$$K_b = \frac{[HB][OH^-]}{[B^-]}$$

而我们得到了一个新的平衡常数，即**碱的电离常数**K_b。

THE HIGHER THE K_b, THE STRONGER THE BASE. THIS IS BECAUSE:

- HIGHER K_b MEANS HIGHER $[OH^-]$, HENCE HIGHER pH.
- K_b MEASURES B^-'S ABILITY TO TAKE A PROTON FROM H_2O.
- K_b IS INVERSE TO K_a. IF HB IS THE CONJUGATE ACID, THEN

BASE B 碱B			K_b
OH^-	HYDROXIDE	氢氧化物	55.6
S^{2-}	SULFIDE	硫化物	10^5
CO_3^{2-}	CARBONATE	碳酸盐	2.0×10^{-4}
NH_3	AMMONIA	氨	1.8×10^{-5}
$B(OH)_4^-$	BORATE	硼酸盐	2.0×10^{-5}
HCO_3^-	BICARBONATE	碳酸氢盐	2.0×10^{-8}

越强的碱有越高的K_b。这是因为：

· K_b越高意味着$[OH^-]$更高，因此pH值也越高。

· K_b表示B^-从水夺取质子的能力。

· K_b是K_a的倒数。如果HB是共轭酸，那么

$$K_a K_b = \frac{[H^+][\cancel{B^-}]}{[\cancel{HB}]} \frac{[\cancel{HB}][OH^-]}{[\cancel{B^-}]} = [H^+][OH^-] = K_w = 10^{-14}$$

Example 范例

WHAT'S THE pH OF A 0.15 M SOLUTION OF AMMONIA, NH_3? CALCULATE AS BEFORE, USING THE REACTION

0.15M的氨（NH_3）溶液的pH值是多少？像以前一样计算，反应式为：

$$NH_3 + H_2O \rightleftharpoons NH_4^+ + OH^-$$

	NH_3	NH_4^+	OH^-
INITIAL CON. 初始浓度。	0.15	0.0	0.0
CHANGE IN CON. 浓度的变化	$-x$	x	x
EQUILIBRIUM CON. 平衡浓度	$0.15-x$	x	x

ASSUMPTION 1: OH^- FROM WATER IS NEGLIGIBLE.

假设1：水形成的OH^-忽略不计。

$$K_b = \frac{[NH_4^+][OH^-]}{[NH_3]} \qquad \frac{x^2}{(0.15-x)} = 1.8 \times 10^{-5}$$

$$\frac{x^2}{0.15} = 1.8 \times 10^{-5}$$

ASSUMPTION 2: x IS NEGLIGIBLE COMPARED TO 0.15.

假设2：x与0.15相比，可以忽略不计。

$$x^2 = 2.7 \times 10^{-6} \qquad x = 1.64 \times 10^{-3}$$

$$[OH^-] = 1.64 \times 10^{-3}$$

$$pOH = 3 - \log(1.64) = 2.78$$

$$pH = 14 - pOH = \mathbf{11.22}$$

Neutralization and Salts 中和及盐

IN WATER, ACIDS GENERATE H^+ AND BASES GENERATE OH^-. WHEN ACIDS AND BASES COMBINE, THESE IONS **NEUTRALIZE** EACH OTHER. FOR EXAMPLE:

在水中，酸产生H^+，碱产生OH^-。当酸和碱结合时，这些离子便相互**中和**。例如：

$$HCl(aq) + NaOH(aq) \longrightarrow Na^+(aq) + Cl^-(aq) + H_2O$$

TWO NASTY CHEMICALS COMBINE TO MAKE AN ORDINARY SOLUTION OF TABLE SALT IN WATER. IF THE WATER EVAPORATES, ONLY SALT CRYSTALS REMAIN.

两个化学物质相结合，产生一个常见的食盐的水溶液。如果水蒸发干，只有盐晶体会存在。

THIS IS TYPICAL, SO TYPICAL, IN FACT, THAT IT'S THE DEFINITION OF A SALT: A **SALT** IS A SUBSTANCE FORMED BY THE NEUTRALIZATION OF AN ACID BY A BASE.

再经典不过的要算盐的定义：**盐**是由酸碱中和形成的一种物质。

BY NEUTRALIZE, WE MEAN THAT THE SALTS ARE MADE FROM EQUIVALENT WEIGHTS OF ACID AND BASE.

中和，是说盐是用等量的酸碱制得的。

AN **EQUIVALENT WEIGHT** OF ACID IS THE AMOUNT THAT WOULD YIELD **ONE MOLE OF PROTONS** IN WATER IF THE ACID IONIZED COMPLETELY.

1 EQUIV HCl = 1 MOL

BUT

1 EQUIV H_2SO_4 = 0.5 MOL

BECAUSE H_2SO_4 CAN GIVE UP TWO PROTONS. SIMILARLY,

1 EQUIV H_2CO_3 = 0.5 MOL

1当量的酸是指酸在水中完全解离产生1摩尔质子所需的量，因此，

1当量 HCl = 1摩尔

但

1当量H_2SO_4 = 0.5摩尔

因为硫酸可以给出两个质子。同样，

1当量H_2CO_3 = 0.5摩尔

AN EQUIVALENT OF BASE IS THE AMOUNT THAT WOULD GIVE UP ONE MOLE OF OH^- IF THE BASE WERE TO IONIZE COMPLETELY. SO

1 EQUIV NaOH = 1 MOL

1 EQUIV $Ca(OH)_2$ = 0.5 MOL

1 EQUIV NH_3 = 1 MOL

BECAUSE

$$NH_3 + H_2O \rightarrow NH_4^+ + OH^-$$

IF IT WERE TO IONIZE COMPLETELY.

1当量的碱是指碱在水中完全解离产生1摩尔OH^-所需的量，因此，

1当量NaOH = 1摩尔

1当量$Ca(OH)_2$ =0.5摩尔

1当量 NH_3 = 1摩尔

这是由于

$$NH_3 + H_2O \rightarrow NH^+_4 + OH^-$$

如果它是完全电离的话。

N当量的酸总需要N当量的碱中和，因为它们各自产生的质子和氢氧根离子的数目相等。

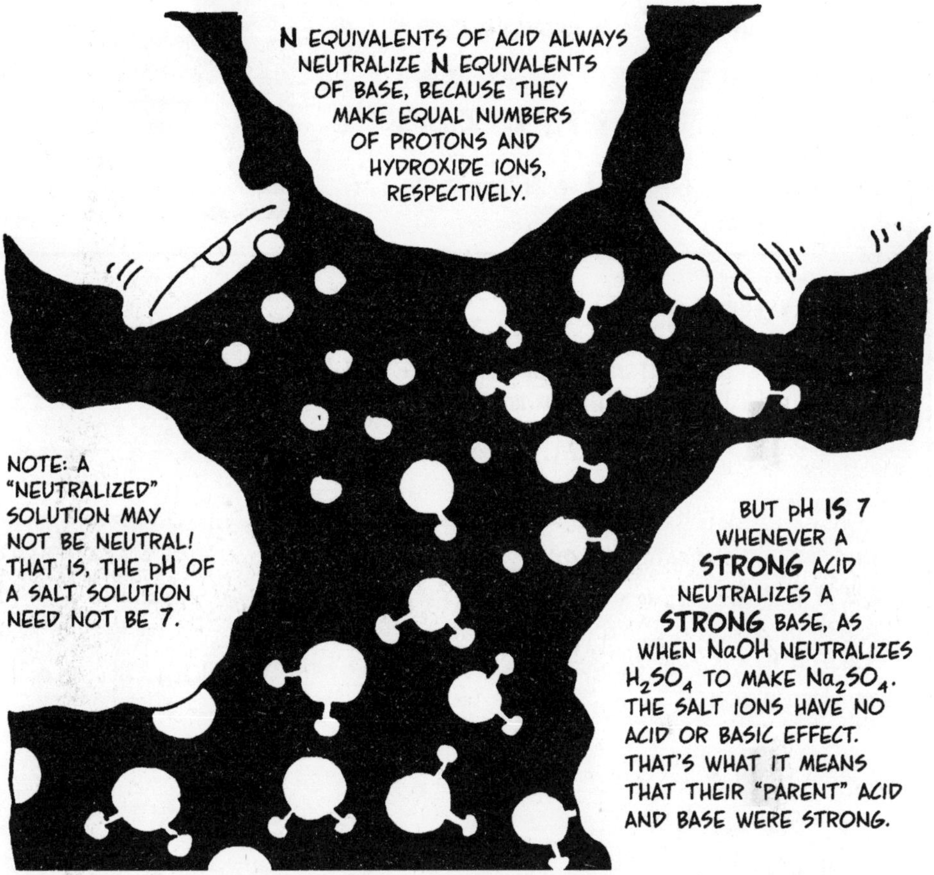

注意：一个“中和”的溶液可能不是中性！也就是说，盐溶液的pH值不一定为7。

但是，强酸与强碱中和时pH为7。就像氢氧化钠中和硫酸时产生硫酸钠。盐离子没有酸性或碱性的效果。这意味着它们的“父母”酸碱是强酸强碱。

WHEN A STRONG ACID NEUTRALIZES A WEAK BASE, THE SOLUTION WILL HAVE pH < 7. CONSIDER AMMONIUM NITRATE, NH_4NO_3, A COMMON INGREDIENT IN FERTILIZER. IT RESULTS FROM THE NEUTRALIZATION OF NH_3 (WEAK BASE) BY HNO_3 (STRONG ACID).

当强酸弱碱中和时。该溶液的pH值<7。例如硝酸铵（NH_4NO_3）——化肥中的常见成分，由NH_4（弱碱）和HNO_3（强酸）中和产生。

$$HNO_3(aq) + NH_3(aq) \longrightarrow NH_4^+(aq) + NO_3^-(aq)$$

NO_3^- HAS NO BASIC EFFECT (BECAUSE HNO_3 IS STRONG), SO WE CAN IGNORE IT. IT'S A "BYSTANDER ION." BUT NH_4^+ IS A WEAK ACID THAT WILL DISSOCIATE, WITH $K_a = 5.7 \times 10^{-10}$.

NO_3^-没有碱性效果（因为硝酸是强酸）。因此，我们可以忽略它。这是一个“旁观者离子”。但NH_4^+是一种弱酸，会产生解离，$K_a = 5.7 \times 10^{-10}$。

$$NH_4^+(aq) \rightleftharpoons NH_3(aq) + H^+(aq)$$

Example 范例

SUPPOSE THE CONCENTRATION OF NH_4NO_3 IS 0.1 M. WHAT IS THE SOLUTION'S pH? WE MAKE THE USUAL TABLE AND COMPUTATION:

假设硝酸铵浓度为0.1M，溶液的pH值是多少？我们像往常一样计算：

	NH_4^+	NH_3	H^+
CONC. BEFORE IONIZATION 电离前的浓度	0.1	0.0	0.0
CHANGE IN CONC. 浓度的变化	−x	x	x
EQUILIBRIUM CONC. 平衡浓度	0.1 − x	x	x

USUAL ASSUMPTION 1: H^+ FROM WATER IS NEGLIGIBLE.

常用假设1：水产生的H^+可以忽略不计。

AT EQUILIBRIUM, K_a IS

在平衡状态，K_a是

$$\frac{[H^+][NH_3]}{[NH_4^+]} = 5.7 \times 10^{-10}$$

MAKING THE USUAL TWO ASSUMPTIONS, WE GET

作完两个常用假设，我们得到

$$\frac{x^2}{0.1} = 5.7 \times 10^{-10}$$

$$x^2 = 5.7 \times 10^{-11} = 57 \times 10^{-12}$$

$$x = [H^+] = 7.55 \times 10^{-6}$$

$$pH = 6 - \log(7.55) = 6 - 0.88$$

$$= \mathbf{5.12}$$

常用假设2：x比0.1少很多，可以忽略不计

USUAL ASSUMPTION 2: X IS MUCH LESS THAN 0.1 AND CAN BE IGNORED.

SIMILARLY, WHEN A STRONG BASE NEUTRALIZES A WEAK ACID, THE RESULTING SALT SOLUTION WILL BE WEAKLY BASIC. FOR EXAMPLE, WHEN NaOH NEUTRALIZES CH_3CO_2H, Na^+ IS A "BYSTANDER ION," WHILE ACETATE, $CH_3CO_2^-$, IS A WEAK BASE. WORK OUT FOR YOURSELF THE pH OF A 0.5 M SOLUTION OF $NaCH_3CO_2$. USE K_b OF $CH_3CO_2^- = 5.7 \times 10^{-10}$.

同样，当一个强碱和弱酸中和时，由此产生的盐溶液会呈弱碱性。例如，当氢氧化钠与CH_3CO_2H中和时，钠离子是一个“旁观者离子”，而醋酸盐（$CH_3CO_2^-$）是一个弱碱。你可以自己算出0.5M $NaCH_3CO_2$溶液的pH值($CH_3CO_2^-$的$K_b = 5.7 \times 10^{-10}$）。

WE CAN SUMMARIZE THE pH OF SALT SOLUTIONS LIKE THIS:

我们可以总结出盐溶液pH值的如下规律：

IF SALT RESULTS FROM NEUTRALIZATION OF 如果盐从中和得到	pH
STRONG ACID, STRONG BASE 强酸，强碱	7
STRONG ACID, WEAK BASE 强酸，弱碱	<7
WEAK ACID, STRONG BASE 弱酸，强碱	>7
WEAK ACID, WEAK BASE 弱酸，弱碱	<7 IF $K_a > K_b$ 7 IF $K_a = K_b$ >7 IF $K_a < K_b$

Titration 滴定

IS THE PROCESS OF NEUTRALIZING AN UNKNOWN SOLUTION BY DRIPPING ("TITRATING") A KNOWN STRONG ACID OR BASE INTO IT.

滴定是把已知强酸或强碱逐滴加入（“滴定”）未知溶液的中和过程。

IF, FOR EXAMPLE, THE UNKNOWN STUFF IS ACIDIC, WE TITRATE IT WITH A STRONG BASE, NaOH, OF KNOWN CONCENTRATION, SAY 0.5 M.

例如，如果未知物是酸性的，我们滴加已知浓度的强碱，NaOH 0.5M。

pH SLOWLY RISES. AT THE **ENDPOINT,** WHEN THE ACID IS NEUTRALIZED, pH RISES RAPIDLY, SIGNALED BY A CHANGE IN COLOR OF AN INDICATOR CHEMICAL.

pH值缓慢上升。在**终点**，当酸被中和时，pH值急剧上升，这将由化学指标剂的颜色变化显示出来。

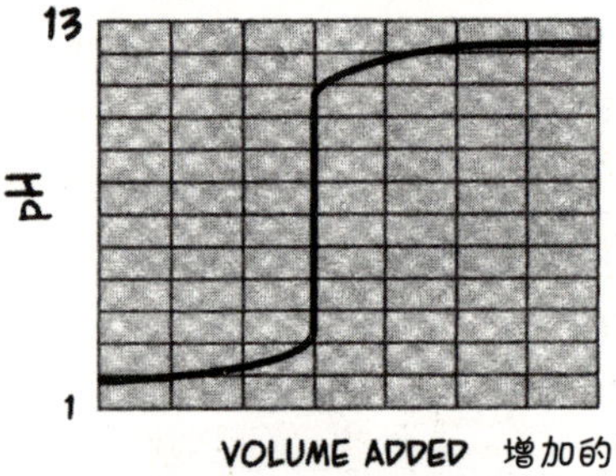

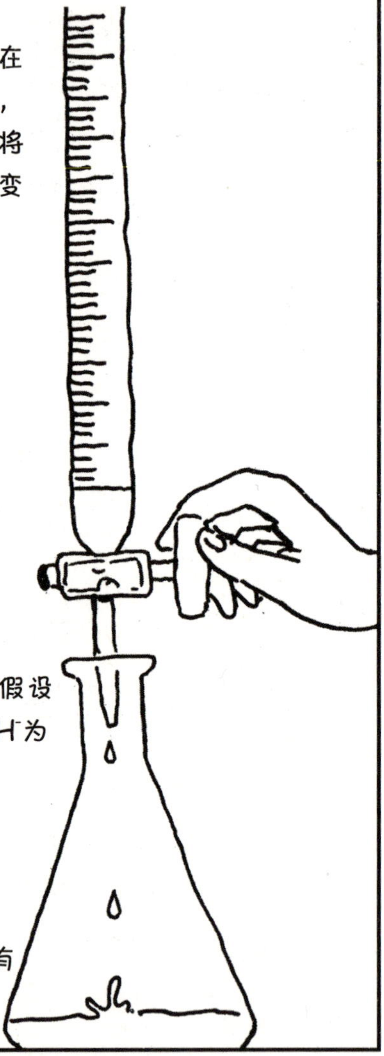

NOW WE CAN FIND HOW MANY EQUIVALENTS WERE IN THE ORIGINAL SOLUTION. SUPPOSE 50 mL OF UNKNOWN NEUTRALIZED 9.3 mL OF NaOH. THEN OH^- CONSUMED WAS

现在，我们可以在原溶液中找到有多少当量的酸。假设50毫升未知溶液用9.3毫升氢氧化钠中和。那么消耗的OH^-为

$$(.0093\ L)(0.5\ mol/L) = 0.0047\ mol.$$

THERE MUST HAVE BEEN 0.0047 EQUIVALENTS OF ACID IN 50 mL OF UNKNOWN, OR 0.094 EQUIVALENTS (.0047 × 1000/50) IN A LITER.

50毫升未知溶液中有0.0047当量的酸，或每升有0.094当量(0.0047 × 1000/50)。

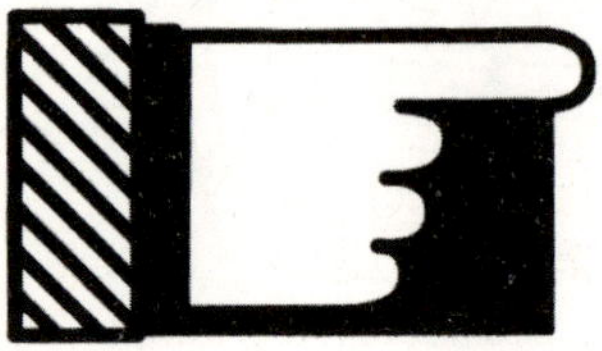

CAUTION: THE pH NEED NOT BE 7 AT THE ENDPOINT! THE TITRATION MAY END WITH A SALT THAT HAS ACIDIC OR BASIC PROPERTIES.

注意：滴定终点的pH值不必是7！滴定结束时可能会形成酸性或碱性的盐。

WHEN SEVERAL IONS GET TOGETHER IN SOLUTION, INTERESTING THINGS HAPPEN...

当几种离子一起出现在溶液中时，有趣的事情将会发生……

Solubility products
溶度积

SOME SALTS ARE VERY SOLUBLE, SOME HARDLY AT ALL. WHEN A SALT SOLUTION REACHES ITS MAXIMUM POSSIBLE CONCENTRATION, WE SAY IT IS **SATURATED.** ANY ADDED SALT JUST FALLS TO THE BOTTOM.

有的盐类易溶，有的几乎根本不溶。当盐溶液浓度达到最大时，我们说它已**饱和**。这时无论添加多少盐都只会落到溶液底部。

SALTS DISSOLVE IN WATER BY IONIZING:

盐通过电离溶于水：

$$H_2O + A_nB_m(s) \rightleftharpoons nA^{m+}(aq) + mB^{n-}(aq)$$

(HERE **A,** THE CATION, HAS OXIDATION NUMBER +m AND **B,** THE ANION, HAS OXIDATION NUMBER −n.) IONS ARE GOING INTO SOLUTION AND FALLING OUT. AT LOW CONCENTRATION, THE FORWARD REACTION DOMINATES. SATURATION IS THE EQUILIBRIUM STATE.

（这里A，阳离子，具有氧化数+*m*；还有B，阴离子，具有氧化数-*n*）。离子正在进入溶液即电离。在低浓度时，正反应占主导地位。饱和时是平衡状态。

HERE IS THE EQUILIBRIUM CONSTANT.

这里是平衡常数。

$$K_{eq} = \frac{[A^{m+}]^n[B^{n-}]^m}{[H_2O][A_nB_m]}$$

THE DENOMINATOR CONTAINS WATER AND THE UNDISSOLVED SALT—BOTH ESSENTIALLY CONSTANT. SO WE IGNORE THEM AS USUAL AND DEFINE K_{sp}, THE **SOLUBILITY PRODUCT:**

分母含有水和未溶解的盐，它们基本都是常数。因此，我们像往常一样忽略它们并对K_{sp}定义，溶度积为：

$$K_{sp} = [A^{m+}]^n[B^{n-}]^m$$

FOR EXAMPLE, A SATURATED SOLUTION OF $CaCO_3$ HAS A CALCIUM CONCENTRATION OF 6.76×10^{-5}M. POSITIVE AND NEGATIVE CHARGES HAVE TO BALANCE, SO THE CARBONATE CONCENTRATION IS ALSO 6.76×10^{-5}M. THEN:

例如，碳酸钙饱和溶液的钙浓度为6.76×10^{-5}M，正电荷和负电荷处于平衡，所以，碳酸根的浓度也是6.76×10^{-5}M，然后

$$\begin{aligned} K_{sp} &= [Ca^{2+}][CO_3^{2-}] \\ &= (6.76 \times 10^{-5})^2 \\ &= 4.57 \times 10^{-9}. \end{aligned}$$

BECAUSE $CaCO_3$ IS SO INSOLUBLE, WE CAN USE Ca^{2+} IONS TO PRECIPITATE DISSOLVED CO_3^{2-} FROM SOLUTION. FOR INSTANCE, WHEN WE MAKE CAUSTIC LYE, NaOH:

由于碳酸钙溶解度很差，我们可以使用钙离子从溶液中沉淀溶解的CO_3^{2-}。举例来说，当我们配出苛性碱液，NaOH：

$$Ca(OH)_2(aq) + Na_2CO_3(aq) \longrightarrow 2NaOH + CaCO_3(s)\downarrow$$

Ca^{2+} AND CO_3^{2-} WILL NOT STAY IN SOLUTION TOGETHER BEYOND WHAT THEIR SOLUBILITY PRODUCT ALLOWS. AS SOON AS THE ADDED Ca^{2+} REACHES A LEVEL THAT MAKES

Ca^{2+}和CO_3^{2-}将不会一起在溶液里超出其溶度积的允许程度。只要添加钙离子达到一定水平，就使得

$$[Ca^{2+}][CO_3^{2-}] = 4.57 \times 10^{-9},$$

CALCIUM CARBONATE BEGINS TO PRECIPITATE OUT.

碳酸钙开始沉淀了。

SOLID 固体的	K_{sp}	SOLID 固体的	K_{sp}
$FePO_4$	1.26×10^{-18}	$BaSO_4$	10^{-10}
$Fe_3(PO_4)_2$	10^{-33}	$PbCl_2$	1.6×10^{-5}
$Fe(OH)_2$	3.26×10^{-15}	$Pb(OH)_2$	5.0×10^{-15}
FeS	5.0×10^{-18}	$PbSO_4$	1.6×10^{-8}
Fe_2S_3	10^{-88}	PbS	10^{-27}
$Al(OH)_3$ (AMORPH) (无定形)	10^{-33}	$MgNH_4PO_4$	2.6×10^{-13}
$AlPO_4$	10^{-21}	$MgCO_3$	10^{-5}
$CaCO_3$ (CALCITE) (方解石)	4.6×10^{-9}	$Mg(OH)_2$	1.82×10^{-11}
$CaCO_3$ (ARAGONITE) (文石)	6.0×10^{-9}	$Mn(OH)_2$	1.6×10^{-13}
$CaMg(CO_3)_2$	2.0×10^{-17}	$AgCl$	10^{-10}
CaF_2	5.0×10^{-11}	Ag_2CrO_4	2.6×10^{-12}
$Ca(OH)_2$	5.0×10^{-6}	Ag_2SO_4	1.6×10^{-5}
$Ca_3(PO_4)_2$	10^{-26}	$Zn(OH)_2$	6.3×10^{-18}
$CaSO_4$ (GYPSUM) (石膏)	2.6×10^{-5}	ZnS	3.26×10^{-22}

K_{sp} CAN HELP US FIND THE EFFECT OF ONE ION ON ANOTHER'S SOLUBILITY. FOR INSTANCE,

溶度积可以帮助我们找到一个离子对另一个离子溶解度的影响。例如：

pH affects solubility.
pH影响溶解度

Example 1. 范例1：

$$Ca(OH)_2 \rightleftharpoons Ca^{2+} + 2OH^-$$

$$K_{sp} = [Ca^{2+}][OH^-]^2 = 5.0 \times 10^{-6}$$

TAKE THE LOGARITHM OF BOTH SIDES:

两边取对数：

$$\log[Ca^{2+}] + 2\log[OH^-] = (\log 5) - 6$$

$$= 0.7 - 6 = -5.3$$

$$\log[Ca^{2+}] - 2pOH = -5.3$$

SUBSTITUTING 代入 $pOH = 14 - pH$,

$$\log[Ca^{2+}] = 22.7 - 2pH$$

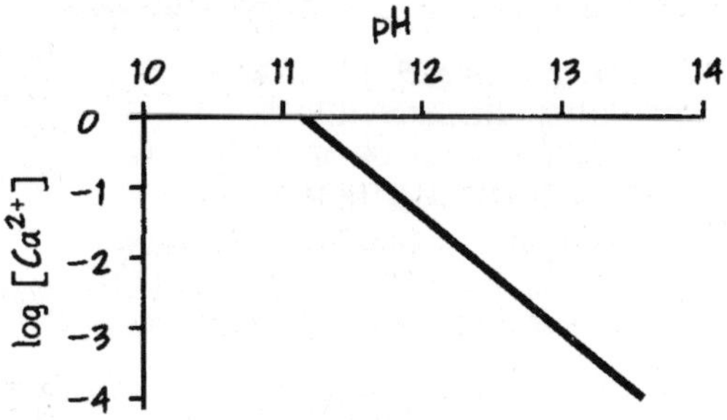

$Ca(OH)_2$ BECOMES HIGHLY SOLUBLE AT pH BELOW 12.

$Ca(OH)_2$在pH值低于12时，溶解度升高

Example 2. 范例2：

$$CaCO_3 \rightleftharpoons Ca^{2+} + CO_3^{2-}$$

WHEN ACID IS ADDED, CO_3^{2-} TAKES UP H^+ TO MAKE HCO_3^- AND H_2CO_3. HAVING THESE TWO DIFFERENT PRODUCTS COMPLICATES THE MATH, BUT ON BALANCE, THE SITUATION IS DOMINATED BY:

$$H^+ + CO_3^{2-} \rightleftharpoons HCO_3$$

加入酸时，CO_3^{2-}得到H^+，成为HCO_3^-和H_2CO_3。这两种不同的产物使计算更复杂，但总体来说，主要因素是：$H^+ + CO_3^{2-} \rightleftharpoons HCO_3$

BY LE CHATELIER'S PRINCIPLE, ADDING H^+ DRIVES THIS EQUATION TO THE RIGHT AND REMOVES CO_3^{2-}. TO MAINTAIN K_{SP}, MORE $CaCO_3$ WILL DISSOLVE.

由勒夏特列原理得知，加入H^+使方程向右反应并去除CO_3^{2-}，为了维持K_{SP}，更多的碳酸钙会溶解。

BOTH EXAMPLES SHOW HOW LOW-pH WATER TENDS TO DISSOLVE MORE Ca^{2+}. THIS IS A GENERAL PATTERN FOR METALS AND EXPLAINS WHY ACIDIFIED LAKES OFTEN HAVE HIGH LEVELS OF DISSOLVED TOXIC METALS.

这两个例子说明pH值低的水怎样更容易溶解Ca^{2+}。这是一个对金属盐的一般模式，并解释为什么酸化湖泊通常会溶解高浓度的有毒金属离子。

Buffers 缓冲溶液

WE CAN USE BASES' PROTON-CAPTURING PROCLIVITIES TO MODERATE THE pH DROP CAUSED BY STRONG ACIDS.

我们可以使用碱的质子捕获能力，以缓和由强酸导致的pH值下降。

FOR EXAMPLE, START WITH A LITER OF .01 M SOLUTION OF SODIUM ACETATE, $NaCH_3CO_2$. THIS IONIZES TO GENERATE .01 mol OF THE WEAK BASE ACETATE, $CH_3CO_2^-$, CONJUGATE TO ACETIC ACID.

例如，0.01M的醋酸钠$NaCH_3CO_2$溶液，电离产生0.01摩尔的弱碱醋酸根（$CH_3CO_2^-$），醋酸的共轭碱。

ADD A LITER OF .01 M HCl, A STRONG ACID. THE ACETATE ION GRABS NEARLY ALL THE PROTONS GIVEN UP BY HCl:

添加1L的0.01M的盐酸，一种强酸。醋酸根离子便几乎捕获了所有盐酸给出的质子：

$$CH_3CO_2^- + H^+ \longrightarrow CH_3CO_2H$$

THE pH OF THE SOLUTION IS THAT OF A .005 M SOLUTION OF ACETIC ACID. (CONCENTRATION IS HALVED BECAUSE WE NOW HAVE TWO LITERS OF LIQUID!) THAT'S pH = **3.53.**

该溶液的pH是一个0.005 M的醋酸溶液的pH。（浓度减半，因为我们现在有两升的液体！）。pH = 3.53。

IF WE HAD ADDED THE HCl TO PURE WATER INSTEAD, THE pH WOULD HAVE DROPPED TO **2.3.** THE ACETATE **MODERATED THE ACIDITY OF THE WATER.**

如果我们用纯水代替，向其中加盐酸，pH值将下降到**2.3**。醋酸中和了水的酸度。

WE SAY THAT THE ACETATE **BUFFERS** THE SOLUTION AGAINST ACIDS.

我们说醋酸**缓冲**了酸对溶液的影响。

WE MAY BE BOTHERED BY THE FACT THAT OUR BUFFER SOLUTION IS MODERATELY ALKALINE, WITH A pH = 8.38.

我们可能感到困扰的是，我们的缓冲溶液是中等碱性，pH = 8.38。

WE COULD LOWER THIS WITH A WEAK ACID, BUT WE DON'T WANT TO GIVE ANY PROTONS TO THE ACETATE IONS. THIS WOULD CUT THEIR BUFFERING ABILITY.

我们可以用弱酸降低它，但我们不希望给任何醋酸离子质子。这将减少它们的缓冲能力。

SO WE **BRILLIANTLY** USE **ACETIC** ACID, CH_3CO_2H. ITS CONJUGATE BASE IS ALREADY ACETATE, SO IT WON'T GIVE UP PROTONS TO THE FREE ACETATE IN SOLUTION.

因此，我们**巧妙**地使用**醋酸**，CH_3CO_2H。其共轭碱是醋酸根，所以它不会把质子让给溶液中的自由醋酸根。

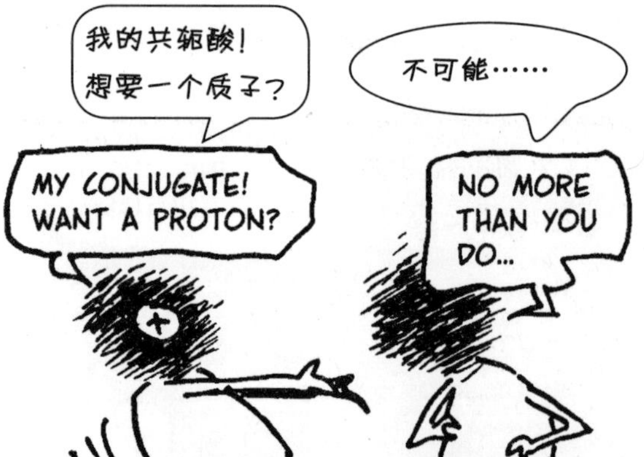

IF WE MAKE A SOLUTION *0.01* M IN ACETATE AND JUST *0.002* M IN ACETIC ACID, THE pH WILL BE 5.5, NOT TOO BAD. (THE CALCULATION IS ON THE PAGE AFTER NEXT.)

如果我们配一个含0.01M的醋酸根和只有0.002M的醋酸溶液，pH为5.5，还不算太坏（计算过程在下两页）。

EVEN BETTER, WE HAVE BUFFERED AGAINST **ACIDS AND BASES SIMULTANEOUSLY!** THE ACETIC ACID WILL GIVE UP ITS H TO A STRONG BASE, WHILE THE ACETATE WILL TAKE PROTONS FROM STRONG ACIDS. pH WILL BE HELD WITHIN A LIMITED RANGE.

更妙的是，我们可以同时**缓冲酸和碱！**醋酸将放弃一个质子给强碱，而醋酸根则会得到强酸的质子，pH值会被控制在有限的范围内。

THIS IS THE TRICK WITH BUFFERS: USE AN ACID AND BASE WITH A **COMMON ION:** COMBINE A WEAK ACID **HB** WITH A SALT THAT IONIZES TO GIVE FREE **B^-**.

这是缓冲溶液的小技巧：使用具有**共同离子**的酸和碱：弱酸**HB**和它的盐混合，电离出自由弱酸根**B^-**。

A BIT OF ARITHMETIC LETS US PREDICT THE pH OF BUFFERS, BOTH BEFORE AND AFTER ADDITION OF ACIDS OR BASES. WE START WITH THE WEAK ACID HB.

简单计算让我们能够预测加入酸或碱时，溶液中的pH值的前后变化。我们从弱酸HB开始。

BY DEFINITION,
根据定义

$$K_a = \frac{[H^+][B^-]}{[HB]}$$

SO 因此，

$$\frac{K_a}{[H^+]} = \frac{[B^-]}{[HB]}$$

TAKING LOG OF BOTH SIDES,

两边取对数，

$$\log K_a - \log [H^+] = \log ([B^-]/[HB])$$

WRITING pK_a FOR $-\log K_a$, THIS BECOMES

pK_a 写作 $-\log K_a$，得出

$$\mathbf{pH - pK_a = \log ([B^-]/[HB])}$$

WHICH IS CALLED THE

Henderson-Hasselbalch Equation.

这就是**亨德森-哈塞尔巴尔赫方程**（Henderson-Hasselbalch equation）。

IN OUR BUFFER SOLUTION, THE SALT CONCENTRATION GIVES $[B^-]$, AND THE CONCENTRATION OF ACID GIVES [HB]. K_a WE KNOW, SO WE CAN SOLVE FOR pH.

在我们的缓冲溶液中，盐的浓度是$[B^-]$，酸的浓度是[HB]。已知K_a，那么我们可以计算出pH。

FOR EXAMPLE, SUPPOSE A BUFFER SOLUTION CONSISTS OF 1 L OF 0.5 M $NaCH_3CO_2$ AND 0.1 M CH_3CO_2H. K_a OF ACETIC ACID IS 1.75×10^{-5}, SO

例如，假设一个缓冲溶液由1升0.5M $NaCH_3CO_2H$和0.1M CH_3CO_2H组成。醋酸的K_a为1.75×10^{-5}，所以

$$pK_a = -\log(1.75 \times 10^{-5})$$
$$= 4.76$$

THEN BY HENDERSON-HASSELBALCH, THE pH OF THE SOLUTION IS

由亨德森-哈塞尔巴尔赫方程得出溶液的pH为

$$pH = pK_a + \log([B^-]/[HB])$$
$$= 4.76 + \log(0.5/0.1)$$
$$= 4.76 + \log 5$$
$$= 4.76 + 0.70 = \mathbf{5.46}$$

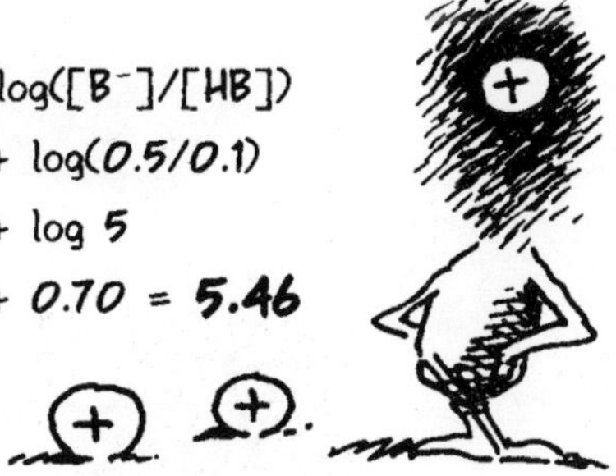

IF A LITER OF 0.05 M HCl IS ADDED, WE ASSUME THAT THE $CH_3CO_2^-$ BINDS WITH ESSENTIALLY ALL THE H^+ FROM HCl:

如果加入1升0.05 M的盐酸。我们假设$CH_3CO_2^-$基本上与所有盐酸给出的H^+结合：

$$CH_3CO_2^- + H^+ \longrightarrow CH_3CO_2H$$

THEN WE MAKE THE USUAL TABLE:
然后，我们像之前一样列一个表：

	CH_3CO_2H	$CH_3CO_2^-$	H^+
ORIG. CON. 原始浓度	0.05	0.25	0.025
CON. CHANGE 浓度变化	0.025	-0.025	-0.025
EQUILIB. CON. 平衡浓度	0.075	0.225	0.0

NOTE THAT CONCENTRATIONS ARE HALVED, BECAUSE WE NOW HAVE **TWO LITERS** OF SOLUTION. THEN HENDERSON-HASSELBALCH SAYS:

注意浓度减半。因为我们现在有**两升**溶液。然后亨德森-哈塞尔巴尔赫提出：

$$pH = pK_a + \log \frac{[CH_3CO_2^-]}{[CH_3CO_2H]}$$
$$= 4.76 + \log(0.225/0.075)$$
$$= 4.76 + \log 3 = 4.76 + 0.48$$
$$= \mathbf{5.24}$$

SEE IF YOU CAN DO THE SAME CALCULATION IF WE HAD ADDED A LITER OF 0.04 M NaOH INSTEAD OF THE HCl.

如果我们加入1升0.04M 的NaOH溶液代替盐酸，看看你能不能做同样的计算。

HENDERSON-HASSELBALCH CAN ALSO GUIDE US WHEN WE WANT TO ADJUST THE pH OF A SYSTEM.

亨德森-哈塞尔巴尔赫还可以指导我们如何调整一个体系的pH。

FOR EXAMPLE, NH_4^+ IS MUCH LESS POISONOUS TO FISH THAN NH_3 BECAUSE THE UNCHARGED MOLECULE CAN PASS THROUGH CELL MEMBRANES EASILY AND INTERFERE WITH METABOLISM. HENDERSON-HASSELBALCH SAYS

例如，NH_4^+对鱼的毒性小于NH_3，因为不带电荷的分子能够轻易通过细胞膜，干扰新陈代谢。亨德森-哈塞尔巴尔赫提出：

$$\log([NH_3]/[NH_4^+]) = pH - pK_a$$

IF, FOR EXAMPLE, WE WANT TO MAKE $[NH_3]/[NH_4^+]$ LESS THAN ONE IN A THOUSAND, I.E., ITS LOG < −3, THEN pH MUST BE LOW ENOUGH THAT

如果，我们想使$[NH_3]/[NH_4^+]$低于千分之一，即它的对数<-3，那么pH值必须足够低，

$$pH - pK_a < -3$$

SINCE pK_a OF NH_4^+ IS 9.3, ANY pH < 6.3 WILL DO.

因为NH_4^+的pK_a为9.3，所以只要pH<6.3即可。

SIMILARLY, WE ADD HOCl TO SWIMMING POOLS TO KILL BACTERIA. THIS MILD ACID PARTLY DISSOCIATES INTO H^+ AND OCl^-. BUT NOW WE **DO** WANT IT TO BE POISONOUS, TO KILL BACTERIA! AGAIN THE NONIONIZED SPECIES HOCl IS THE POISONOUS ONE, SO WE ADJUST POOL pH TO LOWER $[OCl^-]/[HOCl]$.

同样，我们在游泳池中加入次氯酸可以杀死细菌。这种弱酸解离成H^+和OCl^-，但现在我们想让它有杀毒灭菌的效果！而非离子化合物次氯酸是有毒的，因而，我们调整游泳池中的pH值低于$[OCl^-]/[HOCl]$。

WE COVERED A LOT IN THIS CHAPTER. WE MET ACIDS AND BASES, MEASURED THEIR STRENGTH, AND SAW HOW THAT STRENGTH IS RELATED TO THEIR IONIZATION IN WATER. WE NEUTRALIZED, TITRATED, AND LOOKED AT THE RESULTING SALTS. WE SAW HOW ACIDS AND BASES AFFECT A SALT'S SOLUBILITY, AND HOW BUFFERS ARE MADE BY COMBINING WEAK ACIDS AND SALTS.

在这一章中我们讨论了很多。我们了解了酸和碱以及如何测量它们的强度，我们最终知道这种强度与其在水中的电离有关。我们进行了中和与滴定，看到结果生成了盐。我们还知道了酸和碱如何影响盐的溶解度以及如何结合弱酸和盐配制缓冲溶液。

AND NOW FOR SOMETHING COMPLETELY DIFFERENT...

下面，让我们再讨论点完全不同的东西……

Chapter 10
Chemical Thermodynamics

第十章
化学热力学

THE REASSURING THEME OF THIS CHAPTER IS: THE UNIVERSE GETS **LESS IMPROBABLE** ALL THE TIME.

本章的主题是：宇宙每时每刻都存在着**奇迹**。

FOR EXAMPLE, A BRICK FLIES THROUGH A WINDOW, AND THE GLASS SHATTERS AND GOES FLYING.

例如，用一块砖头砸中一扇窗户，那么，玻璃就会碎成一片一片的。

OR: SOME AIR IS LET INTO A VACUUM CHAMBER AND QUICKLY FILLS UP THE SPACE.

或者，空气进入一个真空箱后会很快充满整个空间。

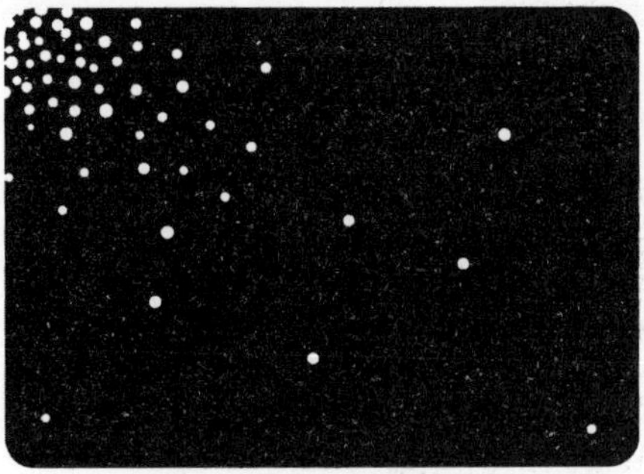

THE REASON IS THE SAME IN BOTH CASES: THERE ARE MANY, MANY, **MANY** MORE WAYS FOR THINGS TO **FLY APART** OR **SPREAD OUT** THAN THERE ARE FOR THEM TO FLY TOGETHER AND GET CONCENTRATED. SPREADING OUT IS VASTLY MORE PROBABLE. IT'S A GENERAL PRINCIPLE OF THE UNIVERSE:

这些事情发生的原因是相同的。物质**飞散**或**扩散**的途径比物质凝聚或集中的途径要多许多。其中扩散法是最有可能的一种。这是宇宙中一种普遍的原则。

YOU NEVER SEE A BRICK HIT A PUDDLE OF GLASS FRAGMENTS AND CAUSE THEM TO FLY UP TO MAKE A WINDOW!

而你永远也不可能看到用砖头去击中玻璃碎片后，这些玻璃碎片会变成一块完整的窗户！

YOU NEVER SEE ALL THE AIR IN A ROOM FLY INTO THE CORNER. (OR IF YOU DO, YOU DON'T LIVE TO TELL THE TALE.)

你也不可能看到在屋子里的空气会自己集中到一个角落里。(如果这件事可能的话，我们就不能活着讲故事了。)

Spontaneous processes tend to spread things out.

自发过程是物质扩散的过程.

YOU MAY OBJECT THAT PICKING UP A BROOM AND SWEEPING THE GLASS SPLINTERS TOGETHER IS A CONCENTRATING PROCESS. AND YOU'D BE RIGHT.

你可能会反对我的观点，并认为拿起扫帚和将玻璃碎片扫在一起是一个集中过程。当然，这样说来你是对的。

BUT I REPLY THAT IN ORDER TO SWEEP, I HAVE TO MOVE MY BODY. MOVING INVOLVES CHEMICAL REACTIONS THAT SPREAD HEAT INTO THE ENVIRONMENT.

但是我为了打扫不得不移动我的身体。而移动身体涉及化学反应，放出的热量会扩散到环境中。

IN FACT, I COULDN'T HAVE MOVED IN THE FIRST PLACE WITHOUT EATING, AND EATING GENERATES WASTE THAT GETS SPREAD AROUND TOO.

事实上，我不可能不吃东西就去运动，而吃东西必然会产生废物，废物也扩散到环境中。

THE FOOD I EAT ULTIMATELY DEPENDS ON SOLAR ENERGY, WHICH SPREADS A TERRIFIC AMOUNT OF MATTER AND ENERGY INTO THE UNIVERSE.

我吃的食物最终源于太阳能，太阳将大量的物质和能量传播到宇宙中.

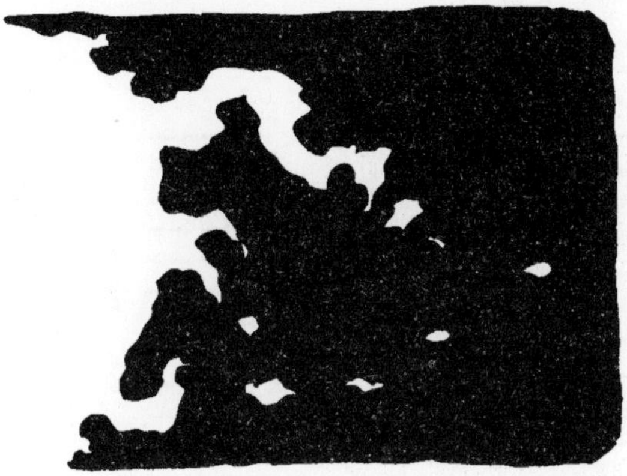

YOU HAVE TO LOOK AT THE BIG PICTURE! ANY PROCESS THAT CONCENTRATES MATTER AND/OR ENERGY IN A SYSTEM IS **MORE THAN OFFSET** BY A GREATER AMOUNT OF SPREADING-OUT ELSEWHERE IN THE UNIVERSE. **THE OVERALL EFFECT IN THE UNIVERSE AS A WHOLE IS TO SPREAD THINGS OUT.**

你不得不看到一幅非常巨大的图景！在宇宙中,一个体系内的任何集中物质或者能量的过程都**远远比不上**宇宙中扩散的过程。作为一个整体，**宇宙中的总效应是向外扩散的。**

IN CHEMICAL SYSTEMS WE CONSIDER THE SPREADING-OUT OF **ENERGY.**

在化学体系中我们考虑的是**能量**的扩散.

IMAGINE A SYSTEM CONSISTING OF SOME TYPICALLY HUGE NUMBER OF MOLECULES, AND LET US CONCENTRATE, FOR THE MOMENT, ON ONE OF THEM.

想象一个由典型的大量分子组成的体系，接着我们暂时把精力集中在一个分子上。

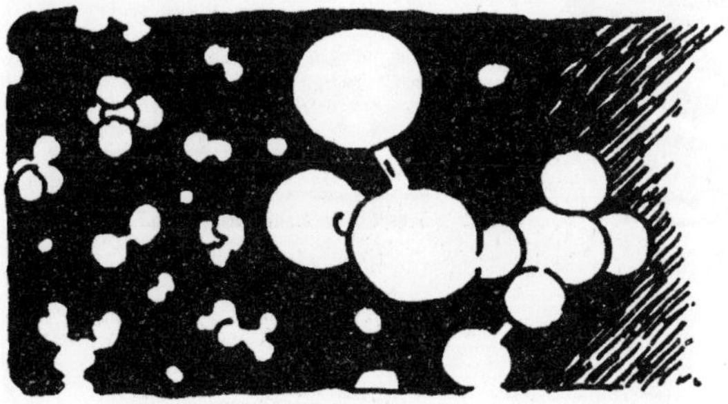

KINETIC ENERGY IS STORED IN A MOLECULE IN THE FORM OF VIBRATION, ROTATION, AND TRANSLATION (I.E., FLYING THROUGH SPACE).

动能以振动、旋转和平动（例如空间飞行）的方式存在于分子之中。

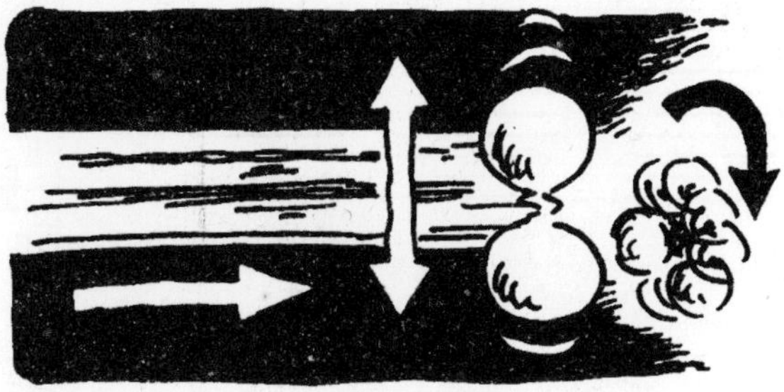

AS WE SAW IN CHAPTER 2, AT THIS SCALE ENERGY IS **QUANTIZED.** ONLY CERTAIN FIXED ENERGY LEVELS ARE ALLOWED.

正如我们在第二章中看到的，这种规模的能量是**量子化**的。只有某些固定能级被允许。

ENERGY IS TAKEN ON OR GIVEN OFF IN PACKETS CALLED QUANTA THAT JUMP THE MOLECULE FROM ONE ENERGY LEVEL TO ANOTHER.

能量以量子的形式被吸收或者被释放，就是这个能量使分子从一个能级跃迁到另一个能级。

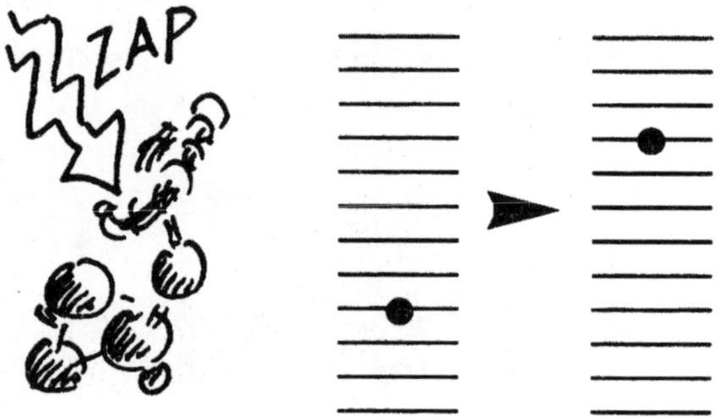

SO THIS IS THE PICTURE: EACH MOLECULE HAS ITS OWN ENERGY LEVELS... AND WE THINK OF THE WHOLE SYSTEM AS ALL THESE ENERGY LEVELS TAKEN TOGETHER, WITH A VAST NUMBER OF QUANTA SPREAD OUT AMONG THEM IN SOME WAY.

所以出现了下面这幅图：每一个分子都有自己的能级……并且我们认为整个系统就是由这些能级整合在一起的，在它们中存在着大量的量子以某种方式进行扩散。

Entropy, S, 熵

MEASURES THE SPREADING OUT OF ENERGY. IT CAN BE DEFINED IN TERMS OF HEAT AND TEMPERATURE:

衡量能量的扩散，可以用热量和温度来定义：

START WITH A SYSTEM AT TEMPERATURE T (MEASURED IN °K) AND ADD A SMALL AMOUNT OF HEAT q.*

系统的初始温度为T（单位：°K），给系统增加一小部分热量q*。

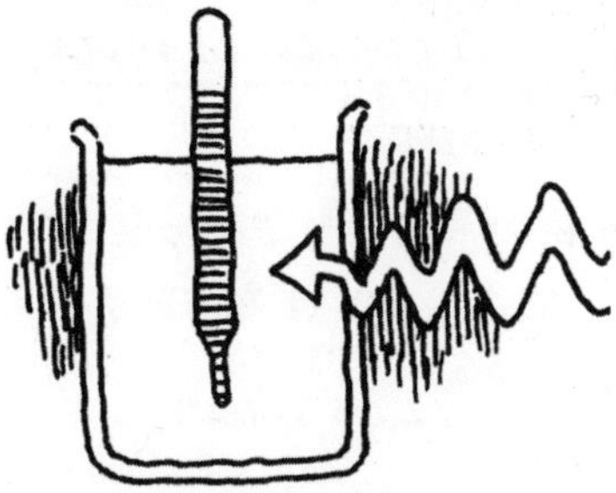

THE **ENTROPY CHANGE** ΔS, IS GIVEN BY

熵的改变量是ΔS定义为：

$$\Delta S = q/T$$

WITH UNITS JOULES/°K.

单位为焦耳/开尔文。

AS THE FOLLOWING DIAGRAMS SUGGEST, ΔS MEASURES THE **EXTRA SPREADING-OUT OF HEAT** IN THE SYSTEM RESULTING FROM THE ADDITION OF q.

正如下文表格显示的，ΔS是用于测量系统由于增加了热量q，而导致的**额外热量的散发**。

SOMETIMES, q CAUSES A SMALL TEMPERATURE INCREASE ΔT. (q = CΔT, WHERE C IS THE SYSTEM'S HEAT CAPACITY.) THE HEAT SPREADS INTO HIGHER ENERGY LEVELS.

有时，热量q引起温度的一部分上升ΔT。（q=cΔT，其中c是系统的比热）。热量以更高的能级扩散。

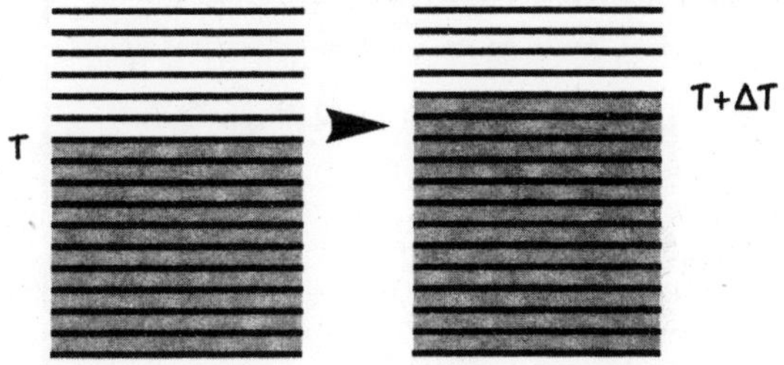

AT OTHER TIMES, q PRODUCES PHASE CHANGE (MELTING, VAPORIZATION). THEN TEMPERATURE REMAINS CONSTANT, BUT MOLECULAR MOTION BECOMES LESS CONSTRAINED, AND MORE LOW-ENERGY LEVELS "OPEN UP." THE HEAT SPREADS INTO THESE ENERGY LEVELS.

也有时候，q发生阶段性的变化（融化、蒸发）。接着温度保持不变，但分子的运动变得相对自由，更多的更低能级被“打开”了。热量就以这些能级扩散。

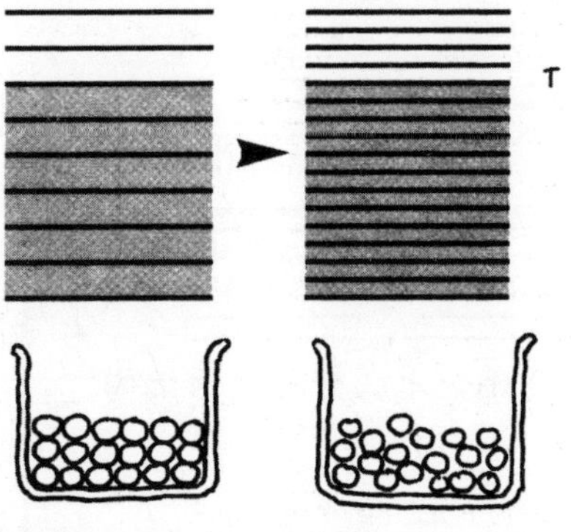

*PHYSICISTS TELL US THAT q MUST BE ADDED **REVERSIBLY**, THAT IS, THE HEAT CAN BE SENT BACK WITHOUT ANY EXTRA EXPENSE OF ENERGY. THIS IS PHYSICALLY IMPOSSIBLE, BUT CAN BE APPROXIMATELY ACHIEVED BY ADDING HEAT IN MANY SMALL STEPS.

* 物理学家告诉我们，q必须可逆增加，也就是说，不需要额外的能量消耗，热量就可以返回传播。这在物理学上是不可能的，但如果逐步一点点地增加热量，却是可以近似做到的。

IT IS NOW POSSIBLE TO CALCULATE THE **ABSOLUTE ENTROPY** OF ANY SUBSTANCE. THIS IS DONE BY ADDING UP ALL THE LITTLE ENTROPY INCREMENTS THAT PILE UP AS THE SUBSTANCE IS HEATED IN SMALL STEPS FROM ABSOLUTE ZERO TO SOME CONVENIENT TEMPERATURE, USUALLY 298°K (ROOM TEMPERATURE, 25°C).

现在计算出任何物质的**绝对熵**是可能的。可以这样做到：把物质逐渐地从绝对零度加热到一个合适的温度，通常是298°K（室温，25°C），熵的增量一点点积累。

AT 298°K, WE WRITE S^o, THE **STANDARD** ABSOLUTE ENTROPY.

298°K时，标准的绝对熵记作S^o

FOR EXAMPLE, FINDING THE STANDARD ABSOLUTE ENTROPY OF WATER INVOLVES THESE STEPS:

例如，要确定水的标准绝对熵需要这些步骤：

CHILL A PERFECT ICE CRYSTAL TO ABSOLUTE ZERO (NOT REALLY POSSIBLE, BUT CAN BE DONE IN THEORY).

把一块冰晶体降温至绝对零度（不完全可能，但理论上是可以做到的）。

SLOWLY ADD SMALL INCREMENTS OF HEAT AND ADD UP ALL THE ENTROPY CHANGES FROM ZERO TO 273°K, THE MELTING POINT (A TRICKY CALCULATION, BUT IT CAN BE DONE!). THIS AMOUNTS TO

慢慢地加热，将从绝对零点到273°K，也就是熔点时，所有熵的变化量相加（这个计算很复杂，但是可以做到的！）：

$$S_{273°} = 47.84 \text{ J/mol°K}$$

MELT THE ICE. WATER'S HEAT OF FUSION IS 6020 J/MOL, AND T= 273°, SO THE ADDED ENTROPY HERE IS

把冰融化成水的热量是6020焦耳/摩尔，T= 273°C，因此，此处增加的熵为

$$\frac{6020}{273} = 22.05 \text{ J/mol°K}$$

HEAT LIQUID WATER FROM 273° TO ROOM TEMPERATURE AND ADD UP THE ENTROPY CHANGES. THEY TOTAL

把冰从273°K加热到室温，把熵的改变量加起来，总计

$$S_{298°} - S_{273°} = 0.09 \text{ J/mol°K}$$

ADD THE THREE SUBTOTALS FOR THE **ABSOLUTE STANDARD MOLAR ENTROPY** OF WATER

把这3项进行相加，得水的绝对标准摩尔熵为：

$$S^o(\text{WATER}) = 47.84 + 22.05 + 0.09$$
$$= \mathbf{70.0} \text{ JOULES/MOL°K}$$

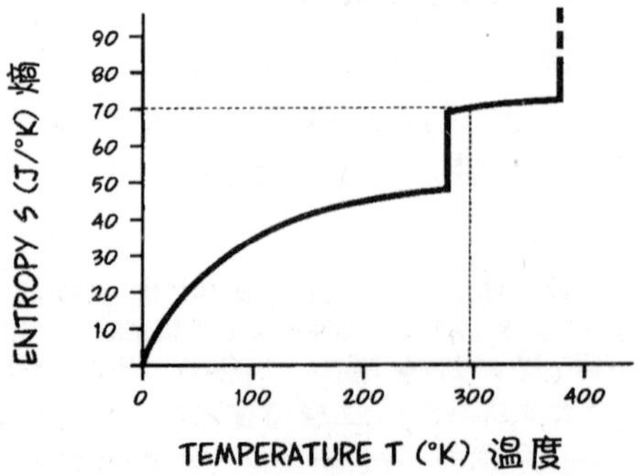

SINCE DIFFERENT SUBSTANCES HAVE DIFFERENT HEAT CAPACITIES AND HEATS OF FUSION AND VAPORIZATION, DIFFERENT AMOUNTS OF HEAT MUST BE ADDED TO RAISE THEIR TEMPERATURES AND CHANGE THEIR STATES. IN OTHER WORDS, EVERY SUBSTANCE HAS ITS OWN CHARACTERISTIC STANDARD ABSOLUTE ENTROPY.

由于不同物质的比热不同，融化蒸发所需的热量不同，因此升高它们的温度和改变它们的状态所需要的热量也就不一样。换句话说，每种物质都有自己的特征性的标准绝对熵。

SUBSTANCE 物质	STANDARD MOLAR ENTROPY 标准摩尔熵 (J/°K-MOL)
ELEMENTAL SOLIDS 基本固体	
C (DIAMOND) 金刚石	2.4
C (GRAPHITE) 石墨	5.7
Fe (IRON) 铁	27.3
Cu (COPPER) 铜	33.1
Pb (LEAD) 铅	64.8
IONIC SOLIDS 离子固体	
CaO	39.7
$CaCO_3$	92.2
NaCl	72.3
$MgCl_2$	89.5
$AlCl_3$	167.2
MOLECULAR SOLID 分子固体	
$C_{12}H_{22}O_{11}$ (SUCROSE) 蔗糖	360.2
LIQUIDS 液体	
H_2O (l)	70
CH_3OH (METHANOL) 甲醇	126.8
C_2H_5OH (ETHANOL) 乙醇	161
GASES 气体	
H_2O (g)	189
CH_4 (METHANE) 甲烷	186
CH_3CH_3 (ETHANE) 乙烷	230
H_2	131
N_2	191
NH_3	193
O_2	205
CO_2	213
CH_3OH (METHANOL, g) 甲醇，气态	240
C_2H_5OH (ETHANOL, g) 乙醇，气态	283

DIAMOND'S AMAZINGLY LOW ENTROPY IS DUE TO ITS HARD, CRYSTALLINE STRUCTURE, WHICH ADMITS VERY LITTLE WIGGLE ROOM. GRAPHITE, MADE OF SHEETS OF ATOMS, HAS MANY MORE ENERGY LEVELS.

金刚石的熵惊人的低是由于它坚硬的结晶体结构，这种结构只允许很小的摆动空间。石墨，由一层层的原子构成，具备更高的能级。

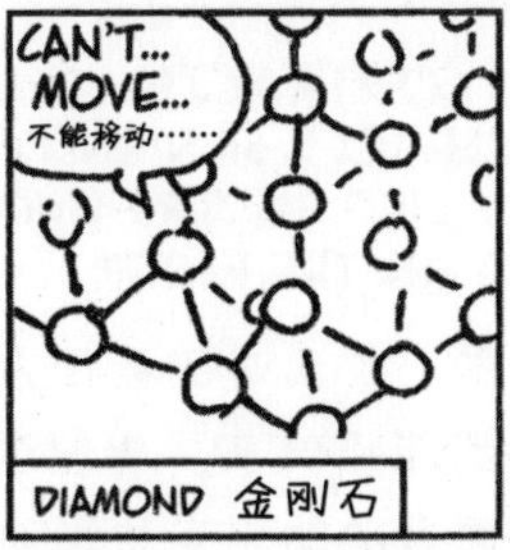

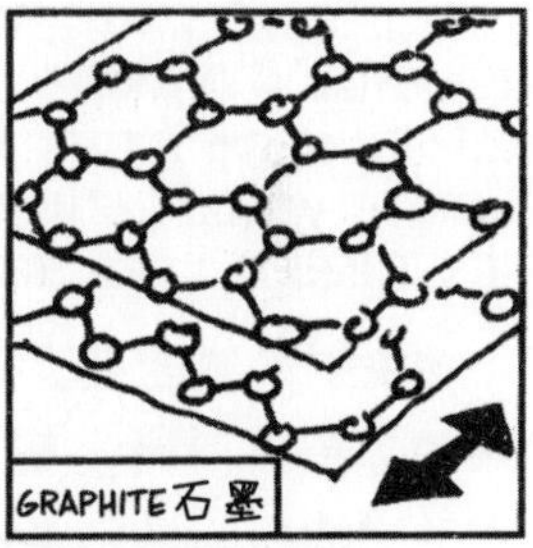

LARGER MOLECULES HAVE HIGHER ENTROPY THAN SMALLER MOLECULES: MORE PARTS TO MOVE.

大分子比小分子具有更高的熵：更大的部分要移动。

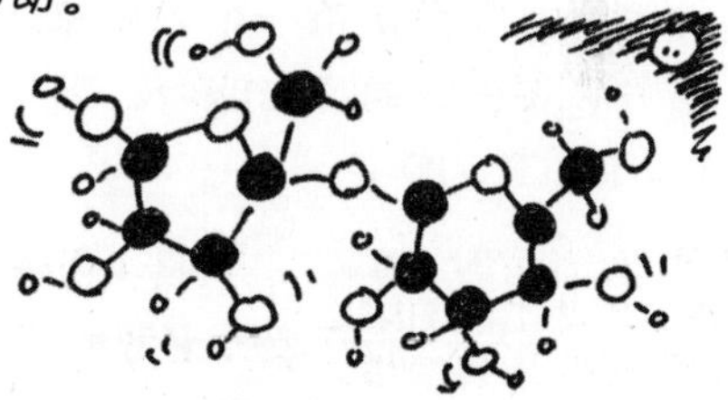

FOR ANY GIVEN SUBSTANCE,

S^o(SOLID) < S^o(LIQUID) < S^o(GAS).

对任一给定的物质，S^o（固态）<S^o（液态）<S^o（气态）

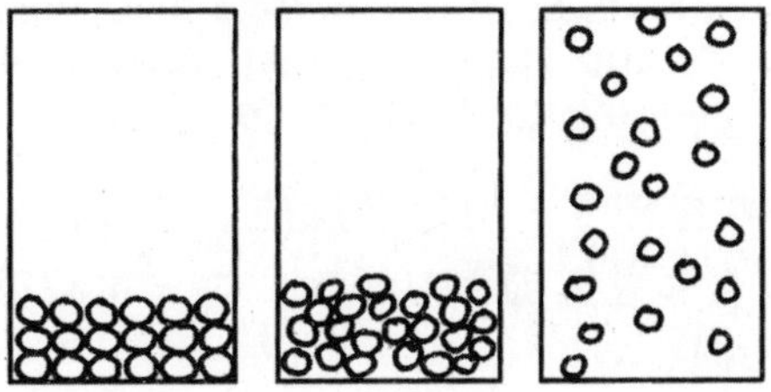

BECAUSE ENTROPY IS RELATED TO SUBSTANCES' COMPOSITION AND INTERNAL STRUCTURE, IT IS POSSIBLE FOR A SYSTEM'S ENTROPY TO CHANGE WITHOUT AN ADDITION OF HEAT. FOR EXAMPLE:

由于熵与物质的组成及内部结构有关，因此，一个系统的熵不需要额外加热就可以改变，这是可能的。例如：

THE **NUMBER** OF PARTICLES IN THE SYSTEM RISES OR FALLS. MORE PARTICLES GENERALLY MEAN MORE ENERGY LEVELS, AND SO ENTROPY GOES UP WITH THE NUMBER OF PARTICLES.

系统内的粒子数增加或减少。更多的粒子通常意味着有更多的能级，因而熵也会随着粒子数而上升。

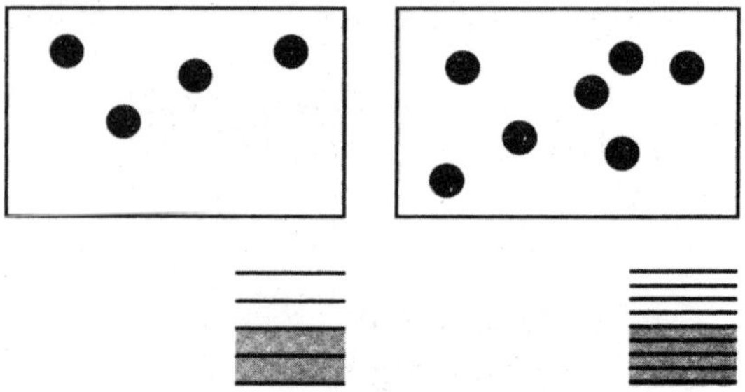

THE SYSTEM **EXPANDS** OR **CONTRACTS**. IT'S A WEIRD QUANTUM-MECHANICAL FACT (TRUST US!) THAT MOLECULES GAIN ENERGY LEVELS WHEN THEY INHABIT A LARGER VOLUME. THEY'RE LIKE DANCERS WHO CAN SHOW OFF MORE MOVES WHEN THERE'S MORE SPACE ON THE FLOOR.

系统扩大或收缩。这是一个离奇的量子力学事实（相信我们！）：当分子位于一个大的体积空间时，获得的能级就高。想象这些分子都是些舞者，当他们在地面上有更大的空间时，他们会有更多的动作来炫耀他们的舞姿。

THIS EFFECT EVEN HAS A FORMULA. IF A GAS EXPANDS AT CONSTANT TEMPERATURE, THEN

这一效果甚至可以用一个公式来表示，当气体在固定温度膨胀时，则

$$\Delta S = R\ln(P_0/P)$$

WHERE P_0 IS THE INITIAL PRESSURE, P IS THE FINAL PRESSURE, AND R IS THE GAS CONSTANT.

其中P_0是初始压强，P是最终压强，R是气体常数。

THE SYSTEM UNDERGOES A **CHEMICAL REACTION**. A CHEMICAL REACTION CHANGES THE NUMBER OF PARTICLES AND THEIR INTERNAL ARRANGEMENTS. THIS IS SO COMPLICATED IT DESERVES ITS OWN SECTION. SO...

系统经历了一次**化学反应**。化学反应改变了粒子的数量和它们的内部排列。这一过程很复杂值得一说。所以……

Entropy and Chemical Reactions
熵与化学反应

THE ENTROPY TABLE IS ONE OF THE CHEMIST'S MOST POWERFUL TOOLS. IT ALLOWS US TO PREDICT WHETHER ANY REACTION WILL GO FORWARD OR NOT (AT STANDARD CONDITIONS).

物质的熵表是化学家最有利的武器之一。它可以预测一个化学反应能否继续（在标准条件下）

ENTROPY RULES THE UNIVERSE. WE'VE ALREADY NOTED THAT THE UNIVERSE GOES TOWARDS MORE PROBABLE, SPREAD-OUT STATES. EXPRESSED IN TERMS OF ENTROPY, THIS BECOMES THE FAMOUS **SECOND LAW OF THERMODYNAMICS,** WHICH SAYS THAT ENTROPY MUST INCREASE. THAT IS, FOR ANY PROCESS WHATSOEVER,

$$\Delta S_{UNIVERSE} > 0$$

熵统治着整个宇宙。我们知道，宇宙可能是朝着发散的状态在变化。用熵的术语来表达，就是著名的**热力学第二定律**：宇宙的熵在增加。也就是任何一个过程$\Delta S_{宇宙} > 0$。

FROM THE STANDARD ENTROPY TABLE, WE CAN FIND THE ENTROPY CHANGE OF THE CHEMICALS INVOLVED IN THE REACTION, WHAT WE WILL CALL ΔS_{SYSTEM}:

从标准的熵表中，我们可以查到化学反应中物质的熵的改变量，我们称作$\Delta S_{系统}$：

$$\Delta S_{SYSTEM} = S^o(\text{PRODUCTS}) - S^o(\text{REACTANTS})$$

$$\Delta S_{系统} = S^o（产物）- S^o（反应物）$$

(S IS A "STATE FUNCTION," I.E., IT DEPENDS ONLY ON THE INITIAL AND FINAL STATE OF THE PROCESS AND NOT ON THE STEPS IN BETWEEN.)

（S是一个"状态函数"，也就是说，它取决于化学反应过程的起始状态和最终状态，而与过程中的阶段性状态无关。）

AS AN EXAMPLE, CONSIDER THE HABER PROCESS AT STANDARD CONDITIONS: SUPPOSE WE HAVE A MIXTURE OF N_2, H_2, AND NH_3... THE PARTIAL PRESSURE OF EACH GAS IS 1 ATM, AND T = 298°K. DOES THE REACTION $N_2 + 3H_2 \longrightarrow 2NH_3$ GO FORWARD?

例如，考虑标准状态下哈伯反应：假设没有气体混合物N_2、H_2和NH_3等，每种气体的分压一个大气压，且T=298°K。那么$N_2+3H_2 \rightarrow 2NH_3$，这个反应会发生吗？

FIRST, COMPUTE THE ENTROPY CHANGE OF THE SYSTEM, I.E., THE MIXTURE OF GASES.

首先，计算系统的变化熵，也就是气体混合物的变化熵。

$$\Delta S_{SYS} = S^0(\text{PRODUCTS}) - S^0(\text{REACTANTS})$$

（系统；生成物；反应物）

$$= 2S^0(NH_3) - S^0(N_2) - 3S^0(H_2)$$

$$= \mathbf{-198}\ \text{J/°K}$$

NOT SO FAST! REMEMBER, IT'S THE ENTROPY OF THE **ENTIRE UNIVERSE** THAT MUST RISE, NOT THE ENTROPY OF THE SYSTEM. WE ALSO HAVE TO CALCULATE THE ENTROPY CHANGE OF THE **SURROUNDINGS.**

不用急！记住，整个宇宙的熵一定是增加的，而系统的熵则不一定。我们还得计算出周围环境的变化熵。

$$\Delta S_{UNIVERSE} = \Delta S_{SYSTEM} + \Delta S_{SURROUNDINGS}$$

（宇宙；系统；周围环境）

BUT 但是

$$\Delta S_{SURR} = \frac{\text{HEAT CHANGE OF SURROUNDINGS}}{T}$$

（环境；环境的热改变量；温度）

THIS HEAT CHANGE IS $-\Delta H$, WHERE ΔH IS THE **ENTHALPY CHANGE** OF THE REACTION. WE SAW THIS IN CHAPTER 5. SO

热的变化是$-\Delta H$，其中ΔH是反应的焓变化量，见第5章。所以

$$\Delta S_{UNIVERSE} = \Delta S_{SYSTEM} - (\Delta H/T)$$

（宇宙；系统）

ΔH FOR THIS REACTION CAN BE READ FROM A TABLE OF ENTHALPIES OF FORMATION. IN FACT, IT'S TWICE ΔH_F OF NH_3 (BECAUSE THERE ARE TWO MOLES PRODUCED):

这个反应中的ΔH可以从反应物的焓表中读出。事实上，ΔH是NH_3的ΔH_F的两倍（因为有2摩尔的NH_3生成）：

$$\Delta H = 2\Delta H_F(NH_3)$$

$$= (2\ \text{MOL})(-45.9\ \text{kJ/MOL})$$

$$= -91.8\ \text{kJ}$$

SO 因此

$$\frac{\Delta H}{T} = \frac{-91{,}800\ \text{J}}{298\ \text{°K}} = \mathbf{-308}\ \text{J/°K}$$

THEN THE **TOTAL** ENTROPY CHANGE ASSOCIATED WITH THIS REACTION IS

于是和这个反应有关的总和的熵变化量为

$$\Delta S_{SYS} - (\Delta H/T)$$

$$= -198\ \text{J/°K} + 308\ \text{J/°K}$$

$$= \mathbf{110}\ \text{J/°K}$$

IT IS POSITIVE! ALTHOUGH THE **SYSTEM'S** ENTROPY FALLS, ENOUGH ENERGY IS SPREAD IN THE **SURROUNDINGS** TO ALLOW THE REACTION TO GO FORWARD!

它是正的！尽管系统的熵在减少，但周围有足够的熵散发，以使反应正常进行。

IT'S ANALOGOUS TO SWEEPING UP BROKEN GLASS. THE PROCESS CONCENTRATES ENERGY WITHIN THE SYSTEM, BUT THE REST OF THE UNIVERSE HAS TO SPREAD OUT ENERGY TO ENABLE IT TO HAPPEN.

打扫碎玻璃也是类似的，在这个过程中，系统位内能量积聚，但宇宙的其他部分在发散出能量以使这个过程发生。

THE SAME APPROACH APPLIES TO **ANY** REACTION AT CONSTANT P AND T. IF ΔH IS THE REACTION'S ENTHALPY, THEN

同样的方法适用于任何一个在不变的压强P和温度T下的反应。如果ΔH表示反应的焓，那么

$\Delta S_{SURROUNDINGS} = -\Delta H/T.$
周围环境

THE TOTAL ENTROPY IS 总的焓为

$\Delta S_{UNIVERSE} = \Delta S_{SYSTEM} + \Delta S_{SURROUNDINGS}$
宇宙　系统　周围环境

WHICH BECOMES 即

$\Delta S_{UNIVERSE} = \Delta S_{SYSTEM} - (\Delta H/T)$
宇宙　系统

THIS IS THE **TOTAL SPREADING OF ENERGY** IN THE UNIVERSE AS A RESULT OF THE REACTION.

这是由反应带来的整个宇宙能量的**总发散量**。

BY THE DEFINITION OF ENTROPY, THE TOTAL AMOUNT OF **ENERGY** SPREAD IS $T\Delta S_{UNIVERSE}$. WE SAY THE REACTION HAS A **FREE ENERGY CHANGE** OF $-T\Delta S_{UNIVERSE}$. THIS LAST EXPRESSION IS CALLED **ΔG,** AFTER THE AMERICAN CHEMIST J. WILLARD GIBBS (1839–1903). MULTIPLYING THE LAST EQUATION BY −T GIVES THIS VALUABLE EXPRESSION FOR ΔG:

根据熵的定义，整个能量发散量是$T\Delta S_{宇宙}$。我们称反应有一个**自由的能量改变量**$-T\Delta S_{宇宙}$。用-T乘以前一个方程，得到一个关于ΔG的有价值的表达式：

$$\Delta G = \Delta H - T\Delta S_{SYSTEM}$$

A REACTION IS **SPONTANEOUS** WHEN $\Delta S > 0$, IN OTHER WORDS, WHEN $\Delta G < 0$. EQUILIBRIUM COMES WHEN $\Delta G = 0$.

当ΔS＞0，也就是ΔG＜0时，反应是**自发**进行的；而当ΔG=0时，反应达到平衡。

NOTE THAT ΔG IS DESCRIBED STRICTLY IN TERMS OF THE SYSTEM, NOT THE SURROUNDINGS.

注意：ΔG*是用来严格描述系统的，而不是描述周围环境的。

*这个表达式被叫做ΔG，是以美国化学家丁·威拉德·吉布斯（1839—1903）的名字命名的。

ΔG REPRESENTS THE NET AMOUNT OF ENERGY THAT CAN POTENTIALLY BE CAPTURED AS WORK WHEN IT SPREADS OUT. IN FACT, YOU CAN THINK OF THE GIBBS FUNCTION AS THE **MAXIMUM AMOUNT OF WORK** THAT CAN BE DONE BY THE REACTION.

ΔG代表潜在的能量发散出来而获得的净能量，事实上，你可以想到将吉布斯函数作为反应的**最大功**。

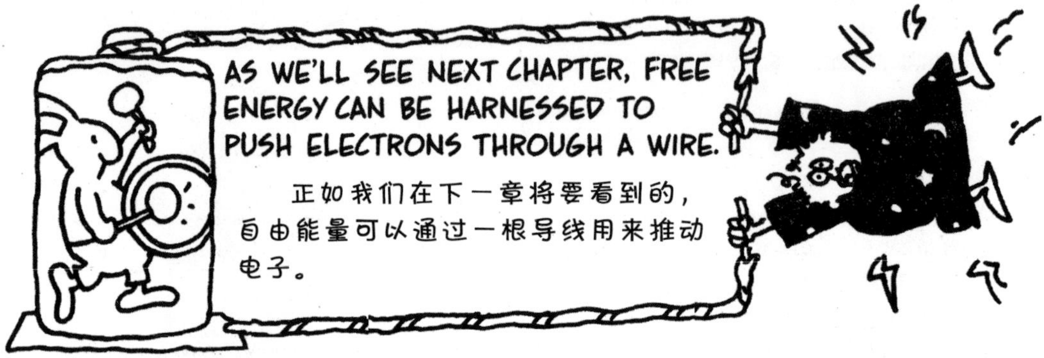

YOU CAN THINK OF THE TWO TERMS IN THE GIBBS FUNCTION GRAPHICALLY:

你可以用图来记住吉布斯函数中的两个术语：

ΔH IS THE CHANGE IN THE GROUND STATE—THE LOWEST ENERGY STATE—BETWEEN REACTANTS AND PRODUCTS. THIS REFLECTS CHANGES IN THE STRENGTH OF CHEMICAL BONDS.

ΔH是反应物与生成物之间基态，也就是最低能级的变化，它反映了化学键强度的变化。

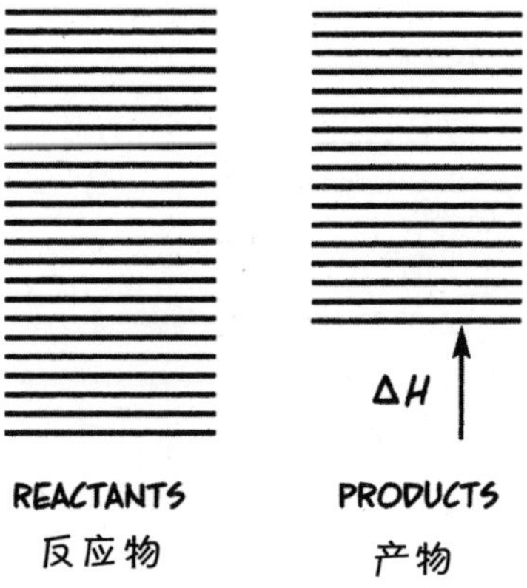

ΔH > 0 MEANS PRODUCTS' GROUND STATE IS HIGHER.

ΔH > 0意味着产物的基态更高。

−TΔS, THE ENERGY ASSOCIATED WITH THE SYSTEM'S ENTROPY CHANGE, REFLECTS CHANGES OF K.E. STATES BETWEEN REACTANTS AND PRODUCTS, I.E., DIFFERENCES OF SIZE, SHAPE, ARRANGEMENT OF MOLECULES, ETC.

−TΔS表示和系统熵的变化量有关的能量，反映了反应物和产物之间动能状态的改变，即大小、形状、分子排列等方面的变化。

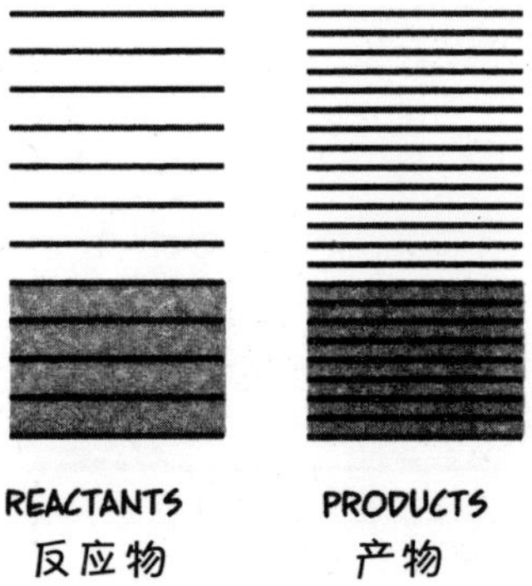

ΔS > 0 MEANS PRODUCTS HAVE MORE ENERGY LEVELS TO FILL.

ΔS > 0意味着产物需要更高的能级去填满。

WHEN IS A REACTION SPONTANEOUS? IT HELPS TO DISTINGUISH AMONG FOUR CASES, DEPENDING ON THE SIGNS OF ΔH AND ΔS (MEANING ΔS_{SYSTEM}).

什么时候反应是自发进行的？这有助于区别四种情况，由ΔH和ΔS（即$\Delta S_{系统}$）的符号来确定。

$\Delta H < 0$ EXOTHERMIC 放热
$\Delta S > 0$ SYSTEM ENTROPY INCREASES 系统熵增加

ΔG IS ALWAYS NEGATIVE. THE REACTION IS SPONTANEOUS AT ANY TEMPERATURE

ΔG总是负的，反应在任何温度下都自发进行。

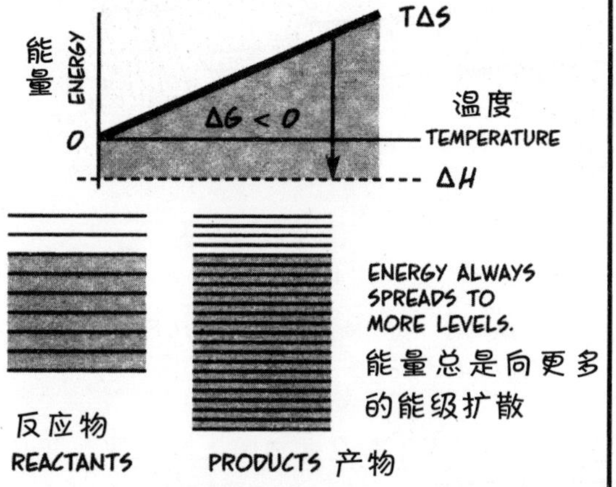

$\Delta H > 0$ ENDOTHERMIC 吸热
$\Delta S < 0$ SYSTEM ENTROPY DECREASES 系统熵减少

ΔG IS ALWAYS POSITIVE. THE REACTION IS NEVER SPONTANEOUS. THE REVERSE REACTION IS ALWAYS SPONTANEOUS.

ΔG总是正的，反应从来不是自发的，而逆反应是自发的。

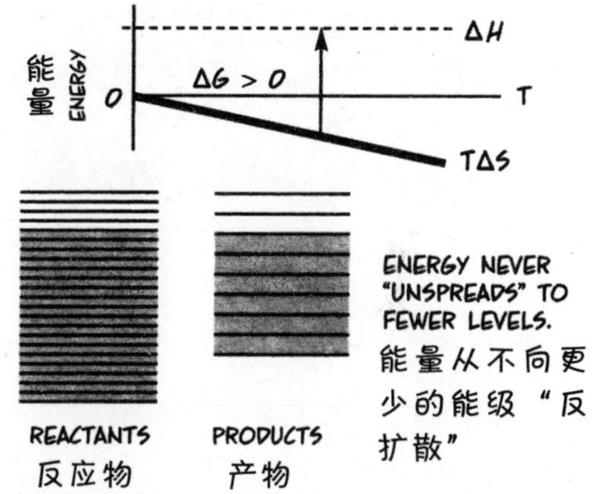

$\Delta H > 0$ ENDOTHERMIC 吸热
$\Delta S > 0$ SYSTEM ENTROPY INCREASES 系统熵增加

$\Delta G < 0$ WHEN $\Delta H < T\Delta S$. $T\Delta S$, THE ENERGY SPREAD OUT BY THE SYSTEM'S ENTROPY RISE, MUST EXCEED ΔH, THE ENERGY DRAWN FROM THE SURROUNDINGS.

当$\Delta H < T\Delta S$，$\Delta G < 0$。$T\Delta S$，通过系统焓增加而散发出去的能量一定超过ΔH，从周围环境吸收能量。

SPONTANEOUS FOR $T > \Delta H/\Delta S$

当$T > \Delta H/\Delta S$，反应是自发的

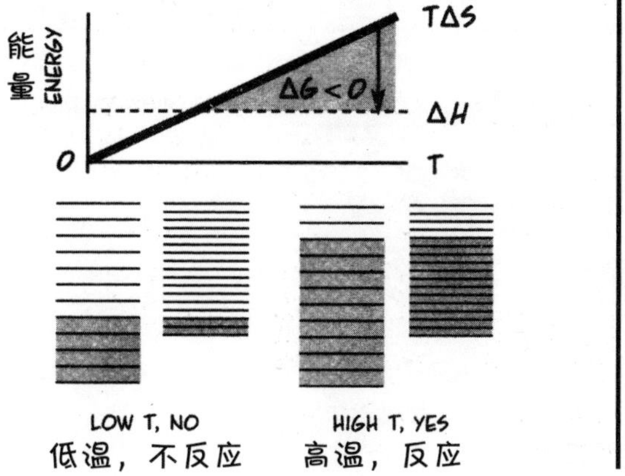

$\Delta H < 0$ EXOTHERMIC 放热
$\Delta S < 0$ SYSTEM ENTROPY DECREASES 系统熵减少

$T\Delta S$ IS THE ENERGY LOST BECAUSE OF THE SYSTEM'S ENTROPY DROP. $\Delta G < 0$ ONLY WHEN THE REACTION RELEASES EVEN MORE ENERGY, I.E., $\Delta H < T\Delta S$, OR WHEN $T < \Delta H/\Delta S$.

$T\Delta S$是由于系统熵减少而丢失的能量，只有当反应放了更高的能量时，$\Delta G < 0$，即$\Delta H < T\Delta S$或$T < \Delta H/\Delta S$.

SPONTANEOUS ONLY FOR LOW T.

只有低温时反应才自发进行

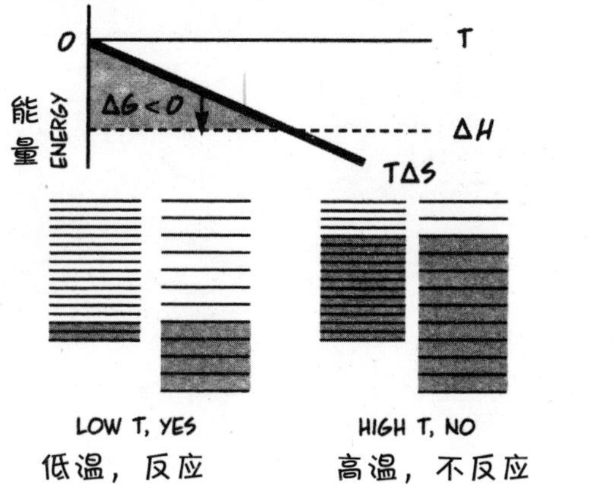

IN OTHER WORDS, THE COMPONENTS OF THE GIBBS FUNCTION, ΔH AND $T\Delta S$, PREDICT THE TEMPERATURE RANGE WITHIN WHICH A REACTION WILL TAKE PLACE SPONTANEOUSLY—PROVIDED THE REACTION HAPPENS AT CONSTANT T AND P.

换句话说，吉布斯函数中的组分ΔH和TΔS，表明了反应自发进行的温度范围——假设反应发生在固定温度T和固定压强P下。

IN THE HABER PROCESS, AS WE SAW, $\Delta S < 0$, $\Delta H < 0$, SO RAISING TEMPERATURE ACTUALLY INHIBITS THE REACTION.* (THE KEY IN THAT CASE, AS LE CHATELIER SAW, WAS TO RAISE THE PRESSURE.)

正如我们看到的在哈伯反应中ΔS＜0，ΔH＜0，所以升高温度实际上阻止了反应的进行*，正如勒夏特烈看到的，此时的关键就是增大压强。

*EVEN SO, IT'S DONE AT FAIRLY HIGH TEMPERATURE BECAUSE OF THE FASTER KINETICS AT HIGH T.

*即便这样，反应还是在高温下进行，因为在高温下动能很大。

TO APPLY GIBBS FREE ENERGY, WE BEGIN WITH A REACTION AT STANDARD CONDITIONS, AND THEN TWEAK THE GIBBS FUNCTION TO REFLECT CHANGES IN PARTIAL PRESSURES OR CONCENTRATIONS.

为了应用吉布斯自由能，我们起初在标准条件下反应，然后调整吉布斯函数，来反映气体的压强或浓度的改变。

EVERY SUBSTANCE HAS A **STANDARD FREE ENERGY OF FORMATION** G_F^o. THIS IS THE FREE ENERGY CHANGE WHEN THE SUBSTANCE IS MADE FROM ITS CONSTITUENT ELEMENTS AT STANDARD CONDITIONS. IN OTHER WORDS, IT IS ΔG OF

ELEMENTS ⟶ SUBSTANCE

NATURALLY, CHEMISTS HAVE COMPILED VAST TABLES OF THESE. HERE IS A LITTLE ONE.

任何物质都有生成标准生成自由能G_F°。当物质是在标准条件下的组成时，G_F°是自由能的变化。也就是说，

它是元素→物质的ΔG。自然的，化学家编排大量的物质的G_F°表格，这儿只是一小部分。

SUBSTANCE 物质	G_F^o (kJ/MOL)
CO_2 (g)	-394.37
NH_3 (g)	-16.4
N_2 (g)	0
H_2 (g)	0
CaO (s)	-604.2
H_2O (l)	-237.18
H_2O (g)	-228.59
O_2 (g)	0
H^+ (aq)	0
OH^- (aq)	-157.29

ONE CAN SHOW (AS WITH ENTHALPY OF FORMATION*) THAT **ANY REACTION TAKING PLACE AT STANDARD CONDITIONS** HAS FREE ENERGY EQUAL TO THE DIFFERENCE BETWEEN THE STANDARD FREE ENERGY OF FORMATION OF THE PRODUCTS AND THE STANDARD FREE ENERGY OF FORMATION OF THE REACTANTS:

可以证明（正如用生成焓*那样），**任何**在标准条件下发生的**化学反应**，其自由能都等于产物的标准生成自由能与反应物的标准生成自由能的差，即：

$$\Delta G = G_F^o(\text{PRODUCTS}) - G_F^o(\text{REACTANTS})$$

$$\Delta G = G_F^\circ(\text{产物}) - G_F^\circ(\text{反应物})$$

*SEE P. 133
*见132页

LET'S WRITE ΔG^o TO INDICATE THAT OUR REACTION TAKES PLACE AT STANDARD CONDITIONS ($T = 298°K$, $P = 1$ ATM). WHAT HAPPENS WHEN WE CHANGE PRESSURE?

用ΔG^o表示标准条件（T=298°k，P=1 ATM）下反应的ΔG。改变压强会发生什么？

WHEN A GAS CHANGES PRESSURE AT CONSTANT T FROM AN INITIAL PRESSURE P_0 TO A FINAL PRESSURE P, THE ENTROPY CHANGE OBEYS THIS EQUATION (OFFERED WITHOUT PROOF—SORRY!):

当一种气体在温度T下，压强从P_0变化到最终的P时，熵的变化满足方程（不加证明地给出——抱歉！）：

$$\Delta S = R \ln(P_0/P)$$ (R THE GAS CONSTANT)

$\Delta S = R\ln(P_0/P)$（R、P气体常数）。

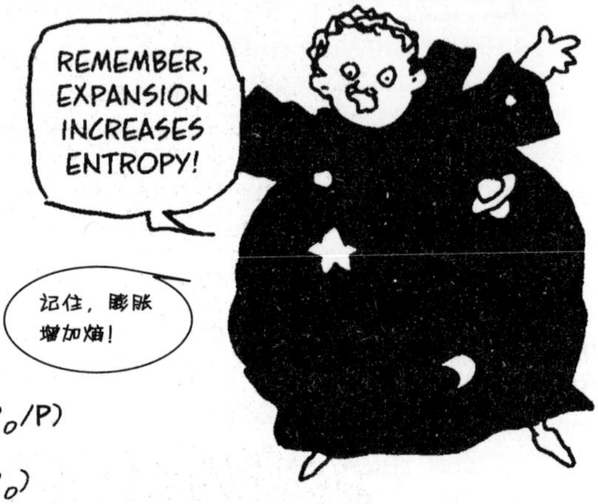

THE PRESSURE CHANGE INVOLVES NO HEAT TRANSFER: $\Delta H = 0$. SO THIS PROCESS (I.E., THE PRESSURE CHANGE) HAS FREE ENERGY:

压强的变化不涉及热量的转移，即$\Delta H=0$，因此，压强变化的这一过程具有自由能量：

$$G_F - G^o_F = \Delta H - T\Delta S = -T\Delta S = -RT\ln(P_0/P)$$

SO 所以

$$G_F = G^o_F - RT\ln(P_0/P) = G^o_F + RT\ln(P/P_0)$$

$$= G^o_F + RT\ln P$$

(BECAUSE $P_0 = 1$ AT STANDARD CONDITIONS).　（因为在标准条件下$P_0=1$）

EXCELLENT! NOW LET P VARY AND CONSIDER REACTIONS AT CONSTANT $T = 298°K$. THEN

好极了！现在让压强P变化而T保持在298°k，那么

$$\Delta G = G_F(\text{PRODUCTS}) - G_F(\text{REACTANTS})$$

　　　产物　　　　反应物

NOW LOOK AT ANY HYPOTHETICAL REACTION WITH BALANCED EQUATION

现在看任一假定的反应，形如平衡方程

$$aA + bB \rightleftharpoons cC + dD$$

AND ASSUME **A, B, C,** AND **D** ARE ALL GASES THAT REMAIN MIXED TOGETHER, WITH PARTIAL PRESSURES P_A, P_B, P_C, AND P_D. THEN

假定A、B、C、D是混合在一起的气体，其各自的压强为P_A、P_B、P_C、P_D，则

$$\Delta G = G_F(\text{PRODUCTS}) - G_F(\text{REACTANTS})$$

$$= G^o_F(\text{PROD}) - G^o_F(\text{REAC}) + RT(c\ln P_C + d\ln P_D - a\ln P_A - b\ln P_B)$$

　　产物　　　反应物

$$= \Delta G^o + RT\ln\left(\frac{P_C^{\,c}P_D^{\,d}}{P_A^{\,a}P_B^{\,b}}\right)$$

Equilibrium Again　再平衡

$$Q = \frac{P_C{}^c P_D{}^d}{P_A{}^a P_B{}^b}$$

IS CALLED THE **REACTION QUOTIENT.** Q IS SMALL WHEN PRODUCTS ARE SCARCE COMPARED TO REACTANTS, AND LARGE WHEN VICE VERSA. IF **A, B, C,** AND **D** ARE DISSOLVED CHEMICALS, WE CAN ALSO WRITE

Q称作**反应商**，当产物相对反应物很少时，Q很小，反之Q很大。如果A、B、C、D是溶解的物质，我们也记：

$$Q = \frac{[C]^c[D]^d}{[A]^a[B]^b}$$

AND IT REMAINS TRUE THAT

以下式子成立：

$$\Delta G = \Delta G^o + RT\ln Q$$

NOTE THAT $\Delta G < 0$ IF Q IS SMALL ENOUGH, AND $\Delta G > 0$ IF Q IS LARGE ENOUGH, THAT IS, IF LOTS OF **C** AND **D** ARE PRESENT.

注意到，Q足够小时，当ΔG＜0，而当Q足够大时ΔG＞0，此时C、D的量足够大。

EQUILIBRIUM OCCURS WHEN $\Delta G = 0$, OR

$$RT\ln Q = -\Delta G^o$$

OR

$$Q = e^{(-\Delta G^o/RT)}$$

当$\Delta G=0$或$RT\ln Q=-\Delta G^o$或$Q=e^{(-\Delta G^o/RT)}$时，达到平衡。

THIS IS A SECOND DERIVATION OF THE EQUILIBRIUM CONSTANT! IT SAYS THAT AT EQUILIBRIUM, THERE IS A CONSTANT K_{eq} SUCH THAT

这可用来推导出平衡常数：在平衡状态，有一个常数K_{eq}满足：

$$\frac{[C]^c[D]^d}{[A]^a[B]^b} = K_{eq}$$

AND SIMILARLY FOR PARTIAL PRESSURES. EVEN BETTER, NOW WE CAN CALCULATE K_{eq} FROM STANDARD FREE ENERGIES OF FORMATION, WITHOUT EVER RUNNING THE REACTION!

对于分压也是类似的。而且，我们可以直接根据生成标准生成自由能K_{eq}计算，而不用进行反应！

$$K_{eq} = e^{(-\Delta G^0/RT)}$$

(AND REMEMBER, IN THIS EQUATION T = 298°K.)

别忘了，方程中的T=298°K

JUST FOR FUN, LET'S SEE IF WE CAN CALCULATE THE IONIZATION CONSTANT OF WATER IN THIS WAY.

娱乐一下，让我们看看能不能用这种方法算出离子方程式的平衡常数。

$$H_2O\ (l) \rightleftharpoons H^+(aq) + OH^-(aq)$$

$$\Delta G^0 = G^0_F(\text{PRODUCTS}) - G^0_F(\text{REACTANTS})$$

产物　反应物

FROM THE TABLE:

从表中可知：

$G^0_F(H_2O\ (l)) = -237.18$ kJ/mol

$G^0_F(OH^-(aq)) = -157.29$ kJ/mol

$G^0_F(H^+(aq)) = 0$

SO 因此，

$$\Delta G^0 = -157.29 - (-237.18) = 79.89 \text{ kJ/mol}$$
$$= 79{,}890 \text{ J/mol}$$

$$K_{eq} = e^{(-\Delta G^0/RT)}$$
$$= e^{(-79{,}890)/(8.3134)(298)}$$
$$= e^{-32.25}$$
$$= 9.9 \times 10^{-15}$$
$$= 10^{-14}$$ OR CLOSE ENOUGH! 或足够接近于10^{-14}！

Chapter 11 第十一章

Electrochemistry 电化学

IN WHICH LIGHTS BLAZE AND BELLS RING, UNTIL THE BATTERY RUNS DOWN...

灯在闪，铃在响，直到电池耗尽……

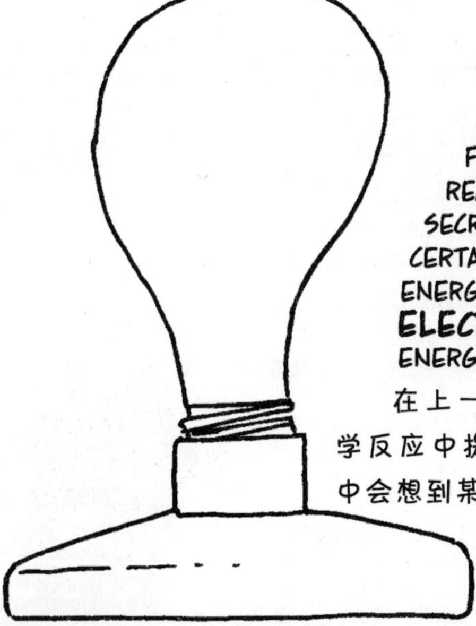

IN THE LAST CHAPTER, WHEN WE SAID ENERGY COULD BE EXTRACTED FROM CHEMICAL REACTIONS, WE SECRETLY HAD A CERTAIN KIND OF ENERGY IN MIND: **ELECTRICAL** ENERGY.

在上一章，当我们说可以从化学反应中提取能量时，我们在脑海中会想到某一种能源：**电能**。

REACTIONS THAT MOVE ELECTRONS AROUND, YOU MAY RECALL FROM CHAPTER 4, ARE CALLED **REDOX REACTIONS.** REDOX REACTIONS TRANSFER ELECTRONS FROM ONE ATOM TO ANOTHER, AND WE WOULD LIKE TO MAKE THAT TRANSFER HAPPEN BY A ROUNDABOUT PATH, PASSING THROUGH A LIGHT BULB, FOR INSTANCE!

第4章中的“转移电子”的反应被称为**氧化还原反应**。在氧化还原反应中，电子从一个原子转移到另一个原子，我们想使该反应通过一种迂回的途径发生，例如，通过一个灯泡！

Redox Redux 氧化还原

REDOX IS SHORT FOR **REDUCTION-OXIDATION.** IN A REDOX REACTION, THE ATOM DONATING THE ELECTRONS IS OXIDIZED, AND THE ONE ACCEPTING THEM IS REDUCED.

氧化还原（Redox）是**氧化还原反应**的简称。在氧化还原反应中，失去电子的原子被氧化，得到电子的原子被还原。

AN ATOM'S **OXIDATION NUMBER** IS THE NUMBER OF EXCESS CHARGES DUE TO THE LOSS OR GAIN OF ELECTRONS. FOR INSTANCE:

氧化数是一个原子得到或失去电子的净电荷数。例如：

$$CH_4 + 2O_2 \longrightarrow CO_2 + 2H_2O$$

C: -4, H: +1; O: 0; C: +4, O: -2; H: +1, O: -2

ON THE LEFT SIDE OF THE EQUATION, OXYGEN'S NUMBER IS ZERO. EACH OXYGEN ATOM TAKES ON TWO ELECTRONS AND SO IS REDUCED TO -2. THESE EIGHT ELECTRONS (2 x 4) COME FROM CARBON AND OXIDIZE IT FROM -4 TO +4. HYDROGEN IS NEITHER OXIDIZED NOR REDUCED.

在等式的左边，氧的氧化数是零。每一个氧原子得到两个电子，所以，氧化数减少到-2。这8个电子（2 × 4）来自于碳，使碳从-4氧化到+4。氢既不会被氧化，也不会被还原。

IN CHAPTER 4, WE SAW OXIDATIONS PERFORMED MOSTLY BY NON-METALS LIKE OXYGEN, BUT REDOX REACTIONS ARE ALSO COMMON AMONG **METALS AND THEIR IONS.** FOR EXAMPLE, ZINC SHEDS ELECTRONS MORE READILY THAN COPPER. WHEN Zn MEETS A Cu^{2+} ION, TWO ELECTRONS JUMP FROM ZINC TO COPPER. Cu^{2+} **OXIDIZES** Zn, AND Zn **REDUCES** Cu^{2+}.

在第四章中，我们看到氧化反应以非金属元素实施居多，如氧，但**金属及其离子**的氧化还原反应也非常普遍。例如，锌比铜更容易失去电子，当锌遇到铜离子时，两个电子从锌转移到铜。Cu^{2+}**氧化**了Zn，Zn**还原**了Cu^{2+}。

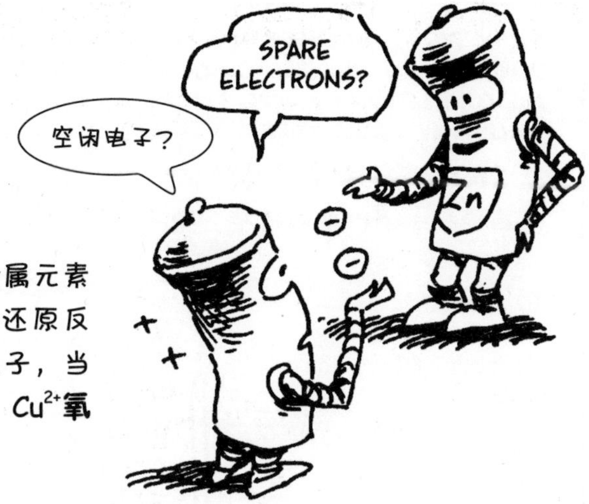

$$Zn + Cu^{2+} \longrightarrow Zn^{2+} + Cu$$

IF A ZINC BAR IS IMMERSED IN A SOLUTION OF COPPER (II) SULFATE,* $CuSO_4$, THE ZINC METAL SLOWLY OXIDIZES AND DISSOLVES, WHILE COPPER IONS PICK UP ELECTRONS AND FALL OUT OF SOLUTION AS PURE METALLIC COPPER.

如果锌棒浸在硫酸铜（$CuSO_4$）溶液中，金属锌会慢慢氧化并溶解，而铜离子得到电子并被还原成金属铜从溶液里析出。

IN THIS REACTION, ELECTRONS MOVE STRAIGHT FROM ONE ATOM OR ION TO ANOTHER. BUT NOW WE DO SOMETHING CLEVER: **SEPARATE** THE OXIDATION FROM THE REDUCTION, BUT CONNECT THE REACTION SITES BY A CONDUCTING WIRE.

在这个反应中，电子从一个原子或离子转移到另一个原子。但是，我们现在要做的更巧妙一些：把氧化反应和还原反应**分离**开来，但是通过一个导线把这个反应连接起来。

*IT'S BLUE, BY THE WAY!

*顺便说一下，它是蓝色的。

A ZINC BAR IS IMMERSED IN A 1M AQUEOUS SOLUTION OF $ZnSO_4$. COPPER IS IMMERSED IN A 1M SOLUTION OF $CuSO_4$. THE TWO BARS—OR **ELECTRODES**—ARE CONNECTED BY A WIRE. ELECTRONS WILL STILL NOT FLOW, HOWEVER, SINCE THEY WOULD CREATE A CHARGE IMBALANCE.

锌棒浸在1M的硫酸锌溶液中。铜棒浸在1M的硫酸铜溶液中。这两个金属棒（或电极）通过电线连接。电子仍不能流动，但是它们将造成电荷不平衡。

TO MAINTAIN CHARGE BALANCE, IONS MUST BE ALLOWED TO FLOW FROM ONE SOLUTION TO THE OTHER.

为了维持电荷平衡，必须让离子从一个溶液流动到另一个溶液中。

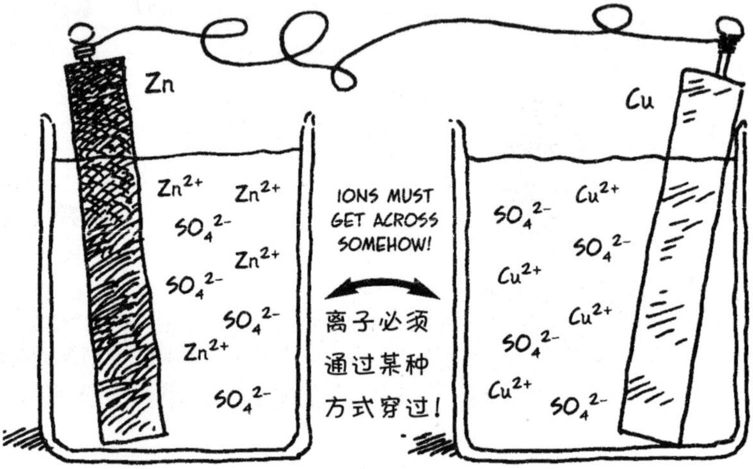

IF WE MAKE A PATH FOR IONS, ELECTRONS WILL MOVE THROUGH THE WIRE. IT'S THE ONLY WAY THEY CAN GET FROM Zn TO Cu^{2+}! DISSOLVED Cu^{2+} IS REDUCED AND DEPOSITED ON THE COPPER ELECTRODE. Zn IS OXIDIZED AND DISSOLVES. SO_4^{2-} MIGRATES TOWARD THE ZINC ELECTRODE. $[Zn^{2+}]$ RISES AND $[Cu^{2+}]$ FALLS.

如果我们为离子建立一条路径，电子会通过导线。这是唯一的出路，Cu^{2+}可以从锌得到电子！溶解的铜离子被还原，沉积到铜电极上。锌被氧化而溶解。SO_4^{2-}向锌电极迁移。$[Zn^{2+}]$升高，$[Cu^{2+}]$降低。

THE ELECTRON SOURCE IS CALLED THE **ANODE.** IT ATTRACTS NEGATIVELY CHARGED **ANIONS** (SO_4^{2-}).

该电子源被称为**阳极**。它吸引带负电荷的**阴离子**（SO_4^{2-}）。

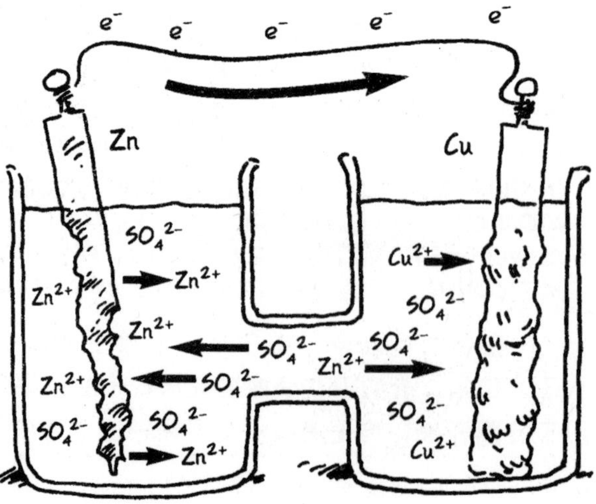

THE ELECTRON SINK, OR **CATHODE,** ATTRACTS POSITIVELY CHARGED **CATIONS** (HERE, MAINLY Cu^{2+} BUT SOME Zn^{2+} TOO).

电子汇，或**阴极**，吸引带正电的阳离子（在这里，主要是Cu^{2+}，但还有少量的Zn^{2+}）。

WHY DO THE ELECTRONS FLOW? BECAUSE FOR THEM IT'S LIKE FALLING DOWNHILL! THE ELECTRONS HAVE A **LOWER POTENTIAL ENERGY** AT THE CATHODE. TO PUT IT ANOTHER WAY, ENERGY WOULD HAVE TO BE ADDED FROM OUTSIDE TO PUSH THE ELECTRONS "UPHILL" FROM CATHODE TO ANODE.

为什么电子会流动？因为对于它们来说这就像是下坡！该电子在阴极有一个**较低的势能**。另一种说法是，不得不增加能量以推动电子从阴极到阳极，这是个“上坡”的过程。

THE REACTION'S "PUSH"—THE ENERGY DROP PER CHARGE—IS CALLED THE **VOLTAGE** OR **ELECTRIC POTENTIAL,** ΔE. ITS UNITS ARE **VOLTS,** ABOUT WHICH MORE SOON. A METER ON THE WIRE SHOWS THAT THE COPPER-ZINC REACTION GENERATES **1.1 VOLTS.** WE CAN HARNESS THIS "ELECTRON SPILLWAY" WITH A LIGHT BULB OR MOTOR OR BELL. THE ELECTRONS **DO WORK.**

该反应的“推力”——单位电荷移动的能量称为**电压**或**电势**，记为ΔE。它的单位是伏特。通过电表显示，铜-锌反应能生成**1.1伏特**。我们可以把一个灯泡、电机或电铃装在“电子溢洪道”上，电子就开始**做功**。

THIS SETUP IS CALLED A **VOLTAIC CELL,** OR LOOSELY SPEAKING, AN ELECTRIC **BATTERY.***

这种装置被称为**原电池**，或者叫**电池组**。*

*STRICTLY SPEAKING, A BATTERY CONSISTS OF SEVERAL CELLS WIRED IN SERIES.

*严格地说，一个电池组由几个电池用线串联组成。

BECAUSE A CHEMICAL CELL PHYSICALLY SEPARATES REDUCTION AND OXIDATION, CHEMISTS LIKE TO THINK IN TERMS OF SEPARATE **HALF-REACTIONS** THAT DESCRIBE THE ELECTRON TRANSFERS. IN THE ZINC-COPPER CELL, THE HALF-REACTIONS ARE:

因为一个化学电池可以分离氧化反应和还原反应，所以，化学家习惯于思考这种**半反应**，用于描述电子转移。在锌铜电池中，半反应是：

OXIDATION: $Zn \longrightarrow Zn^{2+} + 2e^-$
氧化反应：

REDUCTION: $Cu^{2+} + 2e^- \longrightarrow Cu$
还原反应：

WHEN HALF-REACTIONS ARE ADDED TOGETHER, ELECTRONS APPEAR ON BOTH SIDES AND CAN BE CANCELLED:

当半反应叠加在一起时，出现在两侧的电子可以抵消：

$$Zn + Cu^{2+} + \cancel{2e^-} \longrightarrow Zn^{2+} + Cu + \cancel{2e^-}$$

MORE (SIMPLE) REDOX REACTIONS IN SOLUTION AND THEIR HALF REACTIONS:

更多在溶液中的氧化还原反应及其半反应如下：

WHEN IRON FILINGS ARE ADDED TO ACID, THEY REDUCE H^+, AND HYDROGEN GAS IS EVOLVED. (THIS IS HOW RECREATIONAL HYDROGEN USERS MADE IT IN THE 18TH CENTURY!)

当铁屑加入酸中时，它们便可还原H^+，产生氢气（这是18世纪的用氢气球做游戏的人们制备氢气的方法！）。

$$2H^+(aq) + Fe(s) \longrightarrow Fe^{2+}(aq) + H_2(g)$$

HALF-REACTIONS:
半反应：

REDUCTION: $2H^+ + 2e^- \longrightarrow H_2$
氧化反应：

OXIDATION: $Fe \longrightarrow Fe^{2+} + 2e^-$
还原反应：

ON THE OTHER HAND, HYDROGEN IS OXIDIZED BY COPPER IONS:

另一方面，氢被铜离子氧化：

$$H_2 + Cu^{2+} \longrightarrow 2H^+ + Cu$$

REDUCTION: $Cu^{2+} + 2e^- \longrightarrow Cu$

还原反应：

OXIDATION: $H_2 \longrightarrow 2H^+ + 2e^-$

氧化反应：

LISTING ΔE FOR EVERY REDOX REACTION WOULD BE TEDIOUS, BUT IT TURNS OUT WE CAN ASSIGN VOLTAGES E_{OX} AND E_{RED} TO THE HALF-REACTIONS AND ADD THEM TOGETHER.

历数氧化还原反应的ΔE将是相当乏味的。但事实上，我们可以确定半反应的电压E_{OX}和E_{RED}，然后加在一起。

$$\Delta E = E_{OX} + E_{RED}$$

THE VOLTAGE OF ANY FULL REACTION IS FOUND BY ADDING UP ITS HALF-REACTION POTENTIALS. MUCH MORE CONVENIENT!

任何完整反应的电压是其半反应的电势之和。这方便多了！

SO, FOR INSTANCE,

所以，举例来说，

$E_{OX}(Zn \longrightarrow Zn^{2+} + 2e^-) = 0.76V$

$E_{RED}(Cu^{2+} + 2e^- \longrightarrow Cu) = 0.34V$

ΔE OF THE WHOLE REACTION IS

整个反应的ΔE是

$0.77 + 0.34 = 1.10V$

WE CAN THINK OF THESE AS THE OXIDIZED SPECIES' TENDENCY TO GIVE ELECTRONS AWAY AND THE REDUCED SPECIES' URGE TO PICK THEM UP.

我们可以把这些看作被氧化的组分失去电子和被还原的组分迫切得到电子。

HOW CAN WE ASSIGN VOLTAGES TO HALF-REACTIONS WHEN HALF-REACTIONS NEVER HAPPEN ALONE?

半反应又不会单独发生，那我们如何分配半反应的电压呢？

THIS IS HOW: FIRST, SINCE VOLTAGE DEPENDS ON CONCENTRATION, PRESSURE, AND TEMPERATURE, WE ASSUME **STANDARD CONDITIONS:** T = 298°K, P = 1 ATM, CONCENTRATION = 1 M. WE CALL OUR HALF-REACTION VOLTAGE A **STANDARD REDUCTION POTENTIAL,** E^0_{RED}, OR SIMPLY E^0.

这样分配：第一，电压由浓度、压强和温度而定，我们假设**标准条件**：T = 298°K，P = 1 ATM，浓度 =1M，我们将半反应的电压称为**标准还原电势**，E^0_{RED}，或者简称E^0。

IT WILL BE A REDUCTION POTENTIAL, BECAUSE FOR CONVENIENCE **WE WRITE ALL HALF-REACTIONS AS REDUCTIONS.** IF A REACTION RUNS LEFT TO RIGHT, IT'S A REDUCTION; IF RIGHT TO LEFT, IT'S AN OXIDATION, AND

这是一个还原电位，为方便起见，我们把半反应都写作**还原反应**。如果反应方向从左到右，是还原反应；如果从右到左，是氧化反应：

$$E_{RED} = -E_{OX}.$$

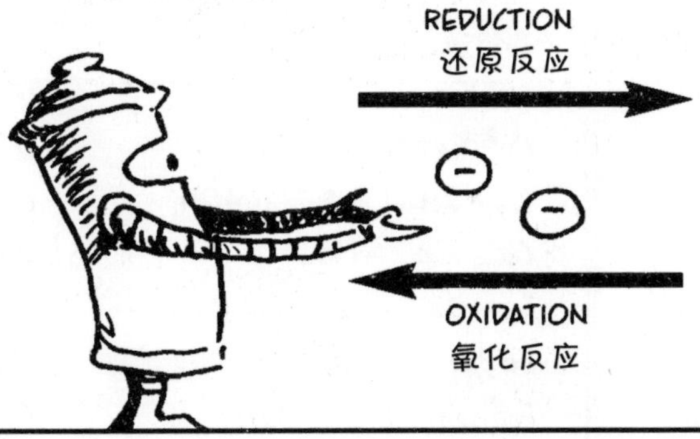

FINALLY, WE MEASURE ALL REDUCTION POTENTIALS AGAINST THAT OF **HYDROGEN,** I.E., THE REDUCTION $2H^+ + 2e^- \longrightarrow H_2$, WHICH IS ASSIGNED A VALUE $E^0 = 0$.

最后，我们用**氢**衡量一切还原电势，也就是，还原反应$2H^+ + 2e^- \rightarrow H_2$，还原电势$E^0 = 0$。

THE HYDROGEN REDUCTION IS DONE BY BUBBLING H_2 AT ONE ATM OVER A CATALYST, PLATINUM DIOXIDE, PtO_2, INTO AN ACID AT pH=0 (AT STANDARD CONDITIONS, $[H^+] = 1$ M).

氢还原反应是在一个标准大气压下，在催化剂二氧化铂（PtO_2）催化的条件下，向PH=0的酸中鼓入H_2来实现的。（在标准条件下，$[H^+] = 1M$）。

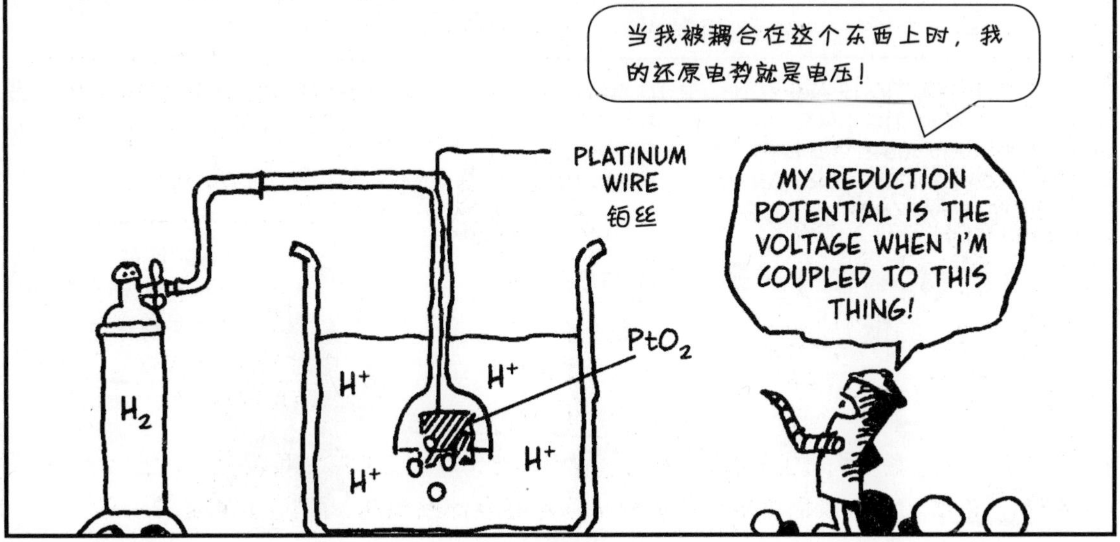

SOME HALF-REACTIONS OXIDIZE H_2 (E.G., $Cu^{2+} + 2e^- \longrightarrow Cu$), WHILE OTHERS ($Fe^{2+} + 2e^- \longrightarrow Fe$) REDUCE H^+. ANYTHING THAT REDUCES H^+ WILL HAVE A **NEGATIVE REDUCTION POTENTIAL.**

有些半反应氧化氢（如：$Cu^{2+} + 2e^- \rightarrow Cu$），而另一些会还原$H^+$（$Fe^{2+} + 2e^- \rightarrow Fe$）。任何还原$H^+$的半反应将产生**负的还原电势。**

HALF-REACTION 半反应	E^0 (V)	HALF-REACTION 半反应	E^0 (V)
$Li^+ + e^- \longrightarrow Li$	-3.05	$Ni^{2+} + 2e^- \longrightarrow Ni$	-0.25
$K^+ + e^- \longrightarrow K$	-2.93	$Sn^{2+} + 2e^- \longrightarrow Sn$	-0.14
$Ba^{2+} + 2e^- \longrightarrow Ba$	-2.92	$Pb^{2+} + 2e^- \longrightarrow Pb$	-0.13
$Sr^{2+} + 2e^- \longrightarrow Sr$	-2.89	$2H^+ + 2e^- \longrightarrow H_2$	0.00
$Ca^{2+} + 2e^- \longrightarrow Ca$	-2.84	$AgCl(s) + e^- \longrightarrow Ag(s) + Cl^-$	0.22
$Na^+ + e^- \longrightarrow Na$	-2.71	$Cu^{2+} + 2e^- \longrightarrow Cu$	0.34
$Mg^{2+} + 2e^- \longrightarrow Mg$	-2.38	$O_2 + 2H_2O + 4e^- \longrightarrow 4OH^-$	0.40
$Be^{2+} + 2e^- \longrightarrow Be$	-1.85	$Cu^+ + e^- \longrightarrow Cu$	0.52
$Al^{3+} + 3e^- \longrightarrow Al$	-1.66	$I_2 + 2e^- \longrightarrow 2I^-$	0.54
$Ti^{2+} + 2e^- \longrightarrow Ti$	-1.63	$Fe^{3+} + e^- \longrightarrow Fe^{2+}$	0.77
$Mn^{2+} + 2e^- \longrightarrow Mn$	-1.18	$Hg^{2+} + 2e^- \longrightarrow Hg$	0.80
$Zn^{2+} + 2e^- \longrightarrow Zn$	-0.76	$Ag^+ + e^- \longrightarrow Ag$	0.80
$Ga^{3+} + 3e^- \longrightarrow Ga$	-0.52	$Ir^{3+} + 3e^- \longrightarrow Ir$	1.00
$Fe^{2+} + 2e^- \longrightarrow Fe$	-0.44	$Br_2(l) + 2e^- \longrightarrow 2Br^-$	1.07
$Cd^{2+} + 2e^- \longrightarrow Cd$	-0.40	$O_2 + 4H^+ + 4e^- \longrightarrow 2H_2O$	1.23
$PbSO_4(s) + 2e^- \longrightarrow Pb(s) + SO_4^{2-}$	-0.35	$PbO_2(s) + SO_4^{2-} + 4H^+ + 2e^- \longrightarrow PbSO_4(s) + 2H_2O$	1.69
$Tl^+ + e^- \longrightarrow Tl$	-0.34	$F_2(g) + 2e^- \longrightarrow 2F^-$	2.87
$Co^{2+} + 2e^- \longrightarrow Co$	-0.27		

IF TWO HALF-REACTIONS ARE COUPLED TO MAKE A WHOLE REACTION, THE HALF-REACTION HIGHER ON THE TABLE RUNS RIGHT TO LEFT, AS AN OXIDATION, AND THE LOWER HALF-REACTION IS THE REDUCTION. THE WHOLE REACTION'S VOLTAGE IS

如果两个半反应组合成整个反应，表中的半反应从右到左变高的是氧化剂，变低的是还原剂。整个反应的电压为：

$$\Delta E^0 = E^0(\text{lower}) - E^0(\text{higher})$$

低　　　　高

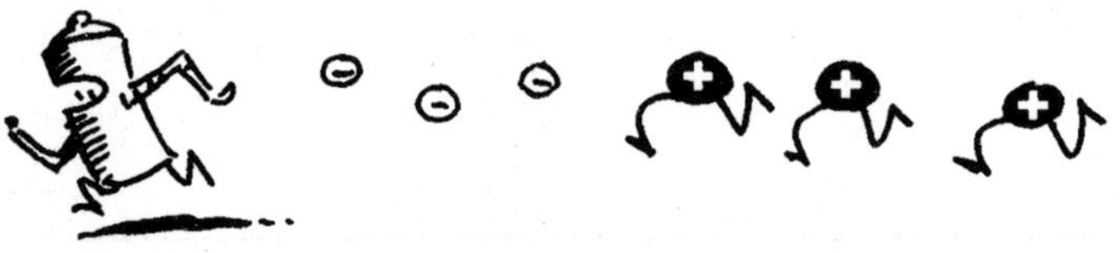

Example: Lead-Acid Battery.
范例：铅–酸蓄电池

IN THE BATTERY UNDER YOUR CAR'S HOOD, THE ANODE IS METALLIC LEAD, Pb(0), OXIDATION NUMBER 0. THE CATHODE IS Pb(+IV), IN THE FORM OF PbO_2. THE ELECTRODES ARE IMMERSED IN STRONG (6M) SULFURIC ACID, H_2SO_4. THE OXIDATION AND REDUCTION CHANGE BOTH ANODE AND CATHODE INTO Pb(+II).

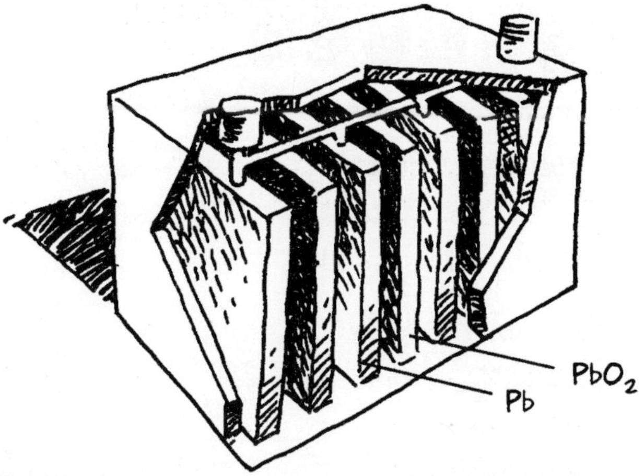

在你汽车的引擎盖下的电池里，金属铅是阳极，Pb(0)，氧化数为0。阴极是Pb(+IV)，以二氧化铅的形式存在。电极都浸在（6M）浓硫酸（H_2SO_4）中。阳极和阴极发生氧化反应和还原反应并同时生成Pb(+II)。

THE HALF REACTIONS ARE 半反应是

OX: $Pb(s) + SO_4^{2-}(aq) \longrightarrow PbSO_4(s) + 2e^-$ $\quad E^0_{RED} = -0.35\,V$

RED: $PbO_2(s) + SO_4^{2-}(aq) + 4H^+(aq) + 2e^- \longrightarrow PbSO_4(s) + 2H_2O$ $\quad E^0_{RED} = 1.69\,V$

THE OVERALL REACTION ADDS UP TO 加起来的总反应是

$$Pb(s) + PbO_2(s) + 2SO_4^{2-}(aq) + 4H^+(aq) \longrightarrow 2PbSO_4(s) + 2H_2O(l)$$

$$\Delta E = 1.69 - (-0.35) = \mathbf{2.04\,V}$$

CAR BATTERIES USUALLY PUT SIX OF THESE CELLS TOGETHER TO ACHIEVE A TOTAL VOLTAGE OF 12V.

通常可以把6个这样的电池组成电池组，共有12V的电压。

LEAD SULFATE IS INSOLUBLE AND BUILDS UP ON THE ELECTRODES WHILE SULFURIC ACID AND THE ELECTRODES ARE CONSUMED. VOLTAGE DROPS...

硫酸铅是不溶的并在电极上析出，而硫酸和电极被消耗。电压下降……

BUT WHEN THE CAR IS RUNNING, THE ENGINE'S MOTION IS CONVERTED TO ELECTRICAL ENERGY BY THE **ALTERNATOR.** THIS PUSHES ELECTRONS BACK TOWARD THE BATTERY'S ANODE, AND THE REACTIONS ARE REVERSED. THE BATTERY **RECHARGES!**

但是，当汽车开动时，发动机运转并通过**交流发电机**产生**电能**。这将推动电子回到电池的正极，反应逆转，电池开始**充电**！

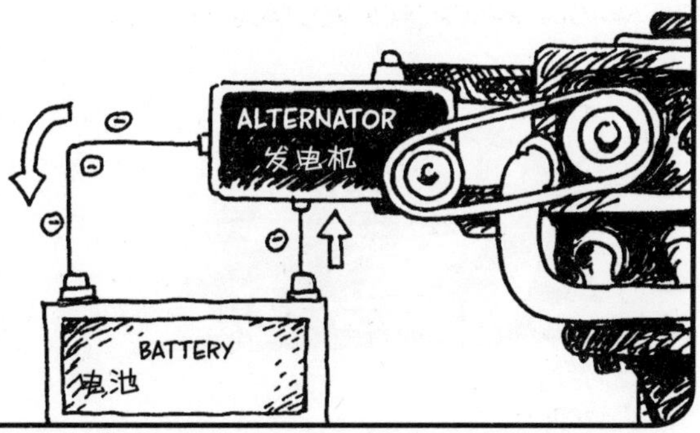

Example: Fuel Cell

范例：燃料电池

A FUEL CELL EXTRACTS ELECTRICAL ENERGY FROM A COMBUSTION REACTION SUCH AS

燃料电池从燃烧反应中获得电能，如

$$2H_2 + O_2 \longrightarrow 2H_2O$$

ONE KIND OF FUEL CELL INTRODUCES HYDROGEN AND OXYGEN ON OPPOSITE SIDES OF A POLYMER (PLASTIC) MEMBRANE. PROTONS CAN PASS THROUGH THE MEMBRANE, BUT IT BLOCKS ELECTRONS.

一类燃料电池在两侧的聚合物（塑料）膜内导入氢气和氧气。质子可以穿过膜，但电子会被膜阻挡。

THE HALF-REACTIONS ARE 半反应为

RED: $O_2 + 4H^+ + 4e^- \longrightarrow 2H_2O$ $\quad E^0 = 1.23\ V$

OX: $H_2 \longrightarrow 2H^+ + 2e^-$ $\quad E^0 = 0$

SO THE TOTAL VOLTAGE OF THE CELL IS—OR SHOULD BE—**1.23 VOLTS.**

因此，电池的总电压大约为**-1.23伏特。**

IN REAL LIFE, A CELL GENERATES LESS THAN 0.9 V. WHY THE DIFFERENCE? ONE REASON IS THAT THE CELL IS NOT 100% EFFICIENT. SOME GASES ESCAPE WITHOUT REACTING, AND THE SYSTEM SUFFERS FROM ELECTRICAL RESISTANCE. AND A FULL 0.2V IS LOST IN OVERCOMING THE REACTION'S **ACTIVATION ENERGY BARRIER.**

在现实生活中，一个电池产生的电压小于0.9V，为什么会有差别呢？原因之一是，电池不是100%有效。有些没有反应的气体逃逸，系统中存在电阻，从而使整整0.2V的电压在克服反应器的**活化能垒**时丢失。

Voltage and Free Energy
电压和自由能

CAN WE PREDICT THE CHANGE IN VOLTAGE WHEN PRESSURES OR CONCENTRATIONS ARE NOT STANDARD? THE ANSWER TURNS OUT TO BE YES, BECAUSE VOLTAGE IS NOTHING BUT **GIBBS FREE ENERGY** IN DISGUISE.

我们是否能预测在非标准压力或浓度时电压的变化呢？答案是肯定的，因为电压是**吉布斯自由能**的另一种表达。

ON P.275, VOLTAGE WAS DEFINED AS ENERGY DROP PER CHARGE, SO TO FIND THE ENERGY CHANGE OF A REACTION, WE MULTIPLY VOLTAGE BY THE AMOUNT OF CHARGE TRANSFERRED:

energy = voltage x charge

SPECIFICALLY, IF ONE VOLT MOVES ONE MOLE OF ELECTRONS, THE TOTAL ENERGY DROP TURNS OUT TO BE 96,485 JOULES.*

1 VOLT-MOL e^- = 96,485 J

在第275页，电压被定义为单位电荷在静电场中由于电势不同所产生的能量差。因此，为了找到一个反应的能量变化，我们用电压乘以电荷转移的量。

能量＝电压×电荷

具体来说，1伏特推动1摩尔电子产生的能量是96485焦耳。*

1伏特-1摩尔e^- = 96485焦耳

*OBVIOUSLY, THE PERSON WHO DEFINED THE VOLT DIDN'T CONSULT WITH ANY CHEMISTS, WHO WOULD PROBABLY PREFER TO MEASURE ΔE IN UNITS OF 1/96,485 VOLT, OR "JOLTS" AND GET RID OF $\mathcal{F}$.

*显然，给伏特下定义的人没有咨询任何化学家，这些化学家会喜欢用1/96 485伏特，或“压特（伏特的谐音）”来作为衡量ΔE的单位，并摒弃F这个常数。

THIS CONVERSION FACTOR, 96.485 kJ/(VOLT-MOL e^-), IS CALLED **FARADAY'S CONSTANT,** AND WRITTEN $\mathcal{F}$. IF A VOLTAGE OF ΔE MOVES n MOLES OF ELECTRONS, THEN

ENERGY DROP = $n\mathcal{F}\Delta E$

THIS REPRESENTS THE MAXIMUM AMOUNT OF WORK THE CELL CAN POTENTIALLY DO.

这种转换因子，96.485千焦耳/(伏特-摩尔 e^-)，被称为**法拉第常数**，书面写作F。1伏特ΔE移动n摩尔的电子，则

能量减少值＝$nF\Delta E$

这代表工作的电池有可能做的最大功。

NOW THE MAXIMUM WORK A **REACTION** CAN DO IS $-\Delta G$, WHERE ΔG IS ITS FREE ENERGY. AND A VOLTAIC CELL IS REALLY A REDOX REACTION! IN OTHER WORDS,

现在一个**反应**可以做的最大功是$-\Delta G$，其中ΔG是它的自由能。而一个伏特电池的确是一个氧化还原反应！换句话说：

$$\Delta G = -n\mathcal{F}\Delta E$$ JOULES, OR 焦耳

$$\Delta E = \frac{-\Delta G}{n\mathcal{F}}$$ VOLTS 伏

THE MINUS SIGN IS AN ARTIFACT OF OUR DEFINITIONS. VOLTAGE IS THE SIZE OF THE ENERGY DROP, WHILE ΔG IS THE ENERGY CHANGE. SO $\Delta E > 0$ WHEN $\Delta G < 0$. THAT IS, **A REDOX REACTION IS SPONTANEOUS WHEN $\Delta E > 0$.**

减号是人为的定义。当能量的变化为ΔG时，电压为能量下降的值。所以当$\Delta G<0$时，$\Delta E>0$。也就是说，**当$\Delta E>0$时，这是一个自发的氧化还原反应。**

IN THE LAST CHAPTER, WE SAW HOW ΔG CHANGES WITH CHANGING CONCENTRATIONS. IF WE HAVE A REACTION

在前一章，我们看到了ΔG如何随浓度改变而变化。如果有一个反应

$$aA + bB \rightleftharpoons cC + dD$$

THEN 接着

$$\Delta G = \Delta G^o + RT\ln Q$$

WHERE Q IS THE REACTION QUOTIENT

其中Q是反应式的商数

$$Q = \frac{[C]^c[D]^d}{[A]^a[B]^b}$$

SINCE $\Delta E = -\Delta G/n\mathcal{F}$ AT ANY CONCENTRATION, WE FIND

由于在任何浓度$\Delta E = -\Delta G/nF$，我们发现

$$\Delta E = \Delta E^o - (RT/n\mathcal{F})\ln Q$$

THIS IS CALLED THE **NERNST EQUATION.** SINCE BALANCED HALF-REACTION POTENTIALS ARE REALLY WHOLE REACTION POTENTIALS MEASURED AGAINST A HYDROGEN ELECTRODE, THE EQUATION IS ALSO TRUE OF REDUCTION POTENTIALS E_{RED}.

这就是**能斯特方程**。由于已平衡的半反应的电势确实为整个反应对氢电极电位的势能，还原电位E_{RED}的方程是

$$E_{RED} = E^o_{RED} - (RT/n\mathcal{F})\ln Q$$

AT EQUILIBRIUM, RECALL, $\Delta G = 0$, SO $\Delta E = 0$ AS WELL. THAT IS, WHEN $Q = K_{eq}$, THE BATTERY GOES DEAD.

回想在平衡状态，$\Delta G = 0$，所以$\Delta E = 0$。也就是说，当$Q = K_{eq}$，电池耗尽。

THERE ARE MANY APPLICATIONS OF THE NERNST EQUATION. WE'LL LOOK AT ONLY ONE, WHEN pH = 7. (AT STANDARD CONDITIONS, REMEMBER, pH = 0!) pH 7 IS WHAT WE FIND IN LIVING ORGANISMS...

还有许多能斯特方程的应用。我们只关注一个，当pH = 7，（在标准条件下，要记住，pH = 0！）我们发现在生物体中的pH值为7……

FOR SIMPLICITY'S SAKE, ASSUME H^+ APPEARS AS A **REACTANT** IN THE HALF REACTION (NOT A PRODUCT), AND ASSUME ALL OTHER SPECIES ARE AT STANDARD 1M CONCENTRATIONS OR CLOSE TO IT. IN THAT CASE WE WRITE THE ADJUSTED VOLTAGE AS $E^{0'}$.

为了简便起见，假设H^+作为一个半反应的反应物（而不是一个产物），并且所有其他组分的浓度等于或接近标准条件1M。在这种情况下，调整电压为$E^{0'}$。

$$E^{0'} = E^0 - (RT/n\mathcal{F})\ln Q$$

IF THE REACTION IS 如果反应

$$hH^+ + aA + bB + \ldots \longrightarrow cC + dD + \ldots$$

AND [A]=[B]=[C]=[D]=1. THEN **ALL FACTORS ARE EQUAL TO ONE** IN THE REACTION QUOTIENT, EXCEPT THE CONCENTRATION OF H^+!

而且[A] = [B] = [C] = [D] = 1。然后，除了H^+浓度外，**所有的因素**在反应商中都等于1！

$$Q = \frac{1}{10^{-7h}} = 10^{7h}$$

SO 因此，

$$E^{0'} = E^0 - (RT/n\mathcal{F})\ln(10^{7h})$$
$$= E^0 - (7hRT/n\mathcal{F})\ln(10)$$

BUT $\ln(10) = 2.3$, SO THIS

但 所以

$$= E^0 - [(2.3)(7)hRT/n\mathcal{F}]$$

NOW ASSUME h = n, THAT IS, A MOLE OF HYDROGEN IS CONSUMED FOR EACH MOLE OF ELECTRONS, WHICH FREQUENTLY HAPPENS IN A NEUTRAL ENVIRONMENT. THEN PLUGGING IN ALL THE CONSTANTS GIVES THIS SIMPLE EQUATION:

现在假设h = n时，也就是每个摩尔电子会消耗1摩尔的氢，这常常会发生在一个中性的环境中。然后，代入所有的常量得到这个简单的等式：

$E^{0'} = E^0 - 0.41$ VOLTS!!!!! 瓦特

Glucose Oxidized 葡萄糖氧化

THE SUGAR GLUCOSE, $C_6H_{12}O_6$, IS THE BASIC FUEL OF LIFE AND A KEY INGREDIENT OF CELLS. IT OXIDIZES BY THIS EQUATION:

葡萄糖（$C_6H_{12}O_6$），是生命的基本燃料和细胞的关键成分。它通过这个方程式被氧化：

$$C_6H_{12}O_6 + 6O_2 \longrightarrow 6CO_2 + 6H_2O$$

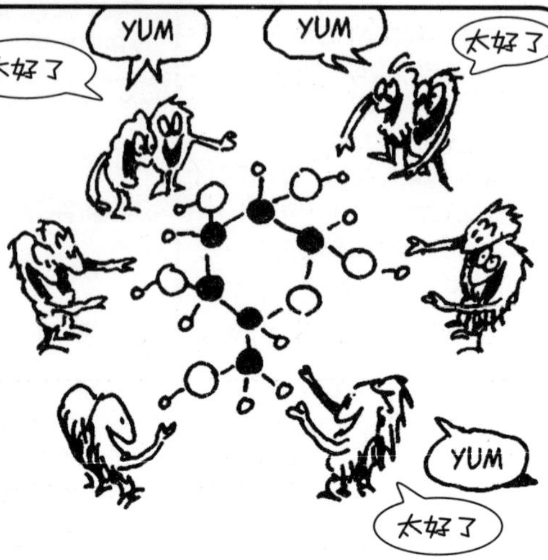

THE HALF-REACTIONS ARE:

半反应是：

$$O_2 + 4H^+ + 4e^- \rightleftharpoons 2H_2O$$

$$6CO_2 + 24H^+ + 24e^- \rightleftharpoons C_6H_{12}O_6 + 6H_2O$$

(WRITTEN AS A REDUCTION AS ALWAYS!)

（一般写成还原反应！）

THE HALF-REACTIONS BOTH HAVE EQUAL AMOUNTS OF H^+ AND e^-, SO WE CAN USE THE FORMULA:

半反应都有同等数量的H^+和e^-，所以，我们可以用以下公式表示：

$$E^{o\prime} = E^o - 0.41$$

OXYGEN'S REDUCTION REACTION IS IN THE TABLE ON P. 280, AND WE CAN WRITE

氧的还原反应在第280页的表中，我们可以写作

$$E^{o\prime} = 1.23 - .41 = \mathbf{0.82}\ \text{V}$$

WE CALCULATE E° OF THE OXIDATION REACTION FROM FREE ENERGY TABLES.

我们从自由能表格计算得到氧化反应的E^o。

SPECIES 组分	G^o_F (kJ/MOL)
$C_6H_{12}O_6(aq)$	−917.22
CO_2	−394.4
H_2O	−237.18

$$\Delta G^o = (-917.22) + (6)(-237.18) - (6)(-394.4) = 26.1\ \text{kJ/mol}$$

$$E^o = -\Delta G^\circ/n\mathcal{F} = -26.1/[(24)(96.485)] = -0.011\ \text{V}$$

$$E^{o\prime} = -0.011 - 0.41 = \mathbf{-0.42}\ \text{V}$$

THEN THE VOLTAGE DROP FOR THE WHOLE REACTION IS

那么整个反应的电压降为

$$\Delta E^{o\prime} = E^{o\prime}(\text{RED}) - E^{o\prime}(\text{OX})$$
$$= 0.82 - (-0.42)$$
$$= \mathbf{1.24\ VOLTS} > 0$$

THE OXIDATION OF GLUCOSE IS SPONTANEOUS!!

葡萄糖的氧化反应是自发的！

WHICH RAISES THE QUESTION: **WHY DON'T WE ALL JUST BURST INTO FLAMES?** THE REASSURING ANSWER IS THAT SPONTANEOUS COMBUSTION IS STOPPED BY THE REACTION'S **ACTIVATION ENERGY.**

这就提出了一个问题：**为什么我们不会因此而着火**？令人欣慰的答案是，自燃能被反应的**活化能**终止。

SO FAR THIS CHAPTER, WE'VE DESCRIBED HOW TO GET ELECTRICITY OUT OF A CHEMICAL REACTION... BUT WE HAVEN'T DISCUSSED HOW TO GET A CHEMICAL REACTION FROM ELECTRICITY.

截至目前，在本章中，我们已经描述了如何获得一个化学反应产生的电力……但是我们还没有讨论如何获得来自电力的化学反应。

ELECTROLYSIS IS WHAT HAPPENS WHEN A SUBSTANCE SPLITS AS THE RESULT OF AN APPLIED ELECTRIC CURRENT.

电解是一种物质被一个外加电流分解产生其他物质的过程。

ALUMINUM, FOR EXAMPLE, IS EXTRACTED FROM ITS ORE ELECTROLYTICALLY...

例如，用电解法从矿石中提取铝……

UNFORTUNATELY, WE DON'T HAVE ROOM FOR THE DETAILS... AND SO ELECTROLYSIS WILL HAVE TO BE LEFT FOR ANOTHER DAY, ALONG WITH A FEW OTHER TOPICS TO BE DESCRIBED IN THE FOLLOWING CHAPTER.

不幸的是，我们没有用大量篇幅讲述细节……还要为下一章节中的其他内容预留空间……电解且听将来的某个日子分解。

Chapter 12 第十二章
Organic Chemistry 有机化学

IT'S ALIVE... OR IS IT? 这是指活着的……还是什么？

OF THE NINETY-TWO NATURALLY OCCURRING ELEMENTS, SOME HAVE COMMANDED MORE OF OUR ATTENTION THAN OTHERS: HYDROGEN, FOR ITS ROLE IN ACIDS; OXYGEN, FOR ITS REACTIVITY AND LOVE OF HYDROGEN; BUT ONLY ONE ELEMENT DESERVES ITS VERY OWN BRANCH OF CHEMISTRY: **CARBON.**

92种天然元素，有的引起我们更多的注意力，例如，氢，它在酸中扮演了重要的角色；氧，具有活性，且对氢极其偏好。但只有一种元素开辟了它自己的化学分支：**碳。**

THANKS TO ITS FOUR OUTER ELECTRONS, CARBON ATOMS CAN BOND WITH EACH OTHER TO FORM LONG CHAINS, WITH OTHER ATOMS ATTACHED TO THE LEFTOVER ELECTRONS. THE SIMPLEST OF THESE CHAINS ARE THE **HYDROCARBONS,** WHICH CONTAIN NOTHING BUT CARBON AND HYDROGEN.

由于碳有4个最外层电子，因此，碳原子可以互相成键形成长链，剩下的电子可以连接到其他原子上。这些最简单的链是**碳氢化合物**，它仅仅包含碳和氢。

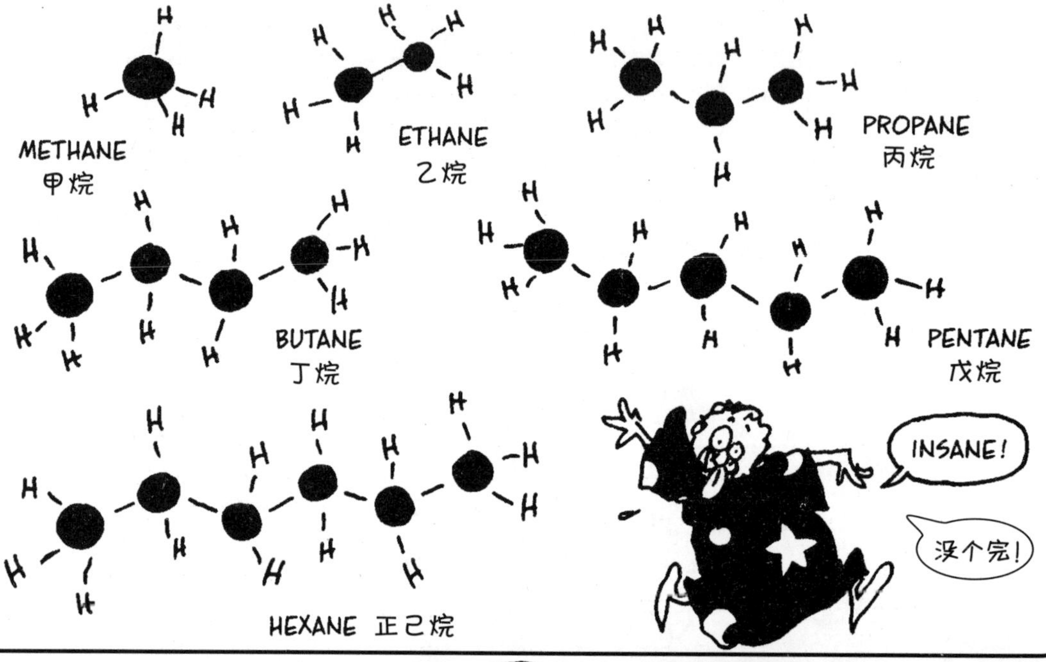

CRUDE OIL IS MADE MAINLY OF HYDROCARBONS. SINCE LONG CHAINS HAVE HIGHER BOILING POINTS THAN SHORT ONES, OIL REFINERIES CAN SEPARATE ("FRACTIONATE") THEM BY LENGTH AND THEN CHEMICALLY "CRACK" THE LONG CHAINS INTO SHORTER ONES. GASOLINE IS A MIXTURE OF CHAINS WITH 5 - 10 CARBONS (OCTANE HAS 8).

原油的主要成分是碳氢化合物。

由于长链化合物比短链的沸点更高，因此，炼油厂可以按长度分离他们（“分馏”），然后用化学方法把长链“裂解”成较短的链。汽油是一种有5—10个碳原子的碳链（辛烷有8个）。

HYDROCARBONS LIKE THOSE ON THE PREVIOUS PAGE, WITH SINGLE BONDS ONLY, ARE CALLED **ALKANES***. A DOUBLE BOND TURNS AN ALKANE INTO AN ALK**ENE**, AND A TRIPLE BOND MAKES IT AN ALK**YNE**. INDIVIDUAL MOLECULES ARE NAMED ACCORDINGLY.

像前面介绍的只有单键的碳氢化合物被称为**烷烃***。烷烃如果有了双键就变成**烯烃**，三键就是**炔烃**。单个分子也相应被命名：

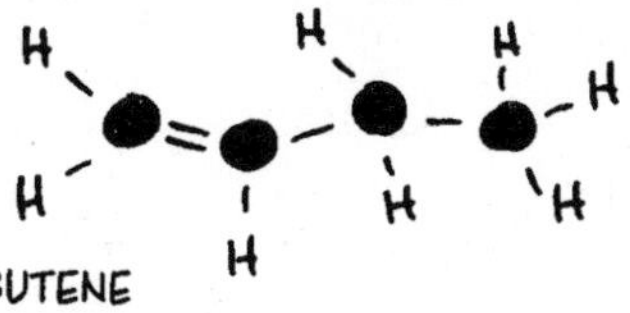

BUTENE
丁烯

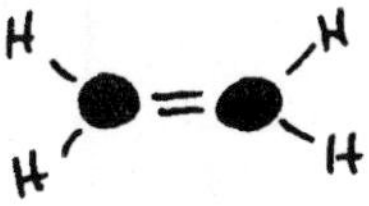

ETHENE
乙烯

ETHYNE
乙炔

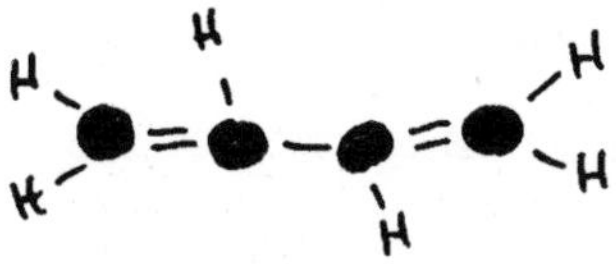

BUTADIENE (TWO DOUBLE BONDS)
丁二烯
（两个双键）

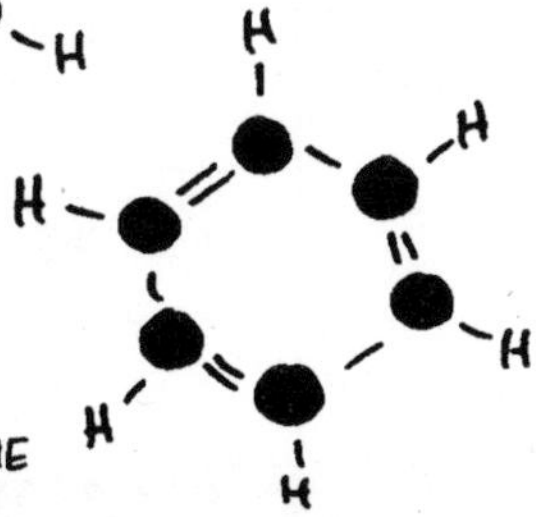

BENZENE
苯

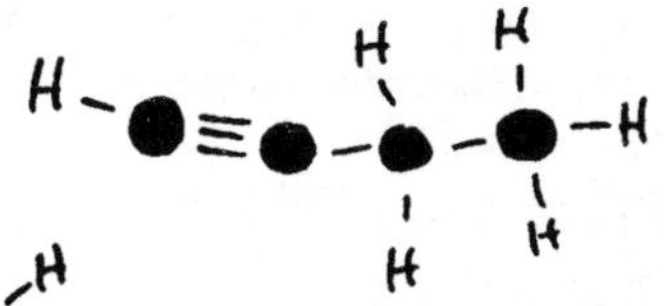

BUTYNE
丁炔

RING-SHAPED STRUCTURES HAPPEN TOO!
环状结构也出现了！

*THEY ARE ALSO CALLED SATURATED HYDROCARBONS, SINCE THEY HAVE THE MAXIMUM POSSIBLE NUMBER OF HYDROGENS. ANYTHING WITH A DOUBLE OR TRIPLE BOND IS CALLED UNSATURATED.

* 它们也被称为饱和烃，因为它们有最大数目的氢。具有双键或三键的化合物被称为不饱和烃。

TO COMPLICATE MATTERS FURTHER, TWO COMPOUNDS WITH THE SAME CHEMICAL FORMULA CAN HAVE DIFFERENT STRUCTURES. VARIANTS OF THE "SAME" MOLECULE ARE CALLED **ISOMERS**.

为了应对未来复杂的问题，两个化学式相同的化合物可以有不同的结构。“相同”分子的变体被称为**异构体**。

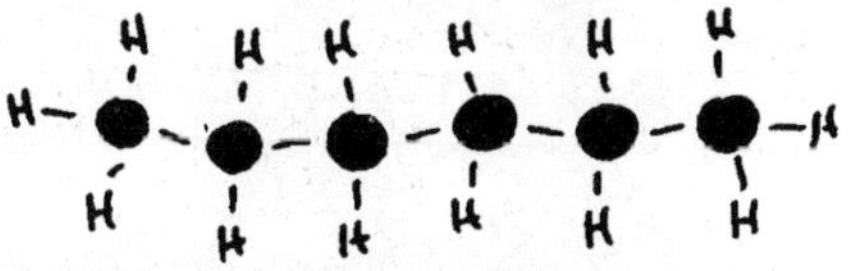

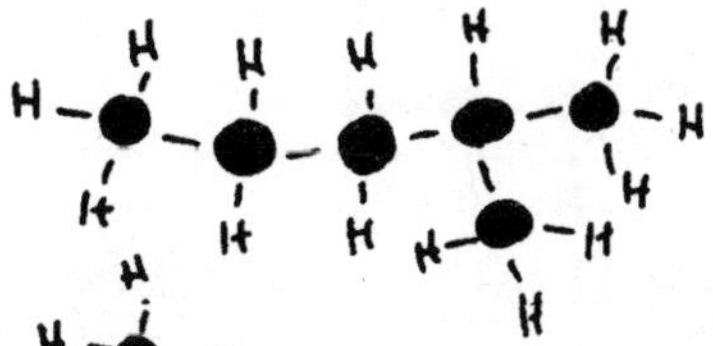

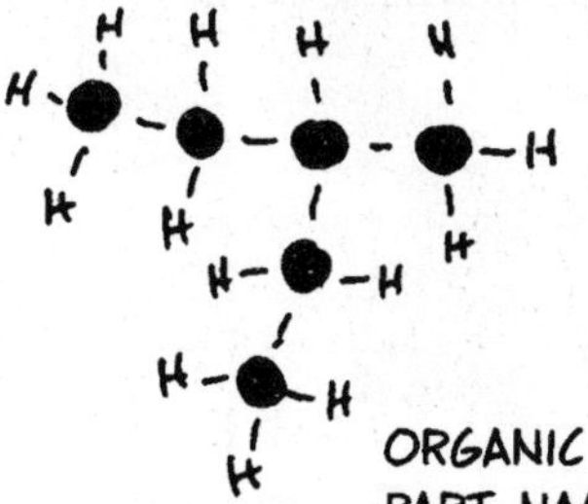

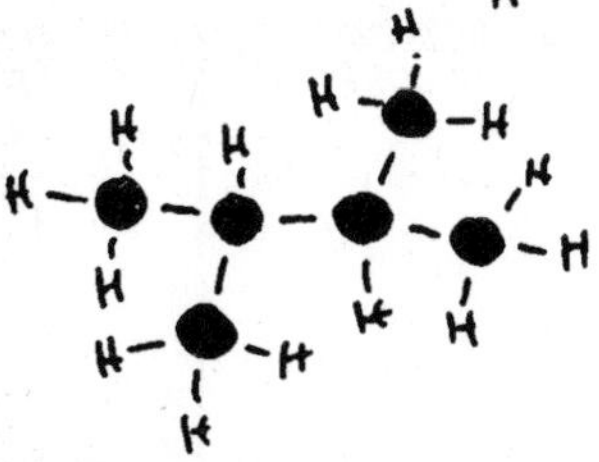

ORGANIC CHEMISTRY IS PART CHEMISTRY, PART NAME GAME, AND PART GEOMETRY!

有机化学是化学的一部分，名称游戏的一部分，几何的一部分！

THINGS ARE EVEN MORE FUN WHEN OXYGEN AND NITROGEN GET INTO THE MIX.

当有氧和氮加入时，事情就更有趣了。

IF A CHAIN HAS AN OH, IT'S CALLED AN **ALCOHOL.**

如果一个链连着一个羟基，这就是**醇**。

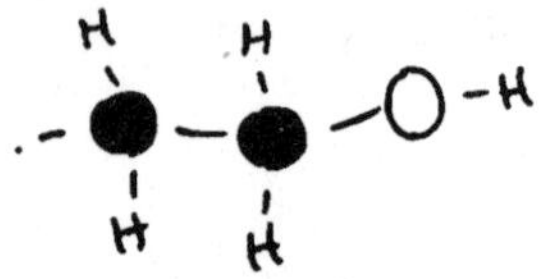

WITH A COOH GROUP, ITS A CARBOXYLIC **ACID.** (ONLY THE HYDROGEN COMES OFF, NOT THE WHOLE OH).

与羧基基团相连，它就成为一个羧酸（只有氢而不是整个羟基脱落）。

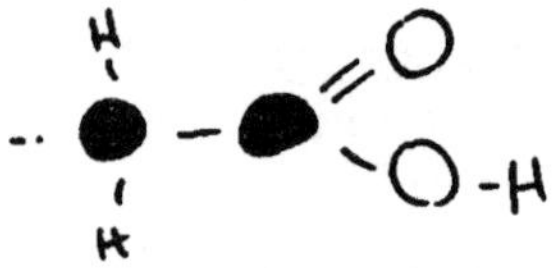

NH_2 MAKES IT AN **AMINE.**

连着氨基就成为**胺**。

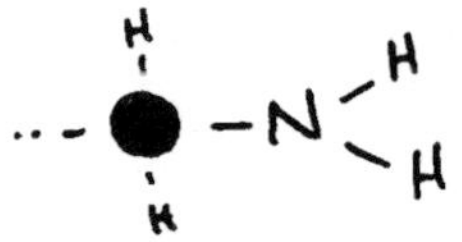

TWO CHAINS LINKED BY OXYGEN FORM AN **ETHER.**

双链用氧相连形成一个**醚**。

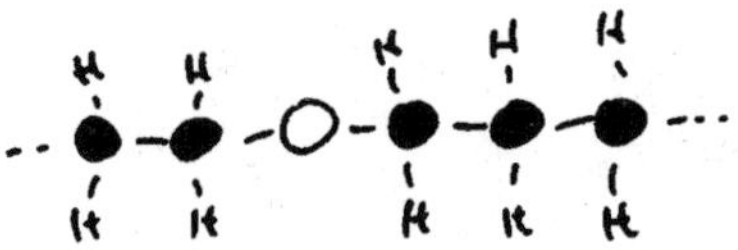

ALDEHYDES LOOK LIKE THIS:

醛像这样：

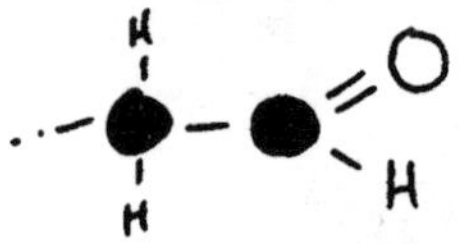

AND THIS IS A **KETONE:**

这是一个**酮**：

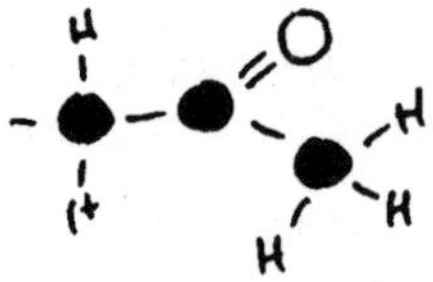

AND DON'T FORGET **ESTERS,** WHICH SMELL NICE.

不要忘了**酯**，闻起来不错。

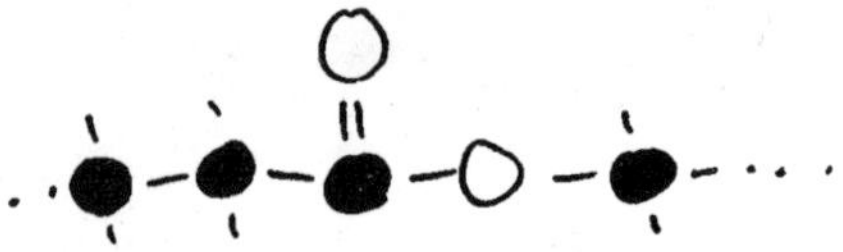

THIS ONE, ETHYL FORMATE, SMELLS LIKE RUM...

其中，甲酸乙酯，闻起来像朗姆酒……

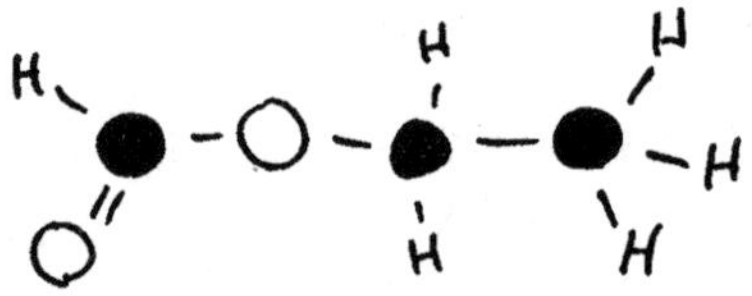

AND PENTYL ACETATE IS "BANANA OIL."

而乙酸戊酯像是"香蕉油"。

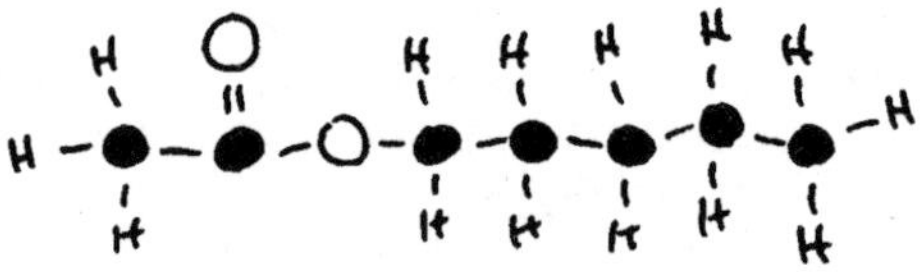

CARBOHYDRATES ("HYDRATED CARBONS") HAVE EXACTLY TWICE AS MANY HYDROGENS AS OXYGENS.* THAT IS, THEIR GENERIC FORMULA IS $C_n(H_2O)_m$. THE SIMPLEST EXAMPLES ARE **SUGARS,** LIKE **GLUCOSE,** $C_6H_{12}O_6$.

碳水化合物（"水合碳"）中的氢正好是氧的两倍*，其通式是$C_n(H_2O)_m$。最简单的例子是**糖**，如**葡萄糖**（$C_6H_{12}O_6$）。

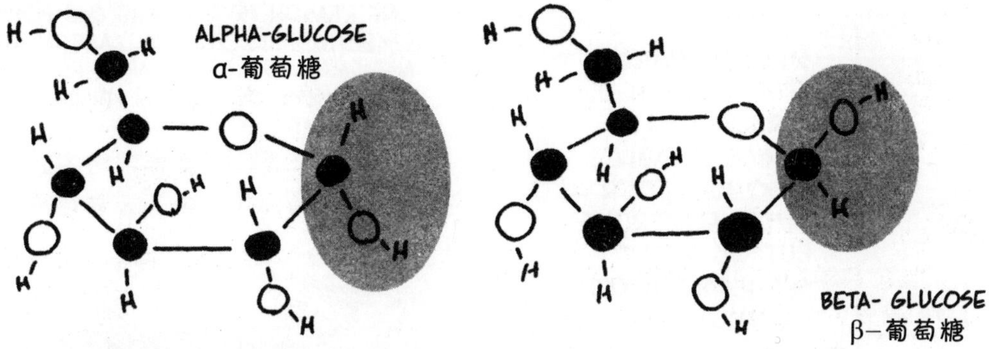

HERE ARE THE TWO MAIN GLUCOSE ISOMERS. IN BETA, THE OH GROUP BESIDE O IS ON THE **SAME** SIDE OF THE RING AS THE SIDE CHAIN. IN ALPHA, OH IS ON THE **OPPOSITE** SIDE FROM THE CHAIN.

这是两个主要的葡萄糖异构体。在β-葡萄糖中，O旁边的OH基团和侧链在环的**同**一侧。在α-葡萄糖中，OH基团和侧链**不**在同一侧。

*THERE ARE EXCEPTIONS. DEOXYRIBOSE IS CONSIDERED A SUGAR, EVEN THOUGH IT IS ONE OXYGEN SHORT.

*也有例外。脱氧核糖仍然被认为是一种糖，即使它少了一个氧。

SINGLE-RING SUGARS ARE CALLED SIMPLE SUGARS OR **MONOSACCHARIDES.** SUCROSE, THE CANE SUGAR YOU BUY AT THE STORE, IS A **DISACCHARIDE** THAT LINKS ALPHA-GLUCOSE TO FRUCTOSE, ANOTHER SIMPLE SUGAR.

单环糖被称为简单糖或**单糖**。你在商店买的蔗糖是**二糖**，一个α-葡萄糖连着一个果糖，这是另一种单糖。

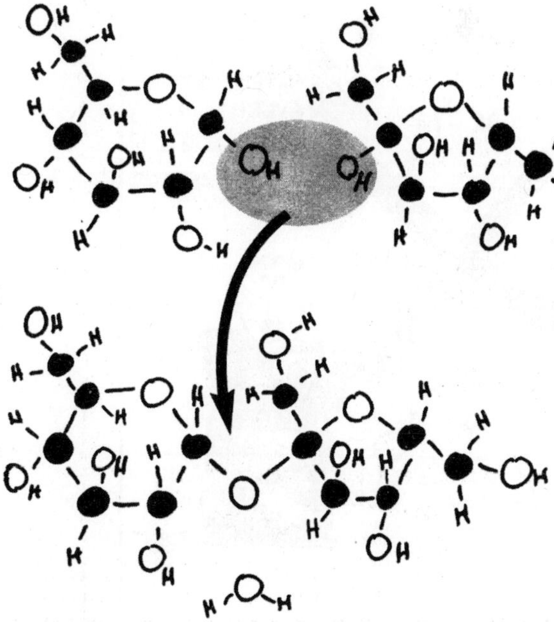

LET'S STOP A MOMENT AND ASK OURSELVES, 让我们稍停一会，问问自己。

Why Carbon and Only Carbon?

为什么会是碳？

WHY IS THIS THE ONE ELEMENT THAT FORMS LONG CHAINS? 为什么只有这一个元素构成长链？

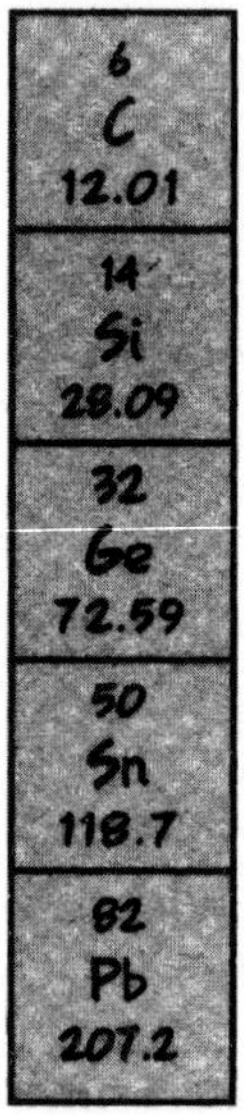

SILICON, WHICH SITS BENEATH CARBON IN THE PERIODIC TABLE, ALSO HAS FOUR OUTER ELECTRONS, BUT WE DON'T SEE HYDROSILICON CHAINS.

硅在化学元素周期表中位于碳的下方，它也有4个外层电子，但我们不曾看到硅碳链。

NOR, FOR THAT MATTER, DO WE SEE CHAINS OF OXYGEN OR NITROGEN.

我们也没有看到过氧链或氮链。

ONE REASON IS THAT THE C–C BOND IS EXCEPTIONALLY STRONG. CARBON ATOMS ARE SMALL, SO THE SHARED ELECTRON CLOUD IS CLOSE TO THE NUCLEI, WHICH ATTRACT IT STRONGLY.

原因之一是C-C键异常结实。由于碳原子很小，所以，共享电子云靠近原子核，产生强吸引力。

HERE ARE SOME BOND STRENGTHS OF INTEREST. (RECALL THAT THE NUMBERS MEAN THE AMOUNT OF ENERGY NEEDED TO BREAK THE BOND.)

下面是一些键的强度（回忆一下，这些数字是破坏键所需能量的大小）。

BOND 键	STRENGTH(kJ/mol) 强度（千焦耳/摩尔）
C–C	347-356*
C=C	611
C≡C	837
C–O	336
C–H	356-460*
Si–Si	230
Si–O	368
O–O	146
O=O	498
N–N	163
N=N	418
N≡N	946

*DEPENDING ON WHAT ELSE IS ATTACHED TO THE CARBON ATOM.

*取决于附着在碳原子上的其他原子是什么。

NOTE THAT THE C—C BOND IS EVEN STRONGER THAN THE C—O BOND. THIS MEANS THAT STABLE CARBON CHAINS CAN FORM IN THE PRESENCE OF OXYGEN.

请注意，C-C键甚至比C-O键更强。这意味着，有氧存在时，也可以形成稳定的碳链。

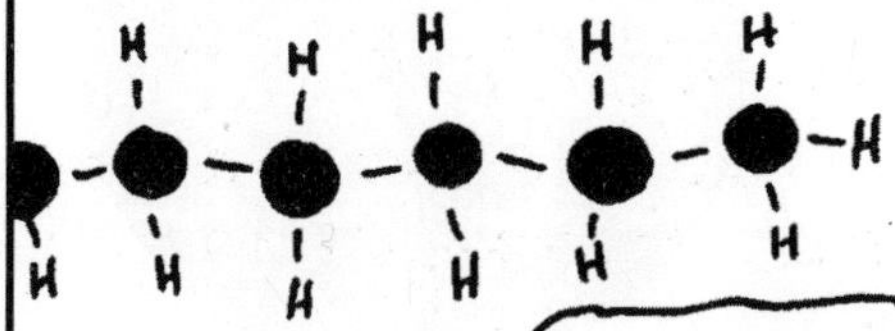

BY CONTRAST, Si—Si BONDS ARE MUCH WEAKER THAN Si—O BONDS. OXYGEN DISRUPTS SILICON CHAINS. MOST SILICON ON EARTH EXISTS AS SiO_2 (SAND) OR SiO_3^{2-} IN SILICATE ROCKS. IN FACT, YOU OFTEN SEE OIL AND SAND SIDE BY SIDE.

相比之下，Si-Si键要弱于Si-O键。氧破坏了硅链。地球上大多数的硅以二氧化硅沙或SiO_3^{2-}的形式存在于硅酸盐岩中。事实上，你可以经常看到油和沙一起出现。

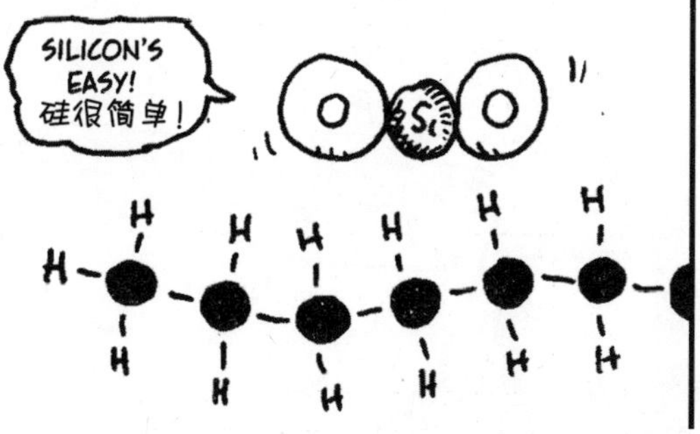

ALSO NOTE THAT TWO C—C BONDS ARE STRONGER THAN ONE C=C BOND. CARBON PREFERS THIS

还要注意的是两个C-C键比C = C键更牢固。碳更喜欢这个

TO THIS:

如果与这个相比的话：

THREE SINGLE BONDS ARE ALSO STRONGER THAN ONE TRIPLE BOND. RESULT: LONG CHAINS ARE PREFERRED OVER SHORT ONES.

三个单键也强于一个三键。结果是长链优先于短链。

BY CONTRAST, OXYGEN PREFERS O=O TO O–O–O, AND NITROGEN PREFERS TO BOND WITH ITSELF AS N≡N. RESULT: NO OXYGEN OR NITROGEN CHAINS!

相比之下，氧更喜欢O=O甚于O-O-O，氮喜欢自己成键为N≡N，结果是没有氧链或氮链！

FINALLY, THE C—H BOND IS STRONG. HYDROCARBONS ARE STABLE AT ROOM TEMPERATURE. OTHER HYDRIDES TEND TO BE UNSTABLE AROUND OXYGEN.

最后，C-H键很牢固。碳氢化合物在室温下保持稳定。其他氢化物在氧气氛围中往往是不稳定的！

IN SUM, ONE OF CARBON'S PREFERRED STATES IS IN LONG SINGLY-BONDED CHAINS, POSSIBLY BRANCHED OR LOOPING BACK ON THEMSELVES AS RINGS, WITH A LOT OF HYDROGEN ATTACHED. **THIS IS TRUE OF NO OTHER ELEMENT.**

总之，碳最稳定的状态之一，就是同很多氢一起形成长单键链，也可能形成支链或者自身成环。**没有别的元素能做到这点。**

BIG, COMPLICATED CARBON MOLECULES FORM THE ESSENTIAL INGREDIENTS OF LIFE... IN FACT, CARBON COMPOUNDS ARE SO INTIMIATELY INVOLVED WITH LIVING SYSTEMS THAT CHEMISTS REFER TO ALL CARBON COMPOUNDS AS **ORGANIC.** CARBON MAKES LIFE POSSIBLE!

大而复杂的碳分子是生命的基本材料……事实上，碳化合物与生命系统密切结合，所以，化学家提出**有机物**就是碳化合物。碳，让生命成为可能！

LUCKILY FOR CHEMISTS, EVEN THE BIGGEST, MOST HORRIBLE ORGANIC COMPOUNDS ARE CHAINS OF SIMPLER SUBUNITS ATTACHED END TO END. THE SIMPLEST EXAMPLE IS POLYETHLENE PLASTIC, $(CH_2)_n$.

化学家很幸运，即使是最大的、最恐怖的有机化合物也是由简单的单元首尾相连成链的。最简单的例子是聚乙烯塑料$(CH_2)_n$。

THE INDIVIDUAL UNITS OF THESE CHAINS ARE CALLED MONOMERS ("SINGLE TYPES"), AND THE WHOLE CHAIN IS A

polymer.

这些链的个体单位称为单体（“单类型”），整个链叫作**聚合物**。

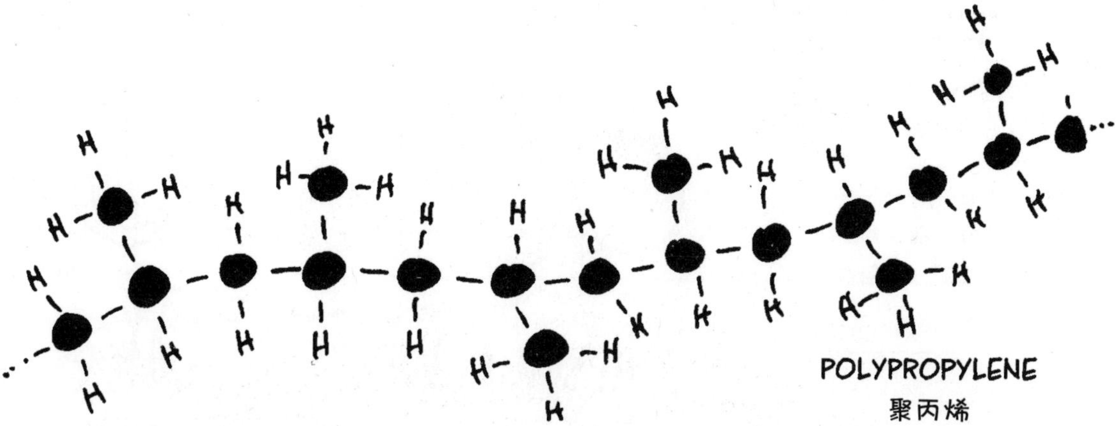

POLYPROPYLENE
聚丙烯

NATURE'S POLYMERS ARE A BIT MORE WHIMSICAL THAN THESE SIMPLE PLASTICS. FOR INSTANCE, **POLYSACCHARIDES** COMBINE MANY SUGARS END TO END. **CELLULOSE** IS FORMED OF REPEATED UNITS OF BETA-GLUCOSE.

大自然的聚合物比这些简单的塑料更多样化。例如，许多单糖首尾相连结合成**多糖**。**纤维素**由β-葡萄糖的重复单元组成。

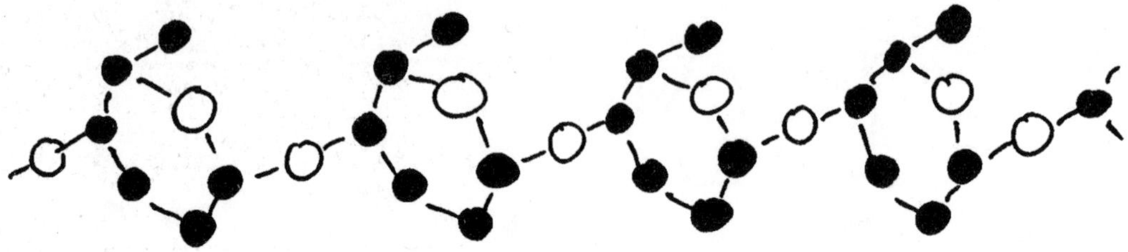

STARCH COMBINES ALPH-GLUCOSE MONOMERS.

淀粉由α-葡萄糖单体组成。

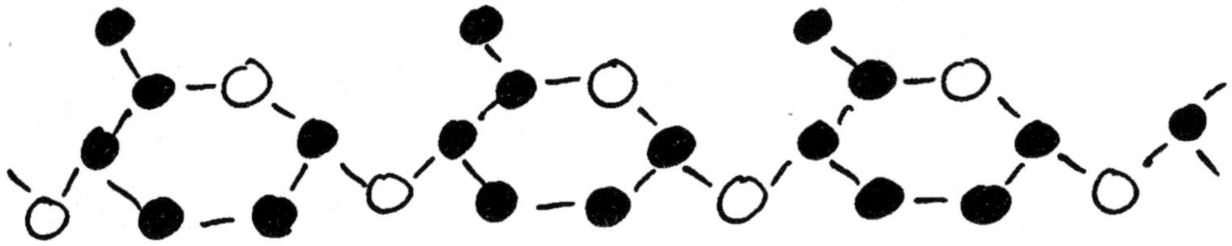

DESPITE THE SEEMINGLY CLOSE SIMILARITY, STARCH AND CELLULOSE ARE VERY DIFFERENT CHEMICALLY. THE STARCH CHAIN IS MORE EASILY BROKEN AND OXIDIZED AS BODY FUEL, WHILE THE TOUGH FIBERS OF CELLULOSE ARE INDIGESTIBLE TO MOST ANIMALS.

尽管看似相同，但淀粉和纤维素在化学上有很大的不同。淀粉链更容易断裂和氧化，主要用作身体的燃料；而纤维素纤维很坚韧，大多数动物消化不了。

Chemicals of Life 生命化学物质

LIVING SYSTEMS TEEM WITH **NON-REPEATING** CHAINS. AMONG THE KEY INGREDIENTS ARE **AMINO ACIDS,** SMALL MOLECULES WITH A BASIC AMINO GROUP (NH_2), AN ACID CARBOXYL GROUP (COOH), AND SOME OTHER GROUP ALL ATTACHED TO THE SAME CARBON ATOM.

生命系统盛产各种**非重复**单元的长链。其主要成分是**氨基酸**，包括一个碱性氨基（NH_2）和一个酸性羧基（COOH）以及另一个官能团，它们都附属于同一个碳原子。

FOR SOME REASON, BIOLOGY FAVORS ONLY TWENTY VARIATIONS ON THIS PATTERN.

出于某种原因，生物学主要选择了以下这20种形式的氨基酸。

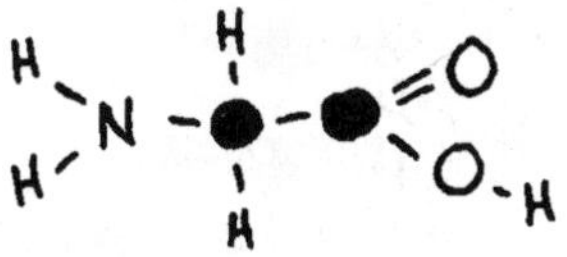

GLYCINE
甘氨酸

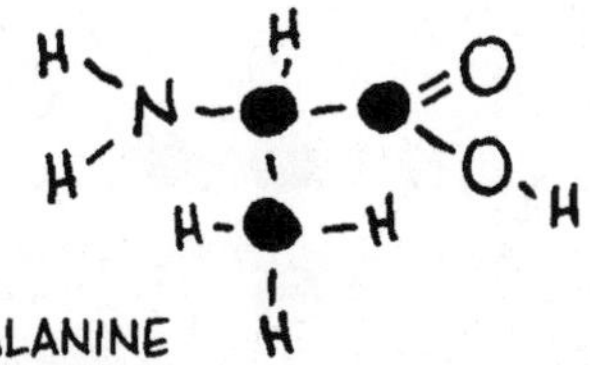

ALANINE
丙氨酸

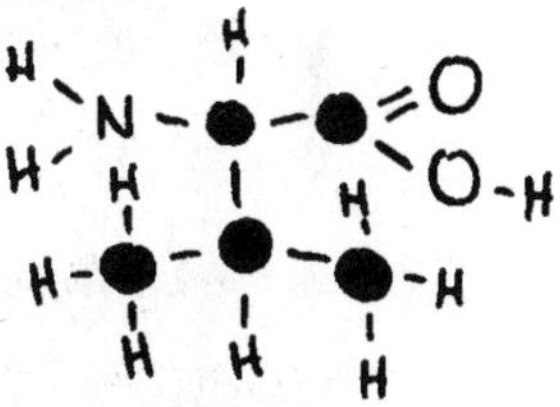

VALINE
缬氨酸

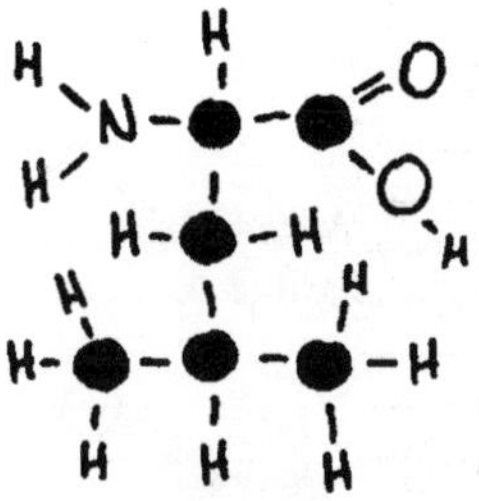

LEUCINE
亮氨酸

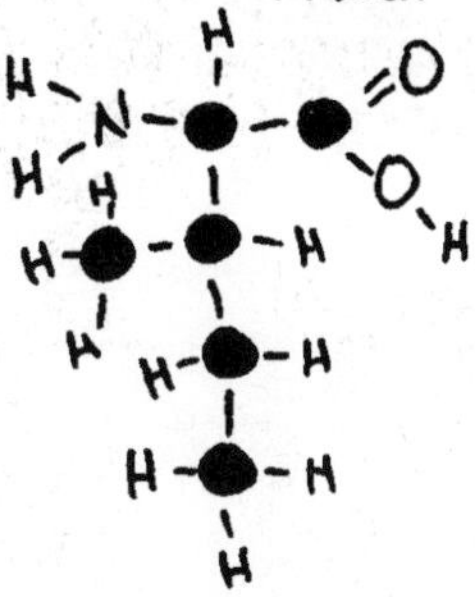

ISOLEUCINE
异亮氨酸

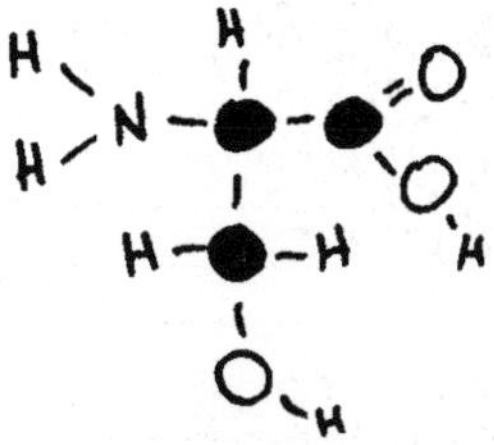

SERINE
丝氨酸

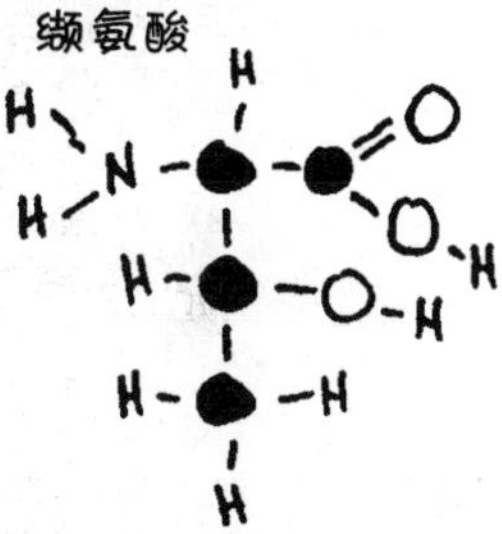

THREONINE
苏氨酸

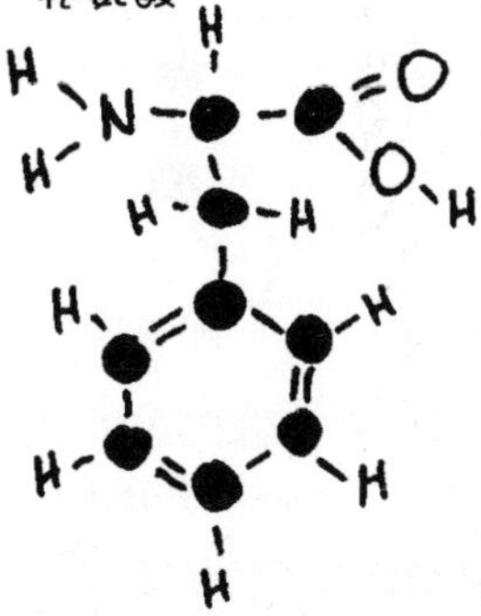

PHENYLALANINE 苯丙氨酸

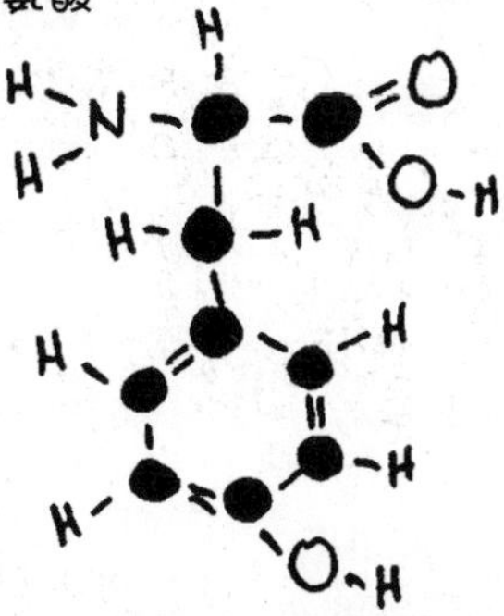

TYROSINE 酪氨酸

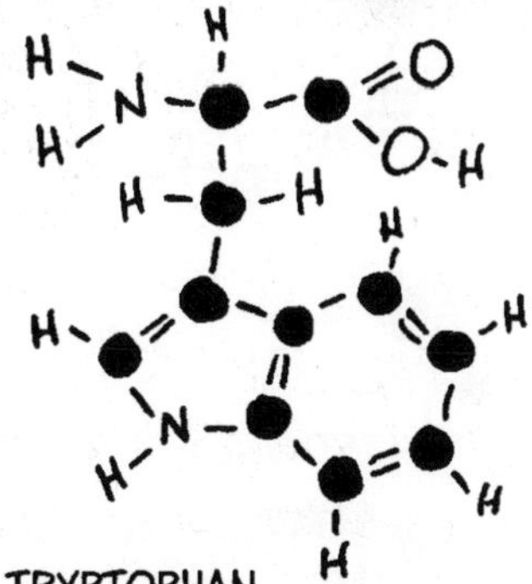

TRYPTOPHAN 色氨酸

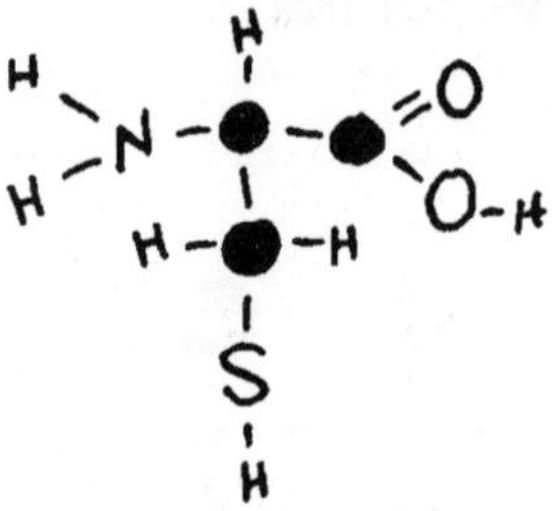

CYSTEINE
半胱氨酸

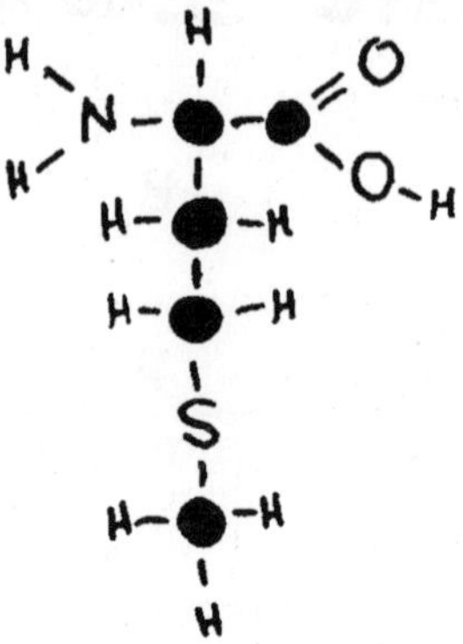

METHIONINE
甲硫氨酸

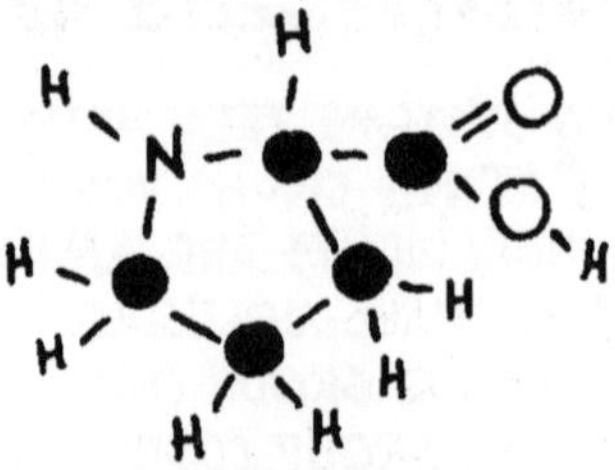

PROLINE
脯氨酸

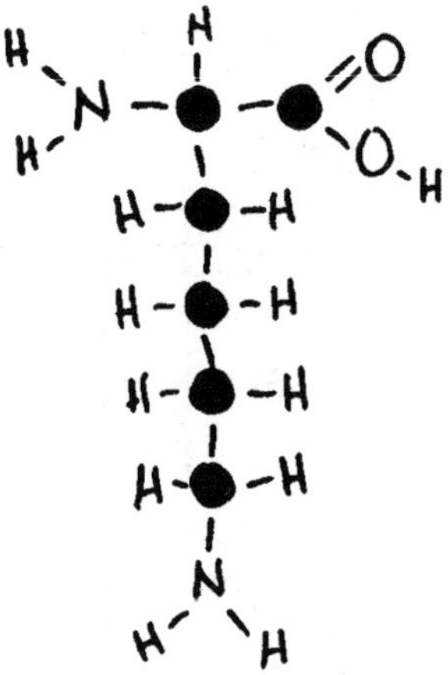

LYSINE
赖氨酸

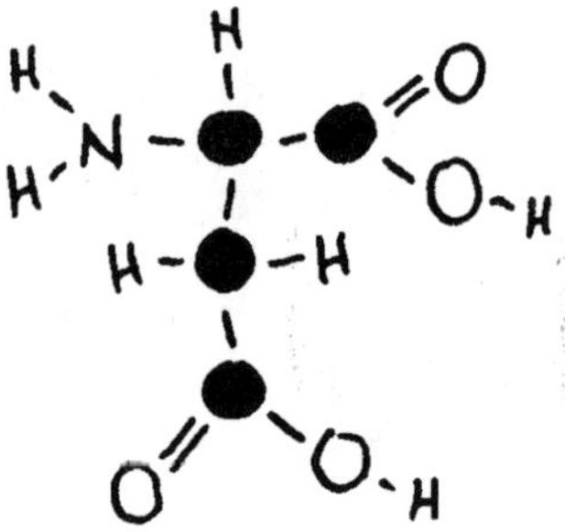

ASPARTIC ACID
天冬氨酸

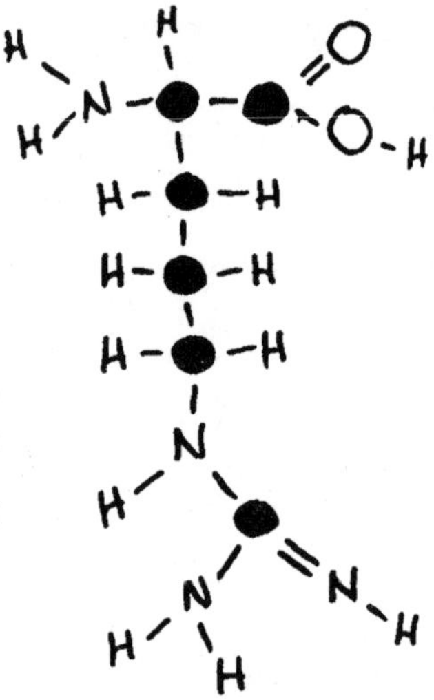

ARGININE
精氨酸

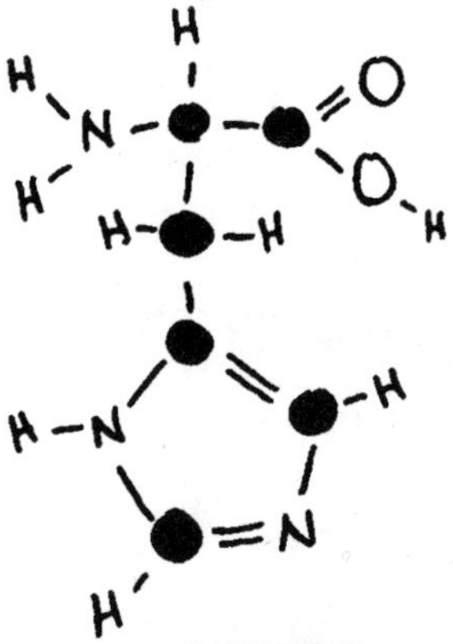

HISTIDINE
组氨酸

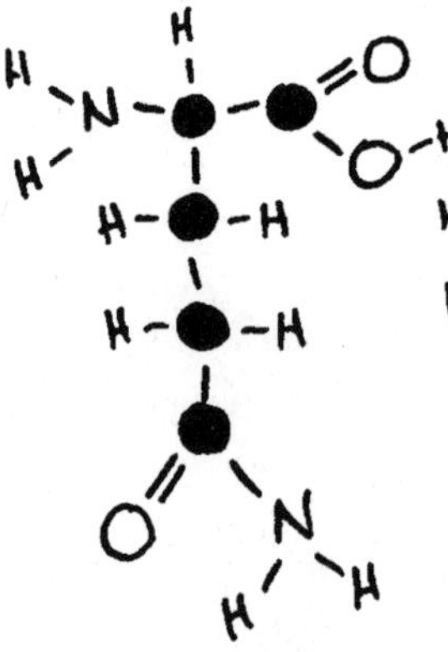

GLUTAMINE
谷氨酰胺

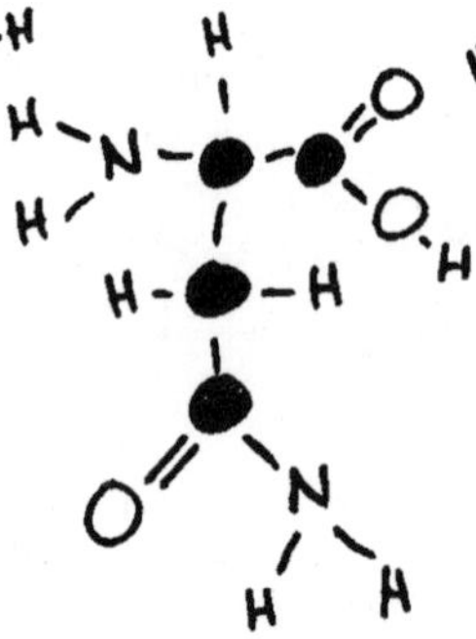

ASPARAGINE
天冬酰胺

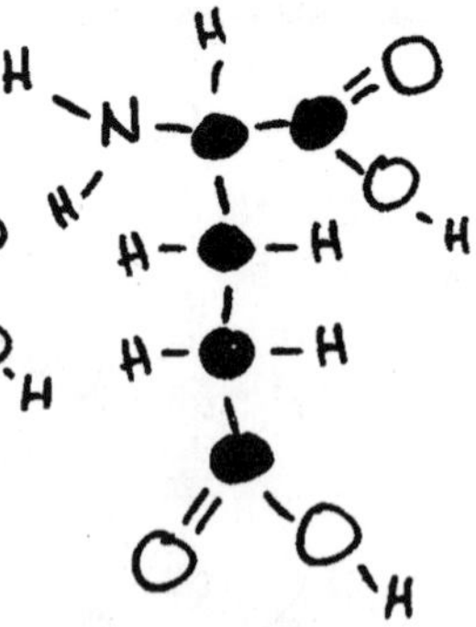

GLUTAMIC ACID
谷氨酸

TWO AMINO ACIDS CAN LINK UP IN A CONNECTION CALLED THE **PEPTIDE BOND.**

两个氨基酸可以以**肽键**连接。

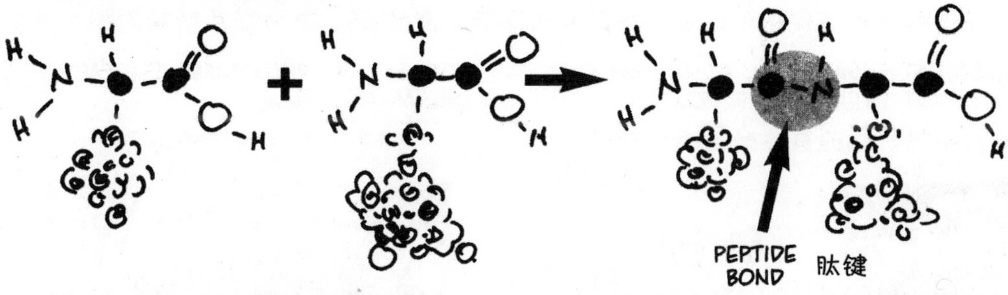

THE RESULTING SHORT CHAIN STILL HAS NH_2 AT ONE END AND COOH AT THE OTHER, SO MORE AMINO ACIDS CAN JOIN TO MAKE A **POLYPEPTIDE** CHAIN.

由此产生的短链仍然在一端有氨基，另一端有羧基，所以还可以有更多的氨基酸加入，形成**多肽**链。

CHARGED OR POLAR SIDE GROUPS ATTRACT OR REPEL EACH OTHER...

带电荷或者极性的基团会互相吸引或排斥……

THE POLYPEPTIDE FOLDS UP, BY A PROCESS THAT IS NOT WELL UNDERSTOOD...

多肽通过一个还未被完全了解的过程折叠起来……

UNTIL IT BECOMES A **PROTEIN.** (IN FACT, PROTEINS SOMETIMES HAVE TWO OR MORE SEPARATE CHAINS WOUND TOGETHER.)

直到它成为一种**蛋白质**（事实上，蛋白质有时会由两个或多个分离的链组合在一起）。

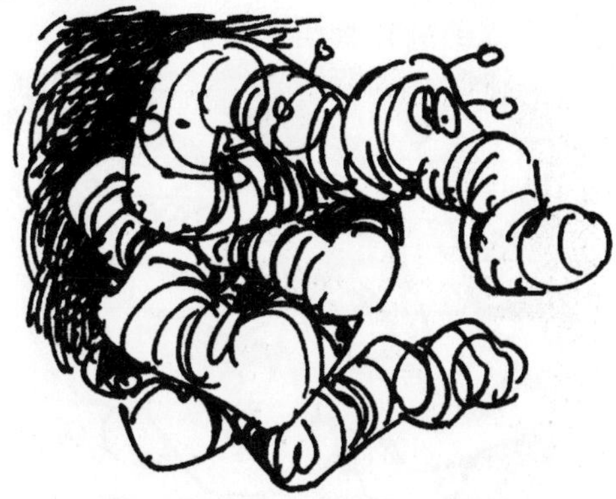

SOME PROTEINS SERVE AS STRUCTURAL MATERIAL, BUT MOST ARE **CATALYSTS** FOR OTHER REACTIONS. CATALYTIC PROTEINS ARE CALLED **ENZYMES.** FOR EXAMPLE:

一些蛋白质是结构材料，但多数是其他反应的**催化剂**。催化蛋白质称为**酶**。例如：

WHEN YOU EAT SUGAR, YOUR BODY MAKES ENZYMES THAT BREAK SUGAR DOWN...

当你吃糖，你的身体产生酶使糖分解……

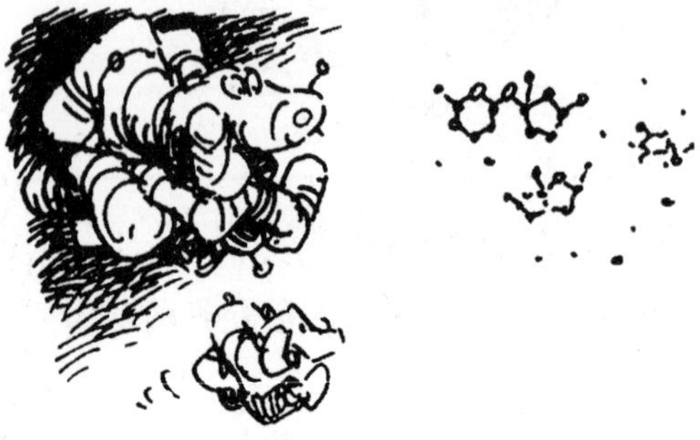

THE ENZYME RECOGNIZES THE PARTICULAR SUGAR MOLECULE...

酶会识别特殊的糖分子……

AND CATALYZES THE REACTION THAT BREAKS IT DOWN INTO SMALLER PIECES.

然后进行催化反应，把糖分解成片段。

THE ENZYME ITSELF IS UNCHANGED IN THE PROCESS.

酶本身在这个过程中保持不变。

MEANWHILE, ANOTHER PROTEIN CALLED **HEMOGLOBIN** TRANSPORTS OXYGEN THROUGH THE BLOOD STREAM TO CELLS, WHERE IT CAN OXIDIZE GLUCOSE AND FREE THE ENERGY YOUR BODY NEEDS TO KEEP GOING.

与此同时，还有另一种蛋白叫作**血红蛋白**，通过血液输送氧气至细胞，它可以氧化葡萄糖并为你的身体提供能量。

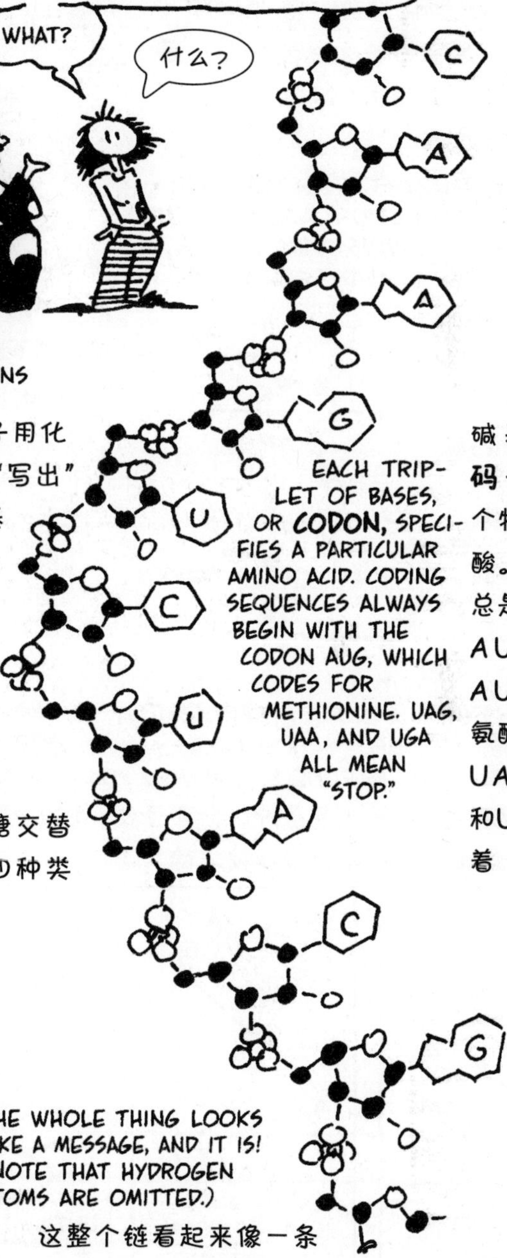

LONG MOLECULES CALLED **NUCLEIC ACIDS** "WRITE" AMINO ACID SEQUENCES FOR PROTEINS IN A CHEMICAL "LANGUAGE."

长链**核酸**分子用化学"语言""写出"蛋白质的氨基酸序列。

EACH TRIPLET OF BASES, OR **CODON,** SPECIFIES A PARTICULAR AMINO ACID. CODING SEQUENCES ALWAYS BEGIN WITH THE CODON AUG, WHICH CODES FOR METHIONINE. UAG, UAA, AND UGA ALL MEAN "STOP."

每个三联碱基，即**密码子**指定一个特定的氨基酸。编码序列总是从密码子AUG开始，AUG是甲硫氨酸的编码。UAG、UAA和UGA都意味着"结束"。

RNA, RIBONUCLEIC ACID, HAS A LONG SPINE OF ALTERNATING PHOSPHATES AND RIBOSE SUGARS, FROM EACH OF WHICH JUTS ONE OF FOUR CHEMICAL BASES KNOWN AS **A, C, G,** AND **U.**

RNA，即核糖核酸，是磷酸和核糖交替排成的长链状分子，每个核糖都连着四种类型的碱基——A、C、G和U之一。

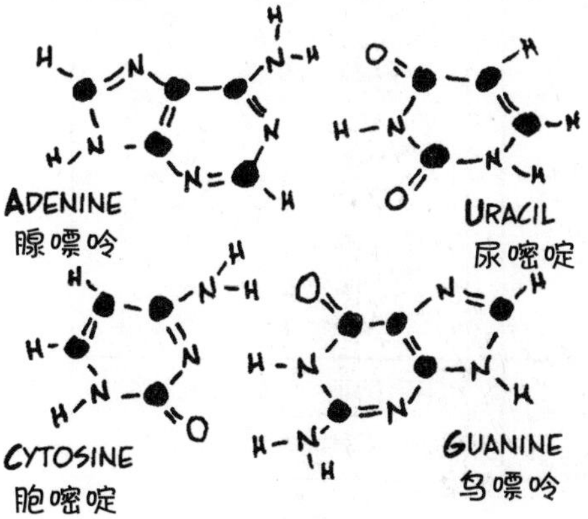

THE WHOLE THING LOOKS LIKE A MESSAGE, AND IT IS! (NOTE THAT HYDROGEN ATOMS ARE OMITTED.)

这整个链看起来像一条信息，实际上它就是信息！（请注意，氢原子被省略。）

THE OTHER NUCLEIC ACID, **DNA,** DEOXYRIBONUCLEIC ACID, HAS TWO STRANDS SIMILAR TO RNA'S WOUND AROUND EACH OTHER. LIKE RNA, DNA USES THE BASES **A, C,** AND **G,** BUT SUBSTITUTES **T** (THYMINE) FOR **U.**

另一个种核酸，即脱氧核糖核酸（DNA），有两个类似的RNA链互相缠绕。像RNA一样，DNA的碱基也包括A、C和G，但用T（胸腺嘧啶）代替了U。

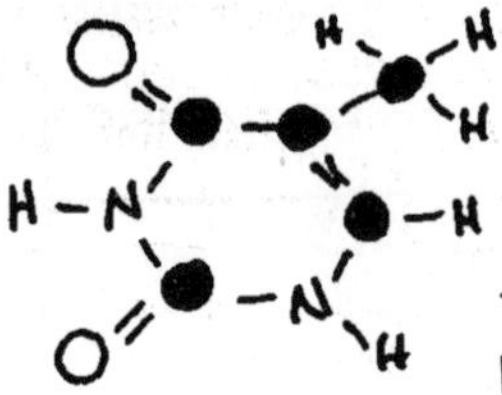

THYMINE
胸腺嘧啶

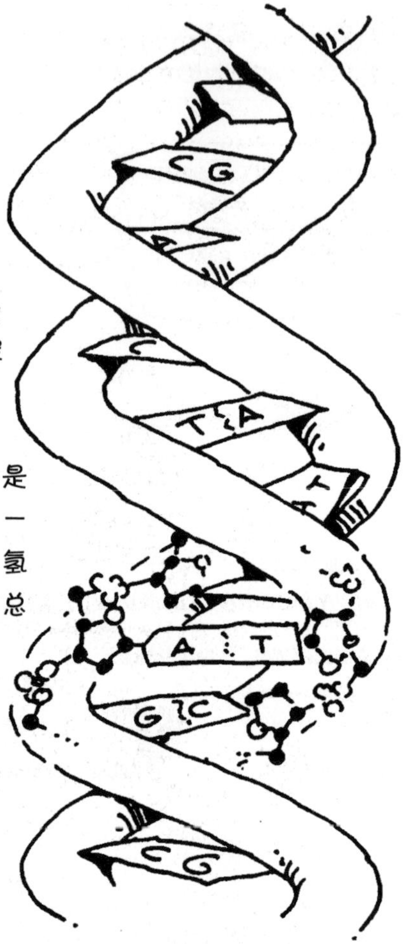

THE TWO STRANDS FIT TOGETHER WITH MIRACULOUS PERFECTION: **A** ALWAYS PAIRS WITH **T,** AND **C** ALWAYS PAIRS WITH **G,** HELD TOGETHER BY HYDROGEN BONDS.

这两条链总是完美地组合在一起：A总是通过氢键与T配对，C总是与G配对。

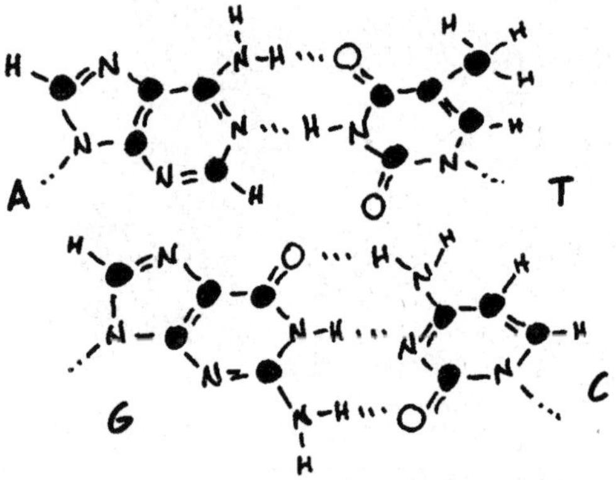

ONE STRAND OF DNA IS THE **COMPLEMENT** OF THE OTHER. IN OTHER WORDS, DNA CARRIES THE INFORMATION NECESSARY TO **REPRODUCE ITSELF!!!** (THE ACTUAL WORK IS DONE BY ENZYMES POWERED BY REDOX REACTIONS.)

一条DNA链与另一条**互补**。换句话说，DNA携带着**复制它本身**所必需的信息！（实际工作是由氧化还原反应驱动酶完成的。）

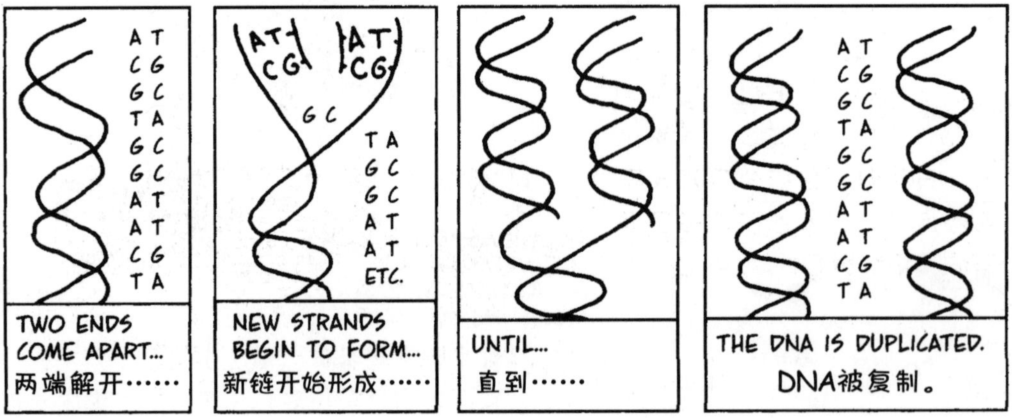

HOW IT DOES SO, AND HOW CODON SEQUENCES ARE TRANSLATED INTO PROTEINS, ARE DETAILS YOU WILL HAVE TO FIND ELSEWHERE. WE SUGGEST **THE CARTOON GUIDE TO GENETICS...**

它是如何做到这些以及密码子序列是如何翻译成蛋白质的，这些细节你需要从别处获得。如有兴趣，我们建议您购买一本**遗传学科学漫画**……

AND THERE ARE A **LOT** OF DETAILS IN ORGANIC AND BIOCHEMISTRY, NO END TO THEM, IN FACT! NOT TO MENTION PHYSICAL, NUCLEAR, ENVIRONMENTAL, NANO-, AND ALL THE OTHER BRANCHES OF CHEMISTRY. YES, READER, THE TIME HAS COME TO REFER YOU TO MORE ADVANCED COURSES, AND TO CONGRATULATE YOU FOR GETTING THROUGH THE BASICS! 'BYE!

在有机和生物化学中的很多细节其实没有尽头！更不要说还有物理化学、核化学、环境化学、纳米化学等化学的其他分支了。是的，读者们，已经到了你们学习更高级课程的时间了，祝贺你们通过了基础知识的学习！再见！

Appendix

附录

Using Logarithms

对数的使用

IN SOME OF OUR CHAPTERS, WE USE A MATHEMATICAL SHORTHAND CALLED LOGARITHMS (OR LOGS, FOR SHORT). THE LOGARITHM IS A CONVENIENT, COMPACT WAY OF WRITING A NUMBER. FOR INSTANCE, INSTEAD OF $[H^+] = 10^{-7}$, WE WRITE pH = 7. pH IS A LOGARITHM.

在本书的几章中，我们使用了一种叫做对数（或简作log）的数学速记法。对数是一种方便简洁的数字书写方式。比如说，我们写pH=7来代替写$[H^+]=10^{-7}$。pH就是对数。

A LOGARITHM IS AN EXPONENT. THE **COMMON LOGARITHM** OF A NUMBER N, log N, IS THE EXPONENT TO WHICH 10 MUST BE RAISED IN ORDER TO EQUAL N:

对数就是幂的指数。一个数N的**常用对数**是这样一个指数：10必须以这个数目自乘才能得到与N相等的幂：

$10^a = N$ IS THE SAME AS $a = \log N$ THAT IS, $10^{\log N} = N$

$10^a = N$ 等同于 $a = \log N$，也就是说，$10^{\log N} = N$

SO $\log 10 = 1$ AND $\log 1 = 0$ AND $\log 100 = 2$ (SINCE $10^0 = 1$, $10^2 = 100$).

所以$\log 10 = 1$，$\log 1 = 0$，$\log 100 = 2$（因为$10^0 = 1$，$10^2 = 100$）。

AND $\log 72.3 = 1.85914$ BECAUSE $10^{1.85914} = 72.3$ (CHECK IT ON YOUR CALCULATOR.)

同理，$\log 72.3 = 1.85914$，因为$10^{1.85914} = 72.3$（请用计算器检验）。

KEY FACT: WHEN NUMBERS ARE **MULTIPLIED**, THEIR **LOGARITHMS ARE ADDED.**

一个关键事实：当几个数**相乘**时，它们的**对数相加**。

$$\log MN = \log M + \log N$$

THIS IS BECAUSE $10^a 10^b = 10^{(a+b)}$. IF $M = 10^a$ AND $N = 10^b$, THEN $MN = 10^a 10^b = 10^{(a+b)}$, SO $a+b = \log MN$. BUT $a = \log M$ AND $b = \log N$.

这是因为$10^a 10^b = 10^{(a+b)}$。设$M = 10^a$而$N = 10^b$，那么$MN = 10^a 10^b = 10^{(a+b)}$，所以$a+b = \log MN$，而$a = \log M$，$b = \log N$。

SIMILARLY

同样

$$\log (M^p) = p(\log M)$$

$$\log \left(\frac{1}{N}\right) = -\log N$$

BECAUSE THIS IS HOW EXPONENTS BEHAVE:

因为它们反映了指数的性质：

$$10^{-a} = \frac{1}{10^a} \qquad 10^{ab} = (10^a)^b$$

log N GIVES US A ROUGH IDEA HOW BIG N IS. THE WHOLE-NUMBER PART OF THE LOGARITHM GIVES N'S ORDER OF MAGNITUDE.

log N可以让我们对N多大有一个粗略的印象。对数的整数部分决定了N的数量级。

$$\log 1{,}234 = 3.0913$$

$$\log 1.234 = 0.0913$$

$$\log 1{,}234{,}000 = 6.0913$$

$$\log (a \times 10^n) = n + \log a$$

THERE'S A GOOD ONLINE CALCULATOR AT

是一个不错的在线计算器。

http://www.squarebox.co.uk/desktop/scalc.html

Natural Logarithms 自然对数

COMMON LOGS HAVE BASE TEN. THEY ARE EXPONENTS OF 10. SOMETIMES, THOUGH, THEY ARE LESS CONVENIENT THAN "NATURAL LOGS," FOR INSTANCE, WHEN A QUANTITY CHANGES AT A RATE PROPORTIONAL TO ITSELF. THAT IS, AT TIME t,

常用对数以10为底，它们是10的指数。但是有时候，它们不如“自然对数”方便，比如说，当一个量以和它自己成比例的速率变化的时候。也就是说，在时间t，

$$r_A(t) = kA_t$$

THEN IT'S POSSIBLE TO SHOW THAT THE QUANTITY A_t IS AT ANY TIME t IS

在这种情况下，任意时间t时的量A_t就可以这样计算：

$A_t = A_0e^{kt}$ WHERE A_0 IS THE INITIAL AMOUNT OF A, AND $e = $ **2.71828**

$At = A_0e^{kt}$，其中A_0是A的初始量，而$e = 2.71828$

THEN $e^{kt} = A_t/A_0$ AND WE WRITE $kt = \ln(A_t/A_0)$, THE **NATURAL LOGARITHM** OF A_t/A_0. THE NATURAL LOG OF ANY NUMBER N IS THE EXPONENT TO WHICH e MUST BE RAISED TO MAKE N.

于是$e^{kt} = A_t / A_0$，我们就可以写$kt = \ln(A_t / A_0)$，也就是A_t / A_0的**自然对数**。任何数N的自然对数都是这样一个指数：e必须以这个数目自乘才能得到与N相等的幂。

$M = \ln N$ MEANS THE SAME THING AS $e^M = N$

$M = \ln N$与$e^M = N$意义相同。

BECAUSE $e^ae^b = e^{(a+b)}$, ETC., THE NATURAL LOGS OBEY THE SAME FORMULAS AS COMMON LOGARITHMS.

因为同样有$e^ae^b = e^{(a+b)}$等性质，自然对数与常用对数遵循相同的公式：

$$\ln MN = \ln M + \ln N$$

$$\ln(1/M) = -\ln M$$

$$\ln(M^n) = n\ln M$$

IN FACT, THE NATURAL LOGARITHM IS A **CONSTANT MULTIPLE** OF THE COMMON LOGARITHM.

事实上，自然对数等于常用对数与一个**常数的乘积**。

$$\ln N = \ln(10^{\log N}) = (\log N)(\ln 10)$$

$\ln 10 = $ **2.302585...**, SO

$\ln N = $ **2.302585** $\log N$

About the Authors

LARRY GONICK IS THE SON AND SON-IN-LAW OF CHEMISTS. HE ONCE CONSIDERED A SCIENTIFIC CAREER, BUT WISELY ABANDONED THE IDEA AFTER BREAKING TWELVE PIECES OF GLASSWARE IN A SINGLE, DISTRESSING THREE-HOUR CHEMISTRY LAB. HE WRITES AND DRAWS NONFICTION COMIC BOOKS AND IS THE STAFF CARTOONIST FOR **MUSE** MAGAZINE. HE LIVES PHYSICALLY IN CALIFORNIA WITH HIS FAMILY AND VIRTUALLY ON THE WEB AT www.larrygonick.com.

拉里·高尼克是化学家的儿子和女婿，他曾梦想成为一名科学家。但在令人厌倦和痛苦的实验中打破了12个玻璃器皿后，他明智地放弃了当科学家的梦想。他开始从事纪实文学的写作与插画绘画工作，并成为*MUSE*杂志的一名专职漫画家。他和家人生活在美国加利福尼亚州。想了解更多关于拉里·高尼克的信息，可登录www.larrygonick.com查询。

CRAIG CRIDDLE IS PROFESSOR OF ENVIRONMENTAL ENGINEERING AND SCIENCE AT STANFORD UNIVERSITY, WHERE HE TEACHES AQUATIC CHEMISTRY AND ENVIRONMENTAL BIOTECHNOLOGY. HE HAS PUBLISHED MANY ARTICLES ON CHEMICALS IN WATER AND WATER CLEANUP, AND HIS TEAM OF GRAD STUDENTS AND RESEARCH ASSOCIATES LIKE TO THINK THEY CAN SOLVE THE WORLD'S WATER CRISIS. PROF. CRIDDLE AND HIS WIFE LIVE IN CUPERTINO, CALIFORNIA, ALONG WITH THEIR DOG AND WHICHEVER OF THEIR FOUR KIDS (MOSTLY GROWN) HAPPENS TO BE HOME. HIS WEB SITE IS www.stanford.edu/group/evpilot/. HE BELIEVES THAT BROKEN EQUIPMENT IS A NATURAL PART OF SCIENCE.

克雷格·克里德尔是斯坦福大学的一名环境工程和科学专业教授，专职教授水环境化学和环境生物技术，发表了诸多关于水资源和水净化方面的文章。他率领研究生团队和研究小组致力于解决全球的水资源危机问题。克雷格和他的妻子、四个已经长大成人的孩子以及饲养的宠物狗居住在美国加利福尼亚州的库比蒂诺。克雷格·克里德尔坚持认为破损的设备是科学的一个天然组成部分。想了解更多关于克雷格·克里德尔的信息，可登录www.standford.edu/group/evpilot/查询。

索引 INDEX

A

absolute entropy 绝对熵 258,259

acids and bases 酸与碱 219,220,222,227

buffers 缓冲溶液 244,246,247

conjugate pairs 共轭酸碱对 220

equivalent weight of 当量 236,237,240

neutralization 中和 234,235,237

activation energy 活化能 200,202,217

air 空气 5,8,11

alchemy 炼金术 6

alternator 交流发电机 281

amino acids 氨基酸 301,303,305

ammonia 氨 78,215,234

amu (atomic mass unit) 原子质量单位 32

anions 阴离子 26,54,62

single-atom 单原子阴离子 63

anode 阳极 24,26,274

Aristotle 亚里士多德 5,16

atmospheric pressure 大气压 8,9,146

atomic mass 原子量 148,167

atomic number 原子序数 31,33,34

atomic size 原子尺寸 52

atomic weight 原子重量 32,33,50

atoms 原子 4,5,17

atomic theory 原子学说 17

atomists 原子论者 4,58

atom building 原子的构造 43

bonds between 原子之间的键 80

electron affinity 电子亲合能 54,56

electronegativity 电负性 61,67,70

ionization energy 电离能 41,52,53

net charge 净电荷 103,272

oxidation number 氧化数 102,103,104

See also electrons 另见电子

attractions 吸引 21,22,23

Avogadro's law 阿伏伽德罗定律 95

Avogadro's number 阿伏伽德罗常数 95

B

balanced equations 配平的方程 92

bases. See acids and bases 碱，见酸与碱

battery 电池组 275,281

boiling point 沸点 56,115,144

carbon chains 碳链 292,296,297

dissolved material 溶解的物质 170,269

heating curve 升温曲线 165,166

helium 氦 35,41,44

bomb calorimeter 弹式量热计 127,128,129

bonds 键 60,65,68

carbon atoms 碳原子的成键 15

potential energy in 键中的势能 123

solvation 溶剂化 172

strength of 强度 142,143,144

See also intermolecular forces 另见分子间作用力 139,141,143

Boyle's law 波义耳定律 148

Brand, Hennig 布兰德，亨尼希 6

buffers 缓冲溶液 244,246,247

bystander ion 旁观者离子 238,239

C

calorimetry 量热测定 126

carbohydrates 碳水化合物 295

carbon 碳 18,19,27

atom 碳原子 27,28,30

hybrid orbital 杂化轨道 79,80

oxidants/reductants 氧化剂 / 还原剂 105,136,280

phase diagram 相图 163,164

valence electrons bonds 价电子成键 51,52,53

carbon chains 碳键 282,296,297

catalysts 催化剂 201,202,203

catalytic converter 催化转化器 202

cathodes 阴极 24,25,274

cations 阳离子 25,26,62

Celsius scale 摄氏温标 114,115

Charles's law 查理定律 148

chemical bonds. See bonds 化学键，见键 85,103,114

chemical reactions 化学反应 1,2,3

activation energy 活化能 200,202,217

alchemy as 作为化学反应的炼金术 6

catalysts 催化剂 201,202,203

defined 定义 31,32,112

electricity from 电化学反应 1,2,3

as energy transfer　作为能量转化的方式　111,119
entropy and　熵和化学反应　257,264
fire as first　作为第一个化学反应的火　7
free energy　自由能　264,267,268
higher-order　多级反应　203
hydrolysis　水解　232
rate of　速率　155,186,187
redox　氧化还原反应　100,101,105
reversible　可逆反应　13,208
solutions and　溶液与化学反应　185
spontaneous　自发反应　253,263,266
collision theory　碰撞理论　192
combination reaction　化合反应　15,90,200
combustion　燃烧　10,11,14
heat of　燃烧热　136
spontaneous　自燃　289
compounds　化合物　14,16,17
concentration　浓度　174,175,176
condensation　凝结　10,90,152
coolants　冷却剂　124
copper　铜　3,24,35
zinc reaction　铜锌反应　273,275,277
corrosion　腐蚀　7,98,99
covalent bond　共价键　71,72,74
strength of attraction　吸引的强度　85
crystalline structures　晶体结构　63,66,69
of carbon　碳的晶体结构　164,259
covalent bonds　共价键　71,72,74
of ice　冰的晶体结构　162
ionic bonds　离子键　62,64,66
metallic bonds　金属键　68,71,75
current, electric　电流　24,69,290

D

Dalton, John　道尔顿，约翰　17
decomposition reaction　分解反应　90,101
Democritus　德谟克利特　4
dipoles　偶极子　140,141
dissolving process　溶解过程　171
acids and bases　酸与碱　222
freezing/boiling points　凝固点 / 沸点　125,182,183
salts in water　盐在水中的溶解过程　172
DNA　脱氧核糖核酸　306
double bond　双键　73,76,77
double-displacement reaction　复分解反应　101
dynamic balance　动态平衡　208

E

elasticity　弹性　145
electric cells　电池　24,185,219
electric potential　电势　40,275,277
electricity　电　21,22,23
attractions/repulsions　吸引 / 排斥　22,303
metal conductors　金属导体　69
See also negetive charge; positive charge
*另见*负电荷；正电荷　22,23,24
electrochemistry　电化学　271
electrodes　电极　24,26,140
electrolysis　电解　24,26,290
electromagnetic radiation　电磁辐射　114
electronegativity　电负性　61,67,70
electrons　电子　25,26,28
affinity　亲合能　54,56
bonds　成键　65,72,73
dipole attraction　偶极吸引　140
ionization energy　电离能　41,52,53
metal　金属电子　51
orbit　轨道　33,37,38
outer　外层电子　46,47,48
paired　成对电子　72,81
particle/wave　粒子波　37
redox reactions　氧化还原反应　100,101,105
rule of eight　八电子规则　58
sharing　共用电子　142
shells　电子层　41,53,73
electropositivity　电正性　61,62,67
electrostatic attraction　静电吸引　62
elementary reactions　电化学反应　271
elements　元素　5,13,14
ancient four　古代的四元素　14
atomic number　原子序数　31,33,34
carbon's uniqueness　碳的独特性　298

charge extremes 电荷端点 82
grouping of 分族 291
isotopes of 同位素 32,33,50
list of 列表 18,35,36
oxidation number 氧化数 102,103,104
periodic table 周期表 50,51,54
empirical formula 实验式 64,89
emulsion 乳化 173
endothermic reactions 吸热反应 131,134,199
energy 能量 33,38,39
activation 活化能 200,202,217
collision 碰撞能 198
conservation law 能量守恒定律 113
electrical 电能 271,281,282
quanta of 能量量子 39
spreading out of 扩散 128,138,253
transfer of 转化 4,113,202
enthalpy 焓 128,129,130
change 焓变 129,130,132
of formation 生成焓 132,133,267
entropy 熵 257,258,259
enzymes 酶 304,306
equilibrium 化学平衡 208,211,214
acids and bases 酸与碱 215
equilibrium constant 平衡常数 212,213,219
pH pH 226,227,228
second derivation of 由其他常数推导 228
solubility product 溶度积 241,242,243
weak ionization 弱电离 228
equivalent weight 当量 236,237,240
evaporation 蒸发 101,152,153
exothermic reactions 放热反应 131,136
explosions 爆炸 7,86,106
explosives 炸药 100,215

F

Faraday's constant 法拉第常数 284
fire 火 1,2,3
first-order reaction 一级反应 191,194
forward reaction 正反应 209,241,269
our basic elements 四大元素 14
Franklin, Benjamin 富兰克林，本杰明 22,23
free energy change 自由能变 267
free radical 自由基 187
freezing point 凝固点 125,182,183
fuel cell 燃料电池 282

G

gases 气体 7,8,9
characteristics of 特性 164
noble 稀有气体 57,59,141
solubility 溶解度 85,177,178
state changes 状态改变 9
temperature and 温度及气体 73
gas laws 气体定律 147,149,167
Gibbs function 吉布斯函数 264,266,267
Gilbert, William 吉尔伯特，威廉 21
glucose 葡萄糖 75,288,289
Guericke, Otto von 居里克，奥托·冯 8,147
gunpowder recipe 黑火药成分 130

H

Haber process 哈伯法 216
half-life 半衰期 189
half-reactions 半反应 276,277,278
halogens 卤素 54,55,74
heat 热量 66,89,90
reaction activation 反应活化热 200,289
See also temperature *另见*温度 13,73,114
heat capacity 热容量 120,121,126
heat change 热量变化 121,122,126
heating curves 升温曲线 165,166
heat of combustion 燃烧热 136
heat of fusion 熔化热 161
heats of formation 生成热 131,132,134
helium 氦 35.41.44
hemoglobin 血红蛋白 304
Henderson-Hasselbalch equation 亨德森-哈塞尔巴尔赫方程 248,249,250
Heraclitus 赫拉克利特 4
Hess's Law 盖斯定律 133
higher-order reactions 多级反应 203
hybrid orbitals 杂化轨道 79,80

hydrocarbons 碳氢化合物 292,293,298
hydrogen 氢 10,16,19
atomic number 原子序数 31,33,34
carbon chains 碳链 292,296,297
electron shell 电子层 41,53,73
heat of combustion 燃烧热 136
pH pH 226,227,228
positive charge 正电荷 22,23,25
redox reaction 氧化还原反应 100,101,105
hydrogen bond 氢键 84,85,123
attraction strength 吸引的强度 142
DNA DNA 306
hydrolysis 水解 232
hydronium 水合氢离子 221,223

I

ice 冰 125,161,162
ideal gas 理想气体 145,149,167
in solution 溶液中 89,101,103
indicator chemicals 化学指示剂 227
intermolecular forces 分子间作用力 139,141,143
internal energy 内能 117,118,120
ion 离子 25,26,41
ionic bonds 离子键 62,64,66
dipole 偶极子 140,141
polarity 极性 82,83,84
strength of attraction 吸引的强度 142
ionic crystals 离子晶体 63,64,66
ionic repulsion 离子排斥 70
ionization 电离 41,52,53
base constant 碱的电离常数 233
equilibrium 平衡 23,94,96
high 强电离 222
ionization energy 电离能 41,52,53
of water 水的电离 212
weak ionization 弱电离 228
isomers 异构体 293,295
isotopes 同位素 32,33,50

J

Jabir 贾比尔 6
Joule, James Prescott 焦耳，詹姆斯·普雷斯科特 121
joules 焦耳 112,121,122

K

Kelvin scale 开尔文温标 114,115
kinetic energy 动能 113,114,118

L

lanthanide series 镧系 49
Lavoisier, Antoine 拉瓦锡，安托万 12,13,14
lead-acid battery 铅-酸蓄电池 281
Le Chatelier's principle 勒夏特列原理 214,244
Lewis diagram 电子式 73,78,80
life 生命 85,203,288
chemicals of 化学物质 89,120,235
glucose oxidation 葡萄糖氧化 288
hydrogen bonding 氢键的形成 306
origin of 起源 203
liquids 液体 3,9,10
boiling point 沸点 56,115,144
evaporation/condensation 蒸发 / 凝结 152
melting point 熔点 56,66,115
phase diagrams 相图 163,164
solubility 溶解度 85,177,178
solutions 溶液 89,98,101
standard molar entropy 标准摩尔熵 258,259
surface tension 表面张力 151
suspensions 悬浊液 173
See also water *另见*水 5,9,12
logarithms 对数 159,191,226
London dispersion force 色散力 141,142,144

M

main-group elements 主族元素 48
mass 质量 14,15,20
mass action, law of 质量作用定律 212
mass-balance table 质量-平衡表 96,107,134
matter 物质 2,4,5
ancient theories of 古代有关物质的理论 5
three types of 三种状态的物质 167
mechanical energy 机械能 112,113
melting point 熔点 56,66,115
heating curve 升温曲线 165,166
Mendeleev, Dmitri 门捷列夫，德米特里 19

metal ions as acids　金属离子，作为酸　68,69,142
metallic bonds　金属键　68,71,85
metals　金属　3,7,8
miscibility　互溶　178,179
molar heat capacity　摩尔比热容　120
molarity　摩尔浓度　176,177,188
mole　摩尔　94,95,96
　Avogadro's number　阿伏伽德罗常数　95
molecules　分子　17,64,71
　attractions between　分子之间的吸引　151
　charged　带电分子　80
　collision theory　碰撞理论　192
　composition　成分　13,25,65
　ionization fraction　离子化分数　212,225
　kinetic energy storage　动能储存　145
　shapes　形状　42,65,76
　solubility　溶解度　85,177,178
　standard entropy　标准熵　258,259
　weight　重量　14,15,19
mullite　莫来石　91,92

N

negative charge　负电荷　22,23,24
　electron　电子　25,26,28
negative reduction potential　负还原电势　280
neon　氖　35,45,57
Nernst equation　能斯特方程　286,287
neutralization　中和　234,235,237
neutrons　中子　30,31,32
noble gases　稀有气体　57,59,141
non-metals　非金属　55,56,58
nonrepeating chains　非重复长链　301
nucleic acids　核酸　305,306
nucleus　原子核　28,30,31

O

orbitals　轨道　33,37,38
organic chemistry　有机化学　291,293,305
oxidants　氧化剂　105,136,280
oxidation　氧化　57,100,101
oxidation numbers　氧化数　102,103,104
oxidation-reduction. See redox reactions
氧化还原，见氧化还原反应　100,101,105
oxygen　氧　11,13,14
　atomic number　原子序数　31,33,34
　carbon chains　碳链　292,296,297
　covalent bond　共价键　71,72,74
　electron shells　电子层　41,53,73
　negative charge　负电荷　22,23,24
ozone　臭氧　187,194,195

P

partial pressure　分压　156,157,160
particles　粒子　4,25,30
　collision of　碰撞　144,145,146
　entropy　熵　257,258,259
　number in mole　以摩尔衡量的数目　95
peptide bond　肽键　303
periodic table　周期表　50,51,54
pH　pH　226,227,228
　buffers　缓冲溶液　244,246,247
　endpoint　终点　240
　Nernst equation　能斯特方程　286,287
　solubility effects　对溶解度的影响　245,251
phase change　相变　159,166
phase diagrams　相图　163,164
photons　光子　114
picometer　皮米　28
plasma　等离子体　168
polarity　极性　82,83,84
polyatomic ions　多原子离子　65,66,81
polymers　聚合物　282,299,300
polypeptide chain　多肽链　303
positive charge　正电荷　22,23,25
　proton　质子　30,31,32
potential energy　势能　113,114,118
pottery　陶器　91,92,96
precipitating　沉淀　89,98,101
pressure　压强　129,145,146
　constant　恒定压强　129
　entropy change　焓变　129,130,132
　external　外部压强　157,158,161
　gas law equation　气体定律方程　149
　gas solubility　气体溶解度　181
　ice melting　冰的融化　258

Le Chatelier's principle 勒夏特列原理 214,244
vapor 蒸气压强 156
Priestley, Joseph 普利斯特列，约瑟夫 9,10,11
properties 性质 2,8,19
metals vs. nonmetals 金属与非金属的对比 55
proteins 蛋白质 303,304,305
Q
quantized energy 量子化的能量 256
quantum mechanics 量子力学 36,37,80
R
radiant energy 辐射能 112,114
rate constant 速率常数 190,191
Razi, al- 拉齐
reactants 反应物 15,89,90
enthalpy of formation 生成焓 132,133,267
mass-balance table 质量-平衡表 134
See also chemical reactions *另见化学反应* 1,2,3
reaction constant 反应常数 149
reaction equations 反应方程 89,92
reaction products 反应产物 89,101,129
reaction quotient 反应商 269,287
reaction rate 反应速率 186,187,190
reaction stoichiometry 反应的化学计量学 93
redox reactions 氧化还原反应 100,101,105
reductants 还原剂 105,280
resonance 共振 80,81
reverse reaction 逆反应 194,209,265
RNA 核糖核酸 305,306
rule of eight 八电子规则 58
S
salt 盐 24,26,55
acid-base neutralization 酸碱中和 235
boiling point 沸点 56,115,144
dissociation in liquid 在液体中的解离 1701,177
solubility products 溶度积 241,242,243
saturation 饱和 178,241,242
second-order reactions 二级反应 192,194,201
soap 肥皂 3,98,99
solids 固体 18,55,89
dissolved 溶解的固体 170
standard molar entropy 标准摩尔熵 258,259
solubility 溶解度 85,177,178
products 溶度积 241,242,243
solutions 溶液 89,98,101
acidity measure 酸度测量 228,245
buffers 缓冲溶液 244,246,247
neutralization 中和 234,235,237
pH pH 226,227,228
reaction rate 反应速率 186,187,190
titration 滴定 7,240,251
weak acid 弱酸溶液 231
solvation 溶剂化 172
specific heat 比热容 120,121,122
spontaneous processes 自发过程 253
starch 淀粉 300
stoichiometric coefficients 化学计量学系数 211
sublimation 升华 161,163
sugars/sucrose 蔗糖 171,259,295
suprefluid 超流体 164
surface tension 表面张力 151
suspensions 悬浊液 173
T
temperature 温度 13,73,114
boiling point 沸点 56,115,144
calorimetery 量热测定 126
critical 临界温度 159
entropy change 熵变 262
gas law equation 气体定律方程 147,149
heat capacity 热容量 120,121,126
melting point 熔点 56,66,115
reaction rate 反应速率 186,187,190
solubility 溶解度 85,177,178
state effects of 对物态的效应 258
thermodynamics 热力学 172,252,261
second law of 第二定律 261
thermometers 温度计 114,122,127
titration 滴定 7,240,251
transition metals 过渡金属 48,51
transition state 过渡态 197

V

valence electrons　价电子　51,52,53
vapor pressure　蒸气压强　156
vinegar　醋　171,173,227
voltaic cell　原电池　275
volts/voltage　电压 / 伏特　275,277,278
　free energy and　自由能与电压　264
volume　体积　16,95,129

W

water　水　5,9,10
　acids/bases　酸 / 碱　222
　boiling point　沸点　56,115,144
　dipole molecule　偶极分子　140,141
　evaporation　蒸发　101,152,153
　freezing expansion　凝固膨胀　162
　ionization　电离　41,52,53
　ionization constant　电离常数　229,230,233
　melting point　熔点　56,66,115
　molecular shape　分子形状　78
　polarity　极性　82,83,84
　specific heat　比热容　120,121,122
　splitting　水解　232
water constant　水常数　225
wavelength　波长　36,37,38
weights　重量　14,15,19
work energy　做功的能量